D1131835

DYNAMICAL THEORY OF DENDRITIC GROWTH IN CONVECTIVE FLOW

Advances in Mechanics and Mathematics

Volume 7

DYNAMICAL THEORY OF DENDRITIC GROWTH IN CONVECTIVE FLOW

JIAN-JUN XU
Department of Mathematics and Statistics, McGill University

Kluwer Academic Publishers
Boston/Dordrecht/London

Distributors for North, Central and South America:
Kluwer Academic Publishers
101 Philip Drive
Assinippi Park
Norwell, Massachusetts 02061 USA
Telephone (781) 871-6600
Fax (781) 871-6528
E-Mail <kluwer@wkap.com>

Distributors for all other countries:
Kluwer Academic Publishers Group
Post Office Box 322
3300 AH Dordrecht, THE NETHERLANDS
Telephone 31 78 6576 000
Fax 31 78 6576 474
E-Mail <orderdept@wkap.nl>

Electronic Services <http://www.wkap.nl>

Library of Congress Cataloging-in-Publication

Xu, Jian-Jun
Dynamic Theory of Dendritic Growth in Convective Flow
ISBN HB 1-4020-7924-9
ISBN E-book 1-4020-7925-7

Printed on acid-free paper.

Printed in the United States of America

Contents

Preface

The first draft of this monograph was written as a set of notes for the series of the lectures on the *Interfacial Wave Theory* of dendritic growth that I delivered in the State Key Laboratory of Crystal Growth at Shandong University, China, from August to October 2001. The targeted audience included the teachers, researchers and graduate students who were interested in the interdisciplinary areas of dynamics of pattern formation, material science, condensed matter physics and applied mathematics. A large portion of the material was later published as an invited review article in *Annual Review of Applied Mathematics and Mechanics* (2002). In writing this monograph I had an opportunity to further refine my previous works and correct some errors in the mathematical details found after their publications. The major revision of the monograph was done in the summer of 2002 during my visit to the State Key Laboratory for Studies of Turbulence and Complex Systems, the Department of Mechanics and Engineering Science at Peking University, where I completed my undergraduate study. I love the school then and even more so now for her ancient oriental academic tradition and the campus for her environmental quietness and natural beauty.

The Interfacial Wave (IFW) theory was first systematically described in my previous monograph "Interfacial Wave Theory of Pattern Formation: Selection of Dendritic Growth and Viscous Fingering in Hele–Shaw Flow" (Springer–Verlag, 1997). In that book, the objects of investigation were a variety of dynamic systems not restricted to dendritic growth. My attention was directed to discussing the common issues arising from various physical systems and exploring the common intrinsic mechanism underlying the phenomena.

Since publication of the monograph, the IFW theory for dendritic growth has further developed due to the extensive work done by my students and me. Some new and profound problems have been analyzed. The present monograph attempts to summarize the new findings on dynamics of dendritic growth with convective flow.

We begin with a description of the macroscopic continuum approach for solidification problems and briefly review, with refined derivations, the IFW theory for the typical dendritic growth system without convection which provides readers with essential background for further study.

The main body of the book is devoted to a systematic study of the interactive dynamics of dendritic growth with convection flow in melt. In particular, it explores the effect of various types of convection flow on the selection and pattern formation of dendritic growth. These subjects have been of great interest to researchers in the broad fields of pattern formation, microgravity research and crystal growth.

This book will be useful for researchers, postdoctoral fellows and graduate students in the fields of condensed matter physics, materials science, microgravity science, theoretical and applied mechanics, chemical engineering, and applied mathematics.

I appreciate the *Spring Sun* Program launched by the Department of Education of China for the promotion of science and technology in some selected significant fields in China. My visit to Peking University was supported by this program. I thank Prof. Huang, Yong-Nian, Prof. Wei, Qing-Ding, and many staff members of the Laboratory for Studies of Turbulence and Complex Systems for their assistance and efforts which made my stay in the Swallow-Garden campus of Peking University most pleasant and productive.

McGill University Jian-Jun Xu

Chapter 1

INTRODUCTION

1. Interfacial Pattern Formations in Dendritic Growth

Dendritic pattern formation at the interface between liquid and solid is a commonly observed phenomenon in crystal growth and solidification process. Fig. 1.1 shows a snow flake, which describes a typical case of dendrite growth (Kobayashi and Furukawa, 1991, Takahashi et al, 1995).

Dendritic growth appears frequently in many forms of material processing, such as alloy casting, metal ingot formation, and welding. The system of dendritic growth is a heterogeneous, complex, dynamical system, which, in general, involves the interactive, co-existing macroscopic transport processes: mass transport, heat transport, momentum transport and phase transition kinetics at the interface. In general, when a liquid is setting in a meta-stable state with a uniform undercooling temperature, a dynamical process can be initiated by some perturbation, such as introducing a tiny seed. Once the process starts, the crystal growth will proceed spontaneously according to physical laws. During this process, a dendritic pattern may occur at the interface between the liquid and solid states.

Dendritic growth is one of the most profound and fundamental subjects in the area of interfacial pattern formation. This is not only due to its underlying vital technical importance in the material processing industries, but also because dendritic growth represents a fascinating class of nonlinear phenomena occurring in heterogeneous dynamical systems.

Theoretical investigation of dendritic growth at the current stage has mainly focused on single free dendritic growth, disregarding the interactions between neighboring dendrites in more complicated multiple den-

Figure 1.1. Photograph of a typical snowflake pattern. From Furukawa (1995)

drite growth. A typical single free dendritic growth is shown in Fig. 1.2, which is the experimental recording curves for two-dimensional dendritic growth from a supersaturated NH_4Br solution, studied by Dougherty and Gollub (Dougherty and Gollub, 1988).

It was first discovered in experiments by Glicksman and his colleagues that once the material properties and growth condition were given, at a later stage of growth the system would automatically select its dendrite-tip growth velocity and the nature of its microstructure at the interface. These characters of limiting state of dendritic growth are very little affected by the details of initial setting. Thus, very naturally, for a long period of time, theoretical investigations of dendritic growth have focused on the following basic and vital issues:

- What mechanism determines the tip growth velocity?
- What is the origin and essence of the microstructure?

These issues have been at the center of research activities in the broad areas of condensed matter physics and materials science, attracting many researchers from various areas (Hurle, 1993, Langer, 1980, Langer and Müller-Krumbhaar, 1978, Langer, 1992, Langer, 1986, Xu, 1987–Xu, 1997, Kessler et al, 1986, Pelce, 1988). The problems remained unsolved for about a half century. For single free dendritic growth with no convection in the melt, these problems are now resolved by the so-called '*Interfacial Wave* (IFW) Theory' (Xu, 1997, Davis, 2001).

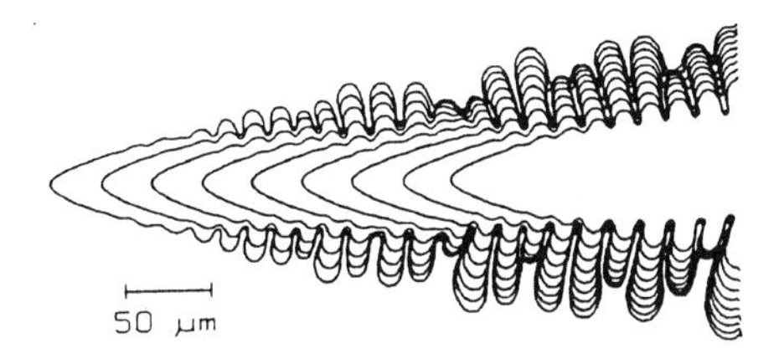

Figure 1.2. Experimental curve of two-dimensional dendritic growth from a supersaturated (NH_4Br) solution. From Dougherty and Gollub (1988)

2. Dendritic Growth Interacting with Convective Flow

It may be presumed that the growth speed of dendrite-tip and the nature of dendritic pattern formation can be affected by external sources, such as fluid flow or an external force field in particular.

Glicksman and Huang first investigated experimentally the effect of buoyancy-driven flow in the melt on dendritic growth (Huang and Glicksman, 1981). They showed that dendrites growing in the gravity field with different directions indeed have different tip velocity and interface shape (see Fig. 1.3). In Fig. 1.4, we show typical dendritic growth in micro-gravity space. It is seen that the without gravity, dendritic growth in its tip region away from the root is rather irrelevant with the growth orientation.

In a realistic dendritic growth system, convection in melt is always unavoidable. Sometimes, it is even artificially generated. The convection generated in the liquid state of a crystal growth system may be summarized in the following categories:

- The convection due to density change during phase transition;
- The convection induced by a force field in the liquid state, such as buoyancy-driven flow in gravity field, or induced by a force field at the interface, such as thermo-capillary flow driven by the gradient of surface tension at the interface;
- The convection induced by external flow.

A realistic system of dendritic growth may contain all of these categories of convection.

The interactive dynamics of convection and solidification is a subject of great interest in the broad areas of materials science and micro-gravity research. It is expected that based on the understanding of the interplay between fluid dynamics and solidification, one might eventually be able to design and develop some advanced new techniques for affecting or controlling the formation of microstructure in material processing in order to improve the electrical or mechanical properties of final materials. In the literature, there is a large number of works contributed to this subject. Lee and his colleagues (Lee, 1991, Lee et al, 1992) conducted experiments with controlled forced convection in the presence of dendritic growth of pure SCN. They also repeated the dendritic growth experiments with natural convection that were first performed by Huang and Glicksman (Huang and Glicksman, 1981). The later development of the investigations on this subject was reviewed by Lee (Lee et al, 1996). Bouissou (Bouissou et al, 1989, Bouissou and Pelce, 1989, Bouissou et al, 1990) performed a series of experiments on influence of an external flow on dendritic growth by using PVA. Emsellem and Tabeling (1995, 1996) performed experiments on dendritic growth from NH_4Br solution when an external flow is imposed. In addition to the interface morphology of dendrite, they measured the concentration around the crystal during its growth.

Dendritic growth with the effect of convection induced by density change was first investigated analytically by McFadden and Coriell (1986). It was also studied by Pines et al (1996) and Xu (1994b) later on. The analytical work on the effect of buoyancy-driven convection on dendritic growth was first given by Canright and Davis (1991). It was also later studied by Sekerka et al (1995, 1996). The effects of uniform external flow on dendritic growth were investigated with different analytical approaches by Ben Amar, Bouillou and Pelce (Amar et al, 1988, Bouissou-89a, Bouissou and Pelce, 1989, Bouissou et al, 1990); Xu and Yu (Xu, 1994a, Yu and Xu, 1999); Lee et al, (Lee, 1991, Lee et al, 1992); Koo et al (1992), Ananth et al (Ananth and Gill, 1989, 1991). Saville and Beaghton (1988) also studied the same problem numerically.

Besides the above investigators, one can find a much longer list of authors who have contributed to this subject.

The above theoretical works are mostly restricted to steady growth with zero surface tension. Without taking surface tension into account, the works so far published have not been able to yield the solution for selection of tip velocity of dendritic growth with convective flow. Nevertheless, studies of the steady dendritic growth with zero surface tension are a quite important initial step, which provides the necessary basis for any further investigations on this subject.

The present monograph is meant to introduce this exciting, interdisciplinary subject to a wide range of readers in the broad fields of theoretical

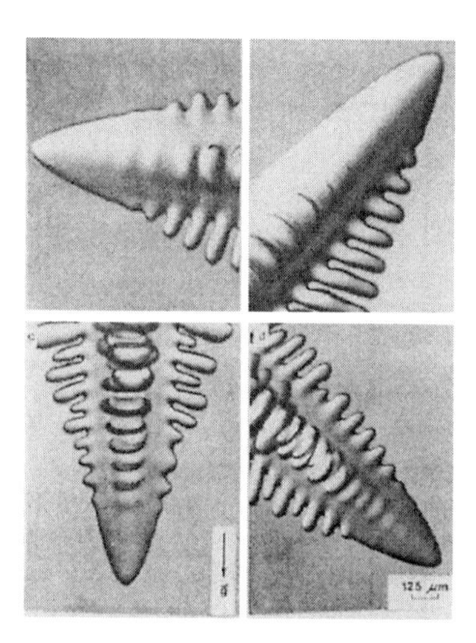

Figure 1.3. Photograph of dendritic growth on ground with different orientations. From Huang and Glicksman (1981)

and applied fluid mechanics and applied mathematics. It shall focus on the topics of a single free dendritic growth from pure melt interacting with convective flow in the liquid phase. In this book, I do not attempt to give a comprehensive review of all previous references, which were all valuable on the road leading to the complete solution of the problems. Instead, I attempt to review and comment on the major analytical works on this subject. In particular, I intend to summarize the results obtained for this subject, in terms of a unified systematic asymptotic approach in the framework of the interfacial wave (IFW) theory.

3. Mathematical Formulation of the General Problem

We shall adopt the macroscopic continuum model and study the behavior of a single free dendritic growth at later stage of growth. This implies that the liquid and solid bulk phases will be treated as continuous media, while the interface is considered as a geometric surface. This continuum model is well applicable to the pattern formation phenomena under investigation.

It is known that at later stage of growth, the average speed of the dendrite-tip is a constant U. Hence, it is appropriate to observe the phenomenon in the moving reference frame fixed at the average location of the growing dendrite-tip. We assume that the liquid phase has the mass density ρ, thermal diffusivity κ_{T}, specific heat c_{p}; while the corresponding thermal characteristics in the solid phase are denoted by ρ_{S}, κ_{TS} and c_{pS}, respectively. We denote the ratio of thermal diffusivities in the solid and liquid by $\alpha_{\mathrm{T}} = \kappa_{\mathrm{T}}/\kappa_{\mathrm{TS}}$. Conventionally, the simplification with $\alpha_{\mathrm{T}} = 1$ is called the symmetric model; while the simplification with $\alpha_{\mathrm{T}} = 0$ is called the one-sided model. In what follows, we shall restrict ourself to the symmetric model.

3.1 Scaling

We choose the growth speed of dendrite-tip, U, as the scale for the velocity. The length scale can be chosen from the intrinsic length scales of the system. There are two intrinsic length scales in the system determined by the macroscopic transport processes: the thermal diffusion length $\ell_{\mathrm{T}} = \kappa_{\mathrm{T}}/U$ and the viscous diffusion length $\ell_{\mathrm{d}} = \nu/U$, where ν is kinematic viscosity of fluid. We shall use the thermal diffusion length ℓ_{T} as the length scale for the growth problem and, accordingly, use ℓ_{T}/U as the time scale. Furthermore, we use $\Delta H/c_{\mathrm{p}}\rho$ as the scale of the temperature, where ΔH is the latent heat of per unit of volume of solid, and define the dimensionless temperature as $T = [T - T_{\mathrm{M0}}]c_{\mathrm{p}}\rho/\Delta H$, where T_{M0} is the melting temperature at a flat interface.

Figure 1.4. Photograph of dendritic growth in micro-gravity space. From Glicksman et al. (1994)

3.2 Macroscopic Transport Equations

With the above scales, in the coordinate system moving with the tip-velocity $\mathbf{U} = U\mathbf{e}_3$, the state of the system is described by the following dimensionless macroscopic fields:

- the temperature field in the liquid phase, $\bar{T}(\mathbf{r}, t)$,
- the temperature field in the solid phase, $\bar{T}_S(\mathbf{r}, t)$,
- the relative velocity field in the liquid phase, $\bar{\mathbf{u}}(\mathbf{r}, t)$.

Here, we use $\mathbf{r}$ and t to denote the dimensionless space vector and time, respectively. We assume that the liquid can be considered as an incompressible Newtonian fluid. So, the governing equations consist of the heat conduction equation and the Navier–Stokes equations. Moreover, for the phenomena under investigation, as the inhomogeneity of density caused by the inhomogeneity of temperature is not important except in terms of a buoyancy effect, the Boussinesq approximation is applicable.

The vector form of the dimensionless governing equations for the dimensionless physical quantities, which are denoted with a bar on the top, can be written as follows:

1. Heat conduction equation in the liquid phase:

$$\left(\frac{\partial \bar{T}}{\partial \bar{t}} + \bar{\mathbf{u}} \cdot \nabla \bar{T}\right) = \nabla^2 \bar{T}. \tag{1.1}$$

2. Heat conduction equation in the solid phase:

$$\left(\frac{\partial \bar{T}_S}{\partial \bar{t}} - \hat{\mathbf{e}}_3 \cdot \nabla \bar{T}_S\right) = \nabla^2 \bar{T}_S. \tag{1.2}$$

3. Continuity equation for the liquid state:

$$\nabla \cdot \bar{\mathbf{u}} = 0. \tag{1.3}$$

4. Momentum equation for the liquid state:

$$\frac{\partial \bar{\omega}}{\partial \bar{t}} + (\bar{\mathbf{u}} \cdot \nabla)\bar{\omega} - (\bar{\omega} \cdot \nabla)\bar{\mathbf{u}} = \mathrm{Pr}\nabla^2 \bar{\omega} + \frac{\mathrm{Gr}}{\mathrm{Pr}|T_\infty|}\nabla \times (\bar{T}\mathbf{e}_g). \tag{1.4}$$

In the above, $\mathbf{e}_g$ is the unit vector along the direction of gravity and we introduce several parameters: $\mathrm{Pr} = \nu/\kappa_T$, the Prandtl number; $T_\infty = -[T_{M0} - (T_\infty)_D]c_P\rho/\Delta H = -\mathrm{St}$, the undercooling parameter, where $(T_\infty)_D$ denotes the dimensional undercooling temperature; the parameter St is sometimes called the Stefan number; $\mathrm{Gr} = g\beta\nu|T_{M0} - (T_\infty)_D|/U^3$ is the Grashof number, where g is the gravitational constant, $\beta = -\rho^{-1}\partial\rho/\partial T|_{T_\infty}$ is the thermal expansion coefficient.

3.3 Interface Conditions

The dimensionless conditions at the interface $\bar{S}(\mathbf{r}, t) = 0$ are:

(i) *Thermodynamic equilibrium for temperature:*

$$\bar{T} = \bar{T}_{\mathrm{S}}\,; \tag{1.5}$$

(ii) *Gibbs–Thomson condition:*

$$\bar{T}_{\mathrm{S}} = \Gamma\{\bar{\mathcal{K}}\}; \tag{1.6}$$

where Γ is the surface tension parameter

$$\Gamma = \frac{\ell_{\mathrm{c}}}{\ell_{\mathrm{T}}} \tag{1.7}$$

and

$$\ell_{\mathrm{c}} = \frac{\gamma c_{\mathrm{p}} T_{\mathrm{M0}} \rho}{(\Delta H)^2} \tag{1.8}$$

is a length scale and usually called the capillary length.

Since ℓ_{c} is determined by the interfacial energy γ, it is sometimes considered as a microscopic length scale. It will be seen later that the parameter Γ, the ratio of the macroscopic length ℓ_{D} and the microscopic length ℓ_{c}, is the most important parameter for the stability of the system.

(iii) *Enthalpy conservation:*

$$(\bar{u}_{\mathrm{I}} + \hat{\mathbf{e}}_3 \cdot \mathbf{n}) = \left[(1+\alpha)\nabla\bar{T}_{\mathrm{S}} - \nabla\bar{T}\right] \cdot \mathbf{n}\,. \tag{1.9}$$

Hereby, we have neglected the variation of the interfacial energy due to stretching or shrinking of the interface and defined

$$\alpha = \frac{\rho_{\mathrm{S}}}{\rho} - 1, \tag{1.10}$$

which measures the density change in the phase transition.

(v) *Conservation of mass:*

$$\bar{\mathbf{u}} \cdot \mathbf{n} + \alpha\bar{u}_{\mathrm{I}} + (1+\alpha)\hat{\mathbf{e}}_3 \cdot \mathbf{n} = 0\,, \tag{1.11}$$

(vi) *Continuity of the tangential component of velocity:*

$$\bar{\mathbf{u}} \cdot \mathbf{e}_{\tau} = -\hat{\mathbf{e}}_3 \cdot \mathbf{e}_{\tau}\,. \tag{1.12}$$

Later, for the sake of simplicity, I shall omit the bar over all dimensionless quantities, as there is no chance of confusion.

To explore dendritic growth with interaction of convection in the liquid state step by step, we must study the following subjects one by one:

- dendritic growth without convection;
- viscous flow past a dendrite with zero growth speed;
- dendritic growth with non-zero speed in a certain type of convection flow.

This book is arranged as follows. Chap. 2 summarizes the results of dendritic growth without convection in the liquid phase. Chap. 3 gives the general mathematical formulation for dendritic growth with convection. Chap. 4 studies viscous flow past a paraboloid. Starting with Chap. 5, we shall investigate steady dendritic growth in various types of convective flow with zero surface tension. In Chap. 5–6, we study dendritic growth in a uniform external flow. In Chap. 7–8, we investigate dendritic growth with buoyancy-driven flow. In Chap. 9 we investigate dendritic growth with the inclusion of interfacial energy. We shall explore the instability of steady solutions and yield the resolution for the problem of limiting state selection of dendritic growth with each type of convection. Finally, in Chap. 10, we give brief concluding remarks.

Chapter 2

INTERFACIAL WAVE THEORY OF DENDRITIC GROWTH FROM PURE MELT WITH NO CONVECTION

Without convection, dendritic growth only involves the thermodynamic process, and is governed just by the heat conduction equation. This subject has been fully explored in the monograph (Xu, 1997). For the sake of self-containment and convenience of the readers, in this chapter we shall give a fairly detailed discussion for this topic with some re-fined derivations.

We adopt the paraboloidal coordinate system (ξ, η, θ), which can be defined through the cylindrical coordinate system (r, θ, z) by (see

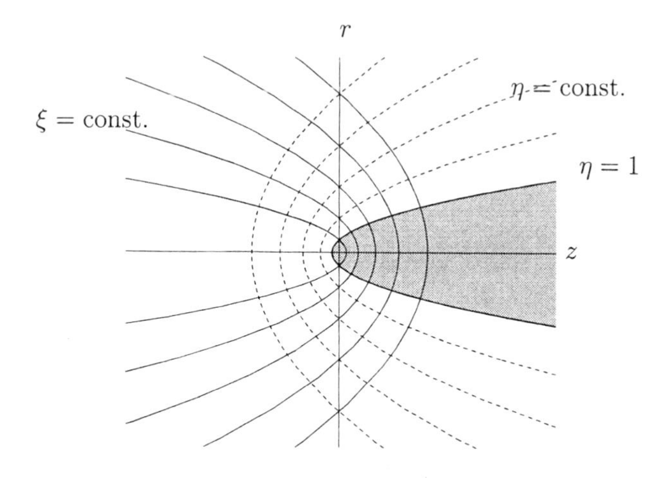

Figure 2.1. The paraboloidal coordinate system (ξ, η, θ) for three-dimensional dendritic growth

Fig. 2.1):

$$\begin{cases} \dfrac{r}{\eta_0^2} = \xi\eta, \\ \dfrac{z}{\eta_0^2} = \dfrac{1}{2}(\xi^2 - \eta^2), \end{cases} \tag{2.1}$$

where the constant η_0^2 is to be determined by setting the location of the dendrite-tip. The dynamical system under investigation is invariant under coordinate translation. One has the freedom to choose the origin of the paraboloidal system (2.1). The constant η_0^2 in (2.1) can therefore be chosen so that the steady interface shape satisfies

$$\eta_s(0) = 1\,. \tag{2.2}$$

It will be seen later that, for a system without convection, the normalization parameter η_0^2 is just the Peclet number of the system with zero surface tension.

In the paraboloidal system, the vector governing equations given in the last chapter are transformed into the following forms:

$$\frac{\partial^2 T}{\partial \xi^2} + \frac{\partial^2 T}{\partial \eta^2} + \frac{1}{\xi}\frac{\partial T}{\partial \xi} + \frac{1}{\eta}\frac{\partial T}{\partial \eta} + \Big(\frac{1}{\xi^2} + \frac{1}{\eta^2}\Big)\frac{\partial^2 T}{\partial \theta^2}$$
$$= \eta_0^2\Big(\xi\frac{\partial T}{\partial \xi} - \eta\frac{\partial T}{\partial \eta}\Big) + \eta_0^4(\xi^2 + \eta^2)\frac{\partial T}{\partial t}\,. \tag{2.3}$$

The boundary conditions are:

1. The up-stream far-field condition:

$$T \to T_\infty = \frac{(T_\infty)_D - T_{\mathrm{M0}}}{\Delta H/(c_{\mathrm{p}}\rho)} < 0 \quad \text{as } \eta \to \infty\,. \tag{2.4}$$

2. The regularity condition:

$$T_{\mathrm{S}} = O(1) \quad \text{as } \eta \to 0\,. \tag{2.5}$$

3. The interface conditions, at $\eta = \eta_s(\xi, t)$:

   (i) the thermodynamic equilibrium condition:

$$T = T_{\mathrm{S}}\,, \tag{2.6}$$

(ii) the Gibbs–Thomson condition with isotropic surface tension:

$$T = -\varepsilon^2 \eta_0^2 \mathcal{K}_3 \left\{ \frac{\mathrm{d}}{\mathrm{d}\xi}, \frac{\mathrm{d}^2}{\mathrm{d}\xi^2}; \frac{\mathrm{d}}{\mathrm{d}\theta}, \frac{\mathrm{d}^2}{\mathrm{d}\theta^2} \right\} \eta_s, \tag{2.7}$$

where the twice mean curvature operator,

$$\mathcal{K}_3 \left\{ \frac{\mathrm{d}}{\mathrm{d}\xi}, \frac{\mathrm{d}^2}{\mathrm{d}\xi^2}; \frac{\mathrm{d}}{\mathrm{d}\theta}, \frac{\mathrm{d}^2}{\mathrm{d}\theta^2} \right\} \eta_s = \frac{1}{\xi^2 \eta_s^2 (\xi^2 + \eta_s^2)^{\frac{3}{2}} (1 + \eta_s'^2 + \eta_{s,\theta}^2)^{\frac{3}{2}}}$$

$$\times \Big\{ (1 + \eta_s'^2)(\xi^2 + \eta_s^2) \Big[-(\xi^2 + \eta_s^2)\eta_{s,\theta\theta} - (\xi\eta_s)'\eta_{s,\theta}^2$$

$$+ 2\xi\eta_{s,\theta} - \xi\eta_s^2\eta_s' + \xi^2\eta_s \Big] - 2(\xi^2 + \eta_s^2)\eta_s'\eta_{s,\theta}(\xi\eta_{s,\theta})'(\eta_s\eta_s' - \xi)$$

$$+ \Big[\xi^2\eta_s^2 + (\xi^2 + \eta_s^2)\eta_{s,\theta}^2 \Big]$$

$$\times \Big[(\eta_s\eta_s' - \xi)(\xi\eta_s)'' + (1 - \eta_s'^2 - \eta_s\eta_s'')(\xi\eta_s)' \Big] \Big\}. \tag{2.8}$$

Here, we introduce the parameter

$$\varepsilon = \frac{\sqrt{\Gamma}}{\eta_0^2}, \tag{2.9}$$

which is called the interfacial stability parameter of isotropic surface tension. We recall that Γ is previously defined and called the surface tension parameter.

(iii) the heat balance condition:

$$\left(\frac{\partial}{\partial \eta} - \eta_s' \frac{\partial}{\partial \xi} \right)(T - T_S) + \eta_{s\theta} \left(\frac{1}{\xi^2} + \frac{1}{\eta_s^2} \right) \frac{\partial}{\partial \theta}(T - T_S)$$
$$+ \eta_0^4 (\xi^2 + \eta_s^2) \frac{\partial \eta_s}{\partial t} + \eta_0^2 (\xi\eta_s)' = 0, \tag{2.10}$$

which, for the axi-symmetric case, is reduced to:

$$\left(\frac{\partial}{\partial \eta} - \eta_s' \frac{\partial}{\partial \xi} \right)(T - T_S) + \eta_0^4 (\xi^2 + \eta^2) \frac{\partial \eta_s}{\partial t} + \eta_0^2 (\xi\eta_s)' = 0, \tag{2.11}$$

Until now, the mathematical formulation remains incomplete. A complete mathematical formulation should also include the boundary conditions that describe the behavior of the solution at the tip of the dendrite

$\xi = 0$, as well as at its root, $\xi = L \gg 1$. In addition to the above boundary conditions, one may also need to impose some initial conditions.

A complete mathematical formulation of the pattern formation problem at late stage of evolution will be given later.

1. Steady State of Dendritic Growth with Zero Surface Tension — Ivantsov's Solution

It is well known that for the case with zero surface tension ($\varepsilon = 0$), arbitrary undercooling, the three-dimensional system (2.3)–(2.11) allows the following steady, axi-symmetric, similarity solution

$$\begin{aligned} T &= T_*(\eta) = T_\infty + \frac{\eta_0^2}{2} e^{\frac{\eta_0^2}{2}} E_1\Big(\frac{\eta_0^2 \eta^2}{2}\Big), \\ T_S &= T_{S*} = T_*(1) = 0, \\ \eta_* &= 1, \\ T_\infty &= -\frac{\eta_0^2}{2} e^{\frac{\eta_0^2}{2}} E_1\Big(\frac{\eta_0^2}{2}\Big), \\ & \quad (0 \leq \xi < \infty), \end{aligned} \tag{2.12}$$

where $E_1(x)$ is the exponential function defined as

$$E_1(x) = \int_x^\infty \frac{e^{-t}}{t} dt \tag{2.13}$$

(see Abramovitz and Stegun, 1964). This solution was first found by Ivantsov in 1946 (Ivantsov, 1947 and Horvay and Cahn, 1961) and is now called the Ivantsov solution.

In the above, the constant η_0^2 is uniquely determined as a function of the undercooling T_∞. The radius of curvature of the parabolic interface $\eta_* = 1$ at the tip $\xi = 0$ is calculated as

$$\ell_t = \eta_0^2 \ell_T. \tag{2.14}$$

One may define the Peclet number as the ratio of the tip radius to the thermal diffusion length, i.e., $Pe = \ell_t/\ell_T$. Obviously, the Peclet number in general is a function of ε as well as T_∞, i.e., $Pe = Pe(\varepsilon, T_\infty)$. Equation (2.14) shows that η_0^2 for the Ivantsov problem is actually the Peclet number with zero surface tension, i.e., $\eta_0^2 = Pe_0 = Pe(0, T_\infty)$. Returning to the dimensional tip radius ℓ_t, formula (2.14) shows

$$\ell_t U = (Pe_0)\kappa_T. \tag{2.15}$$

The Ivantsov solution describes a steadily growing, axi-symmetric, smooth needle crystal, whose interface has no microstructure. Moreover, as a similarity solution, it represents a continuous family of dimensional, physical solutions, with arbitrary tip velocity for given growth conditions and material properties. However, experimental observations show that the interface of growing dendrite always exhibits micro-structure (see Fig.1.2) and, for given material and growth condition, in the limit $t \to \infty$ the dendrite-tip velocity always approaches a certain constant, which appears to be unaffected by the details of initial settings of growth (Glicksman et al, 1975, 1976).

The experimental results raised the questions:

- What is the mechanism which determines the tip growth velocity? From the dynamic point of view, this problem can be translated to: what is the selection criterion for the limiting state of dendritic growth as $t \to \infty$?

- What is the origin and essence of the microstructure?

These problems are fundamental subjects in the field of condensed matter physics and materials science and have been at the center of broad theoretical and experimental research activities during the past decades.

It is noted that although the Ivantsov solution with zero surface tension provides no information on the selection of dendrite-top velocity, it does describe the shape of the dendrite tip with a high accuracy, once the tip velocity is determined correctly through experiments. Therefore, the Ivantsov solution has been one of the most significant results on this subject. It provides an important background to all further research work.

The key to resolving the above-imposed problems of the selection of dendrite-tip velocity and microstructure formation is understanding the role of surface tension at the interface. The surface tension at the interface between the solid and liquid phases is usually a very small quantity. However, it is precisely this extremely small quantity that plays a vital role in interfacial pattern formation phenomenon. It is now realized that the above two problems are related to each other. To resolve them, one needs to study the **basic state** of the system with nonzero surface tension and its linear stability.

2. The Basic State for Dendritic Growth with Nonzero Surface Tension

To study the long-term behavior of a dynamic system, one needs to first find the basic state of the system, then perform the stability analysis for the basic state. The basic state of a dynamical system, for some cases, may be given by a fully time-independent, steady solution; while for other cases, it may be given by a nearly steady solution with a slow time-dependency. How to specify the basic state in dendritic growth with nonzero surface tension is an important subject, which must be approached with great caution.

In the literature, many researchers attempted to find a classic, steady needle solution for the basic state of dendritic growth. Such efforts were not successful for the case of isotropic surface tension. The so-called classic, steady needle crystal solution, like the Ivantsov solution, is fully time-independent, with both a smooth tip and an infinitely long smooth, non-oscillating tail. Given the fact that the system with $\varepsilon = 0$ allows a steady needle solution, what about nonzero surface tension? The question of whether or not the system with $\varepsilon \neq 0$ still allows a steady needle crystal solution is not trivial. It involves the subtle mathematical issue of how to catch the exponentially small terms missed by the regular asymptotic expansion. This issue is sometimes called *asymptotics beyond all orders* (Kruskal and Segur, 1991).

At an early stage of research on dendritic growth, most researchers thought that when a small isotropic surface tension is included, the steady needle solution would still persist with a small perturbation from the Ivantsov needle solution in the whole infinite region. Furthermore, at infinity, since the curvature of the Ivantsov needle tends to zero, most researchers thought that the surface tension effect would vanish. Hence, the steady needle solution with nonzero surface tension should approach the Ivantsov needle. It was with this idea that Nash and Glicksman formulated the needle crystal growth problem (Nash and Glicksman, 1974). Their problem formulation was then adopted by many researchers in the field without question. It is now recognized that the above ideas are incorrect. In fact, out of the expectation of these researchers, it was found that with any small amount of surface tension, the system only allows a steady solution as a small perturbation in a **finite region** behind the tip. It does not allow any classic needle solution in the whole infinite region.

It was even more unexpected that the effect of surface tension is more important in the root region of the needle than in the tip region! With these results, one can determine two types of so-called non-classic needle solutions of dendritic growth:

- The steady solution with a long, but finite dendrite-stem. Namely, dendrite has a root with a fixed, large coordinate $\xi = \xi_{\max} = O(1/\varepsilon)$, and the steady conditions at the root.
- The nearly steady solution with a growing, finite long stem. In this case, the length of dendrite-stem is continuously growing, and the root conditions may be changing with time.

The exact forms of these non-classic needle solutions, of course, will be different, if the detailed root conditions are specified differently. However, when the material properties and other growth conditions are fixed, the difference between all these solutions in a finite region behind the tip is exponentially small, as the stability parameter $\varepsilon \to 0$. Thus, these non-classic needle solutions are actually indistinguishable in the experiments in the region not close to the root. Moreover, for a given fixed ξ, η, as $\varepsilon \to 0$, the above two types of solution have the same regular perturbation expansion, which for the case of axi-symmetric dendritic growth can be written in the form

$$\begin{aligned} T &= T(\xi,\eta) = T_0(\eta) + \varepsilon^2\eta_0^2 T_1(\xi,\eta) + \cdots, \\ T_S &= T_S(\xi,\eta) = T_{S0} + \varepsilon^2\eta_0^2 T_{S1}(\xi,\eta) + \cdots, \\ \eta_s(\xi) &= 1 + \varepsilon^2\eta_1(\xi) + \cdots . \end{aligned} \tag{2.16}$$

To understand the selection mechanism and essence of pattern formation, we are mostly interested in the dynamics of dendritic growth in the region away from the root. Since in this region all the above defined non-classic needle solutions are actually indistinguishable, we may treat them as one solution and define it as the basic state of dendritic growth.

In the this monograph, we only consider axi-symmetric, basic state and express it as $\{T_B(\xi,\eta), T_{BS}(\xi,\eta), \eta_B(\xi)\}$.

3. Regular Perturbation Expansion of Axi-symmetric, Basic State of Dendritic Growth

The interfacial stability parameter ε is, in practice, very small. Its numerical magnitude is about ≈ 0.1–0.2. Thus, in order to examine the effect of the isotropic surface tension, it is very natural to consider a regular perturbation expansion (RPE) of the basic state (2.16) in the limit $\varepsilon \to 0$ (Xu and Yu, 1998, Xu, 1990a).

We substitute (2.16) into the system (2.3)–(2.11) and equate coefficients of like powers of ε to zero, then we can derive the approximations at each order of ε.

3.1 $O(\varepsilon^0)$

The zeroth-order approximation solution is the Ivantsov solution.

3.2 $O(\varepsilon^2)$

In the first-order approximation, we derive

$$L\{T_1\} = \left\{ \frac{\partial^2}{\partial \xi^2} + \frac{\partial^2}{\partial \eta^2} + \left(\frac{1}{\xi} - \eta_0^2 \xi\right)\frac{\partial}{\partial \xi} + \left(\frac{1}{\eta} + \eta_0^2 \eta\right)\frac{\partial}{\partial \eta} \right\} T_1 = 0, \quad (2.17)$$

with the boundary conditions:

1. As $\eta \to \infty$,

$$T_1 \to 0 \quad \text{(exponentially)}. \quad (2.18)$$

2. As $\xi \to \infty$,

$$T_1 \to 0 \quad \text{(algebraically)}. \quad (2.19)$$

3. As $\eta \to 0$,

$$T_{S1} \qquad \text{regular}. \quad (2.20)$$

4. At $\eta = 1$,

$$T_1 = T_{S1} + \eta_1 \,, \quad (2.21)$$

$$T_{S1} = -\mathcal{K}_0(\xi) = -\frac{\xi^2 + 2}{(1+\xi^2)^{\frac{3}{2}}} \,, \quad (2.22)$$

$$\frac{\partial}{\partial \eta}(T_1 - T_{S1}) + (2 + \eta_0^2)\eta_1 + \xi \frac{d\eta_1}{d\xi} = 0 \,. \quad (2.23)$$

5. The tip-regularity condition, at $\xi = 0$,

$$\eta_1'(0) = 0 \quad (2.24)$$

and

$$\eta_1(0) = 0. \quad (2.25)$$

By the method of separation of variables, it is derived that the general solution for the temperature in the solid phase is

$$T_{S1}(\xi,\eta) = \sum_{n=0}^{\infty} \frac{\alpha_n}{L_n\left(-\frac{\eta_0^2}{2}\right)} L_n\left(\frac{\eta_0^2\xi^2}{2}\right) L_n\left(-\frac{\eta_0^2\eta^2}{2}\right), \tag{2.26}$$

while the general solution for the temperature in the liquid is

$$T_1(\xi,\eta) = \sum_{n=0}^{\infty} \beta_n L_n\left(\frac{\eta_0^2\xi^2}{2}\right) \frac{e^{-\frac{\eta_0^2\eta^2}{2}} U(n+1,1,\frac{\eta_0^2\eta^2}{2})}{e^{-\frac{\eta_0^2}{2}} U(n+1,1,\frac{\eta_0^2}{2})}. \tag{2.27}$$

Furthermore, one can expand the function $\eta_1(\xi)$ in the Laguerre series

$$\eta_1(\xi) = \sum_{n=0}^{\infty} \gamma_n L_n\left(\frac{\eta_0^2\xi^2}{2}\right). \tag{2.28}$$

Thus the problem is to determine the coefficients $\{\alpha_n, \beta_n, \gamma_n, (n = 0, 1, 2, \ldots)\}$ such that the boundary conditions (2.21)–(2.23) can be satisfied. In what follows, we shall give the analytical forms for these coefficients.

(1) It follows from the boundary condition (2.21) that

$$\beta_n = \alpha_n + \gamma_n. \tag{2.29}$$

(2) The boundary condition (2.22) gives

$$\sum_{n=0}^{\infty} \alpha_n L_n\left(\frac{\eta_0^2\xi^2}{2}\right) = -\mathcal{K}_0(\xi) = -\left[\frac{1}{(1+\xi^2)^{\frac{1}{2}}} + \frac{1}{(1+\xi^2)^{\frac{3}{2}}}\right]. \tag{2.30}$$

Due to the orthogonality of the functions $L_n(x)$ $(n = 0, 1, 2, \ldots)$, the coefficients α_n in the above expansion can be determined from the integral

$$\alpha_n = -\int_0^{\infty} e^{-x} L_n(x) \mathcal{K}_0(\sqrt{2x}/\eta_0) dx. \tag{2.31}$$

(3) We are now going to apply the boundary condition (2.23). From (2.26) and (2.27), one can write

$$\frac{\partial T_{S1}}{\partial \eta}(\xi, 1) = \sum_{n=0}^{\infty} b_n L_n\left(\frac{\eta_0^2\xi^2}{2}\right) \tag{2.32}$$

and

$$\frac{\partial T_1}{\partial \eta}(\xi, 1) = \sum_{n=0}^{\infty} a_n L_n\left(\frac{\eta_0^2\xi^2}{2}\right), \tag{2.33}$$

where

$$a_n = 2\beta_n A_n \,, \qquad b_n = 2n\alpha_n B_n \,, \tag{2.34}$$

$$\begin{cases} A_n = (n+1)^2 \dfrac{U(n+2,1,\frac{\eta_0^2}{2})}{U(n+1,1,\frac{\eta_0^2}{2})} - (n+1) - \dfrac{\eta_0^2}{2}, \\ B_n = 1 - \dfrac{L_{n-1}(\frac{-\eta_0^2}{2})}{L_n(\frac{-\eta_0^2}{2})}. \end{cases} \tag{2.35}$$

Thus, from the boundary condition (2.23), we have

$$\sum_{n=0}^{\infty}(a_n - b_n)L_n\Big(\frac{\eta_0^2\xi^2}{2}\Big) + (2+\eta_0^2)\sum_{n=0}^{\infty}\gamma_n L_n\Big(\frac{\eta_0^2\xi^2}{2}\Big)$$
$$+2\sum_{n=0}^{\infty} n\gamma_n\Big\{L_n\Big(\frac{\eta_0^2\xi^2}{2}\Big) - L_{n-1}\Big(\frac{\eta_0^2\xi^2}{2}\Big)\Big\} = 0 \tag{2.36}$$

and

$$(a_n - b_n) + (2+\eta_0^2+2n)\gamma_n - 2(n+1)\gamma_{n+1} = 0, \tag{2.37}$$
$$(n = 0,1,2,3,\ldots).$$

From (2.34) and (2.37), we find that

$$\gamma_{n+1} = g_n\gamma_n + f_n\alpha_n, \quad (n = 0,1,2\ldots), \tag{2.38}$$

where

$$g_n = (n+1)\frac{U(n+2,1,\frac{\eta_0^2}{2})}{U(n+1,1,\frac{\eta_0^2}{2})}, \tag{2.39}$$

$$f_n = \left\{(n+1)\frac{U(n+2,1,\frac{\eta_0^2}{2})}{U(n+1,1,\frac{\eta_0^2}{2})} - 1 - \frac{\eta_0^2}{2(n+1)}\right\} - \frac{nB_n}{(n+1)\eta_0^2}$$
$$= \left\{\frac{1-\frac{\eta_0^2}{2}}{n+1} - 2 + \frac{n}{n+1}\frac{L_{n-1}(-\frac{\eta_0^2}{2})}{L_n\Big(-\frac{\eta_0^2}{2}\Big)} + g_n\right\}. \tag{2.40}$$

For any given γ_0, the recurrence formula (2.38) allows us to generate the series

$$\{\gamma_0, \gamma_1, \gamma_2, \cdots \gamma_n \cdots\}.$$

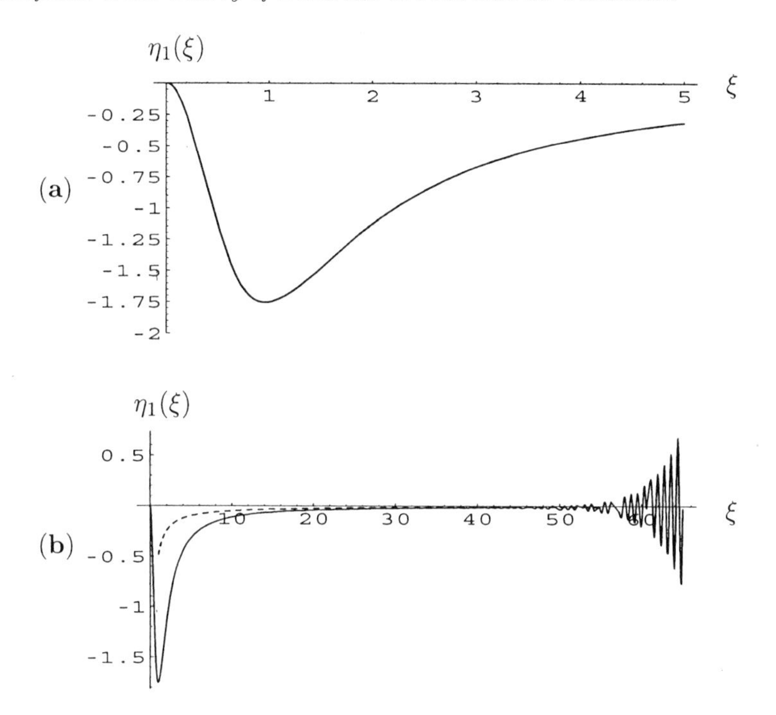

Figure 2.2. The solution $\eta_1(\xi)$ for $\frac{\eta_0^2}{2} = 0.005$: (**a**) in the region of $0 \leq \xi < 5$; (**b**) in the region of $0 \leq \xi < 65$. The dashed line represents the asymptotic solution in the far field, as $\xi \to \infty$

Thus, the function $\eta_1(\gamma_0, \xi)$ can be evaluated as

$$\eta_1(\gamma_0, \xi) = \sum_{n=0}^{\infty} \gamma_n L_n\Big(\frac{\xi^2}{2}\Big). \tag{2.41}$$

The tip-regularity condition (2.24) is automatically satisfied. The value of γ_0 is then uniquely determined by the tip condition (2.25),

$$\eta_1(\gamma_0, 0) = \sum_{n=0}^{\infty} \gamma_n = 0. \tag{2.42}$$

In the above, through (2.29), (2.30), and (2.38)–(2.42), we have obtained an analytical solution for the problem.

3.3 The Asymptotic Behavior of the Regular Perturbation Expansion Solution as $\xi \to \infty$

The Laguerre polynomial expansion for $\eta_1(\xi)$ obtained in the last section, (2.41) is valid for $0 \leq \xi < \infty$. However, numerically, it can only

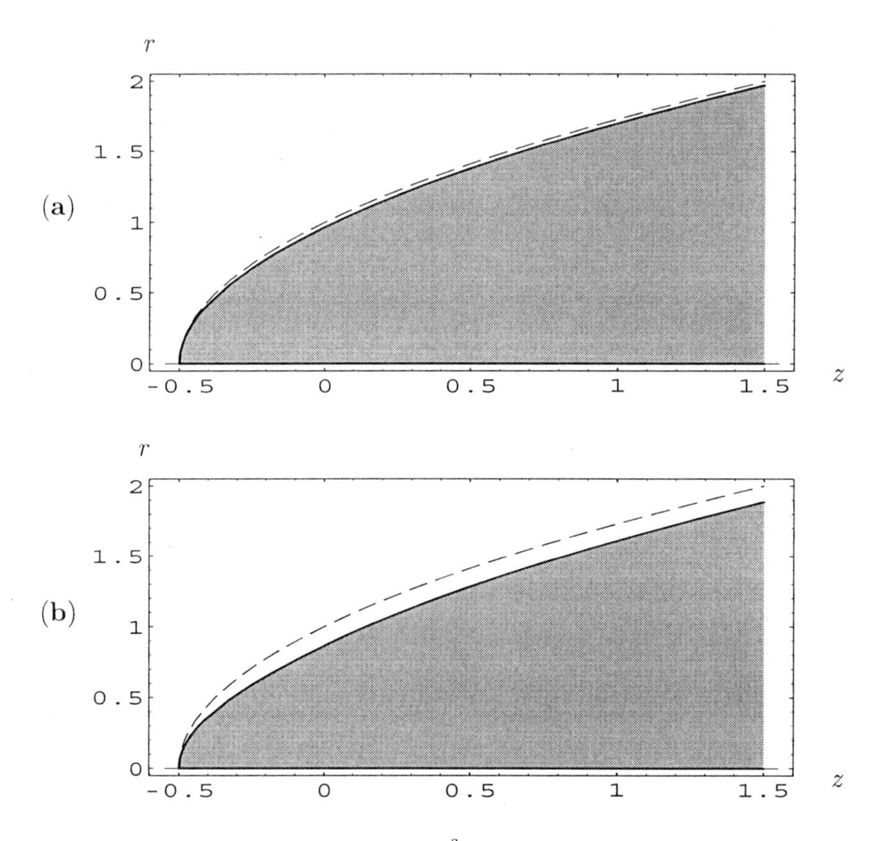

Figure 2.3. The shape of dendrite with $\frac{\eta_0^2}{2} = 0.005$: (**a**) for $\varepsilon = 0.1$; (**b**) for $\varepsilon = 0.2$

be used in the region near the tip ($\xi \leq \xi_{\max}$). Its partial summation with any large number of terms, N, starts to rapidly oscillate when $\xi \gg 1$, as the sum becomes dominated by the last term $\gamma_N L_N(\xi^2/2)$ (see Fig. 2.2). To describe the asymptotic behavior of the solution in the limit $\xi \to \infty$, we expand the solution, which is an analytic function at $\xi = \infty$, in the Taylor series

$$\begin{aligned}
T_1(\xi,\eta) &= \frac{A_1(\eta)}{\xi} + \frac{A_2(\eta)}{\xi^2} + \cdots, \\
T_{S1}(\xi,\eta) &= \frac{A_{S1}(\eta)}{\xi} + \frac{A_{S2}(\eta)}{\xi^2} + \cdots, \\
\eta_1(\xi) &= \frac{C_1}{\xi} + \frac{C_2}{\xi^2} + \cdots .
\end{aligned} \tag{2.43}$$

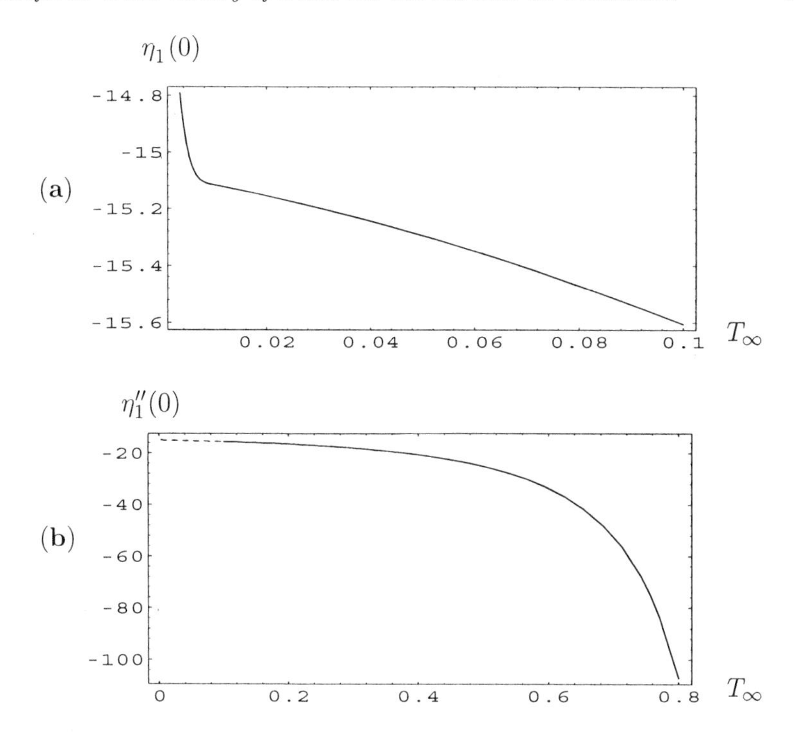

Figure 2.4. The variations of $\eta_1''(0)$ with the parameters T_∞: (**a**) for the small undercooling temperature regime $0 < T_\infty < 0.1$; (**b**) for the large undercooling temperature regime $0.1 < T_\infty < 0.8$

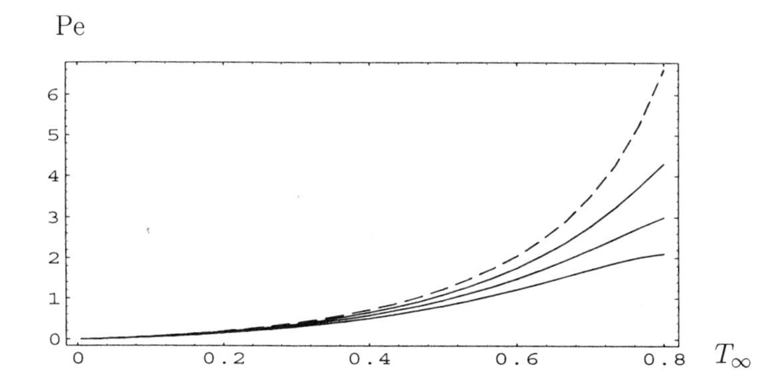

Figure 2.5. The variations of the Peclet number $\mathrm{Pe} = \ell_t/\ell_T$ with the parameters T_∞ and ε for the cases: $\varepsilon = 0.0, 0.1, 0.15$ and 0.2 from top to bottom. The dashed line is the Ivantsov solution with $\varepsilon = 0$

The first terms in the Taylor series are subject to the system

$$\left\{\frac{d^2}{d\eta^2}+\left(\frac{1}{\eta}+\eta_0^2\eta\right)\frac{d}{d\eta}+\eta_0^2\right\}\begin{pmatrix}A_1\\ A_{S1}\end{pmatrix}=0 \tag{2.44}$$

with the boundary conditions at $\eta=1$:

$$A_1(1) = A_{S1}(1)+C_1\,, \tag{2.45}$$

$$A_{S1}(1) = -1\,, \tag{2.46}$$

$$\frac{d}{d\eta}(A_1-A_{S1})+(1+\eta_0^2)C_1=0\,. \tag{2.47}$$

It is solved that

$$A_1(\eta)=\sqrt{\pi}\hat{A}_1 e^{-\frac{\eta_0^2\eta^2}{2}}M\left(\frac{1}{2},1,\frac{\eta_0^2\eta^2}{2}\right)=\hat{A}_1 e^{-\frac{\eta_0^2\eta^2}{4}}K_0\left(\frac{\eta_0^2\eta^2}{4}\right), \tag{2.48}$$

$$A_{S1}(\eta)=\sqrt{\pi}\hat{A}_{S1} e^{-\frac{\eta_0^2\eta^2}{2}}U\left(\frac{1}{2},1,\frac{\eta_0^2\eta^2}{2}\right)=\hat{A}_{S1} e^{-\frac{\eta_0^2\eta^2}{4}}I_0\left(\frac{\eta_0^2\eta^2}{4}\right). \tag{2.49}$$

Here and hereafter, the functions $K_\nu(x)$ and $I_\nu(x)$ are νth-order modified Bessel functions. The boundary conditions (2.45)–(2.47) uniquely determine the three unknown constants $\{\hat{A}_1,\hat{A}_{S1},C_1\}$. The results are as follows:

$$\hat{A}_1 = \frac{I_1+(1-a_0)I_0}{K_1+(1-a_0)K_0}\hat{A}_{S1}\,, \tag{2.50}$$

$$\hat{A}_{S1} = -\frac{1}{I_0}e^{\frac{\eta_0^2}{4}}, \tag{2.51}$$

$$C_1 = \frac{I_0K_1-K_0I_1}{I_0\left\{(1-a_0)K_0+K_1\right\}}, \tag{2.52}$$

where

$$K_{0,1}=K_{0,1}\left(\frac{\eta_0^2}{4}\right),\quad I_{0,1}=I_{0,1}\left(\frac{\eta_0^2}{4}\right),\quad a_0=2+\frac{2}{\eta_0^2}. \tag{2.53}$$

The above procedure can be continued to higher-order approximations with no difficulty. The asymptotic solution for $\xi\to\infty$ is shown by the dashed line in Fig. 2.2.

3.4 Some Numerical Results of the Interface Shape Correction

Note that from a numerical point of view, the Laguerre series solution (2.28) can be easily calculated in the tip region, the numerical calculation becomes more and more difficult as ξ becomes bigger and bigger, because the convergence of the Laguerre series expansion becomes increasingly slow. As a consequence, to guarantee numerical accuracy, more and more terms in the Laguerre series must be included. On the other hand, the far-field asymptotic expansion (2.43) is quite accurate and can be easily calculated for larger and larger ξ.

With the above results, to carry out the numerical calculations for the solution $\eta_1(\xi, \gamma_0)$, one may divide the interval $0 \leq \xi < \infty$ into two sub-intervals: $(I) : (0 \leq \xi \leq \xi_{\max}$ and $(II) : (\xi_{\max} \leq \xi < \infty)$. In sub-interval (I), one may calculate the solution with the Laguerre expansion form (2.41), while in sub-interval (II) with the asymptotic form (2.43). By matching both solutions numerically at $\xi = \xi_{\max}$, one can determine the unknown constant γ_0.

The solutions depend on the number η_0^2, which is related to T_∞. For $\frac{\eta_0^2}{2} = 0.005$, we obtain $\gamma_0 = -0.1681$. The solution $\eta_1(\xi)$ is displayed in Fig. 2.2. Thus, with $\varepsilon > 0$, we have

$$\eta_s(\xi, \varepsilon) \approx 1 + \varepsilon^2 \eta_1(\xi, \eta_0^2). \tag{2.54}$$

The dendrite shapes with $\varepsilon = 0.1$ and 0.2 are shown in Fig. 2.3(a) and Fig. 2.3(b), respectively, in which the dashed lines represent the Ivantsov paraboloid with $\varepsilon = 0$.

The mean curvature of the interface at the tip is

$$\mathcal{K}(0) \approx 2 - \varepsilon^2 \eta_1''(0). \tag{2.55}$$

On the other hand, the dimensionless tip radius, $\mathrm{Pe} = \ell_t/\ell_T$ can be calculated with the formula of curvature, namely

$$\frac{\mathcal{K}(0)}{\eta_0^2} = \frac{2}{\mathrm{Pe}}. \tag{2.56}$$

Therefore, we get

$$\mathrm{Pe} = \frac{\ell_t}{\ell_T} = \frac{\eta_0^2}{1 - \varepsilon^2 \eta_1''(0)/2}. \tag{2.57}$$

The derivatives $\eta_1''(0, \eta_0^2)$ can be numerically computed as

$$\eta_1''(0) = \lim_{\xi \to 0} \frac{2\eta_1(\xi)}{\xi^2}. \tag{2.58}$$

The variations of the Peclet number $\mathrm{Pe} = \ell_t/\ell_T$ with the parameters T_∞ and ε are shown in Fig. 2.5. This steady correction for the Ivantsov needle solution due to the isotropic surface tension is of practical interest. Such a correction can be observed experimentally.

Up to this point, we have derived the approximate analytical form of the basic state in the order ε^2. We do not attempt to continue the above procedure further to determine its even higher-order approximations. For practical purposes, there is no need to do so.

4. Global Interfacial Wave Instability

We now turn to study global linear stability of the non-classic needle solutions q demonstrated in the last section and referred to as the basic states. For global linear stability, one needs to investigate the evolution of infinitesimal perturbations around the basic state. We shall adopt the normal mode approach, by assuming that the perturbations are in the form of quasi-stationary waves and then investigate their evolution by solving an eigenvalue problem under a certain set of boundary conditions. The eigenfunctions obtained with the normal mode approach are so-called global modes. The eigenvalue corresponding to a global mode gives both the growth rate of the amplitude of the perturbation and the frequency of the oscillation.

The subject under study obviously is a phenomenon with multiple length scales. To find the global mode solutions, we shall naturally apply the Multiple Variables Expansion (MVE) method (Kevorkian and Cole, 1996) with the matching procedure. It was first found in 1989 that in the system there exists a special simple turning point in the complex plane that plays a crucial role in understanding the dynamics of dendritic growth (Xu, 1989, 1991a, 1991b, 1996, 1997). Due to the presence of such a singular point for the asymptotic expansion solutions, the complex plane must be further divided into three regions: the outer region, the inner region near the singular point, and the tip inner region. The outer solutions in the outer region will be derived by the MVE method. These outer solutions can be interpreted as some special interfacial travelling waves, propagating along the interface of the basic state. The local dispersion relationship for these interfacial waves is obtained in the zeroth-order approximation, while the amplitude functions of the waves are determined in the first-order approximation. It is the first-order approximation that identifies the turning-point type of singularity in the outer solutions. Near the turning point and the leading edge of the dendrite tip, the outer solutions are invalid. Different asymptotic expansions for the exact solutions are needed. Hence one must choose proper new length scales and derive the inner solutions in the inner regions of the turning point and dendrite tip, respectively. The inner equation in the vicinity of the turning point can be reduced to the Airy equation with complex coefficients, whereas the inner solutions in the vicinity of the tip can be expressed by Hankel functions. Finally, all asymptotic expansion solutions must be matched in the intermediate regions. The global mode solutions and a quantization condition for the eigenvalues are then obtained.

It is found that given $\varepsilon > 0$, the system permits a discrete set of complex eigenvalues $\sigma_n = (\sigma_R - i\omega)_n$ $(n = 0, 1, 2, \ldots)$ and corresponding global wave modes (Xu, 1991a). The global instability mechanism discovered here is called the *global trapped-wave* (GTW) instability, and its presence explains the origin and persistence of the pattern formation in the solidification process. In a certain sense, the global instability mechanism discovered in dendritic growth, is similar to the so-called *over-reflection mechanism* explored in the critical layer theory of shear flow of fluid dynamics (Lin, 1955) and in the density wave theory of galactic dynamics (Lin and Lau, 1979).

Though the basic state under study is axi-symmetric, the perturbed states may be non-axi-symmetric. Thus, the general global modes may have m-th fold symmetry $(m = 0, 1, 2, \ldots)$. However, it is derived that the system only allows global modes of $m = 0, 1, 2$, and the axi-symmetric $(m = 0)$ global modes are the most dangerous modes. Hence when ε equals a critical number ε_*, the system permits a uniquely determined axi-symmetric *global neutrally stable mode*; its corresponding eigenvalue σ has a zero real part. This global neutrally stable mode will be selected at the later stage of growth. Therefore the stability criterion $\varepsilon = \varepsilon_*$ is also the selection criterion for the tip speed.

It is also found that up to the approximation of $O(\varepsilon)$, the interface correction term $\eta_1(\xi)$ in the basic state solution induced by the isothermal surface tension does not affect the eigenvalues σ, as well as the eigen-functions of the perturbed states. From here, one can conclude that for the first order approximation of perturbed state solutions, one may simply use the Ivantsov's solution to replace the basic state solution without introducing additional errors.

5. Three-Dimensional, Linear Perturbed States Around the Axi-symmetric Basic State of Dendritic Growth

We write the basic state in the form

$$\begin{aligned} T_{\rm B}(\xi,\eta,\varepsilon) &= T_*(\eta)+O(\varepsilon^2),\\ T_{\rm SB}(\xi,\eta,\varepsilon) &= O(\varepsilon^2),\\ \eta_{\rm B}(\xi,\varepsilon) &= 1+\varepsilon^2\eta_1(\xi)+O(\varepsilon^4), \end{aligned} \tag{2.59}$$

and separate the general unsteady solutions into two parts:

$$\begin{aligned} T &= T_{\rm B}+\tilde{T}(\xi,\eta,\theta,t,\varepsilon),\\ T_{\rm S} &= T_{\rm SB}+\tilde{T}_{\rm S}(\xi,\eta,\theta,t,\varepsilon),\\ \eta_{\rm s} &= \eta_{\rm B}+\tilde{h}(\xi,\theta,t,\varepsilon)/\eta_0^2\,. \end{aligned} \tag{2.60}$$

Here, we consider initially infinitesimal, general three-dimensional perturbations around the basic state. The linearized perturbed system is a homogeneous system as follows:

$$\begin{aligned} &\frac{\partial^2\tilde{T}}{\partial\xi^2}+\frac{\partial^2\tilde{T}}{\partial\eta^2}+\frac{1}{\xi}\frac{\partial\tilde{T}}{\partial\xi}+\frac{1}{\eta}\frac{\partial\tilde{T}}{\partial\eta}+\Big(\frac{1}{\xi^2}+\frac{1}{\eta^2}\Big)\frac{\partial^2\tilde{T}}{\partial\theta^2}\\ &\quad=-\eta_0^2\Big(\xi\frac{\partial\tilde{T}}{\partial\xi}-\eta\frac{\partial\tilde{T}}{\partial\eta}\Big)+\eta_0^4(\xi^2+\eta^2)\frac{\partial\tilde{T}}{\partial t}\\ &\quad(0\le\xi\le\xi_{\max}=O(1/\varepsilon);\ 0\le\eta<\infty), \end{aligned} \tag{2.61}$$

with the boundary conditions

1. As $\eta\to\infty$

$$\tilde{T}\to 0\,. \tag{2.62}$$

2. As $\eta\to 0$

$$\tilde{T}_{\rm S}=O(1)\,. \tag{2.63}$$

3. The interface conditions: making the Taylor expansions around the interface of the basic state, it follows that, for the case of isotropic surface tension, at $\eta=\eta_{\rm B}(\xi,\varepsilon)$:

   (i)

$$\tilde{T}-\tilde{T}_{\rm S}=-\Big\{\frac{\partial T_{\rm B}}{\partial\eta}-\frac{\partial T_{\rm SB}}{\partial\eta}\Big\}\frac{\tilde{h}}{\eta_0^2}\,, \tag{2.64}$$

(ii)

$$\tilde{T}_{\mathrm{S}} = \frac{\varepsilon^2}{\hat{S}(\xi)} \left\{ \frac{\partial^2 \tilde{h}}{\partial \xi^2} + (1 + \frac{1}{\xi^2}) \frac{\partial^2 \tilde{h}}{\partial \theta^2} + \frac{1 + 2\xi^2}{\xi \hat{S}^2(\xi)} \frac{\partial \tilde{h}}{\partial \xi} - \frac{2\hat{S}}{\xi} \frac{\partial \tilde{h}}{\partial \theta} - \frac{\tilde{h}}{\hat{S}^2(\xi)} \right\}$$

$$+(\text{higher-order terms}) \,, \tag{2.65}$$

(iii)

$$\frac{\partial}{\partial \eta}\left(\tilde{T} - \tilde{T}_{\mathrm{S}}\right) + \eta_0^2 \hat{S}^2(\xi) \frac{\partial \tilde{h}}{\partial t} + \xi \frac{\partial \tilde{h}}{\partial \xi} + \tilde{h} + \left\{ \frac{\partial^2}{\partial \eta^2} (T_{\mathrm{B}} - T_{\mathrm{SB}}) \right\} \frac{\tilde{h}}{\eta_0^2}$$

$$- \frac{1}{\eta_0^2} \frac{\partial \tilde{h}}{\partial \xi} \frac{\partial}{\partial \xi} (T_{\mathrm{B}} - T_{\mathrm{SB}}) - \frac{\eta_{\mathrm{B}}' \tilde{h}}{\eta_0^2} \frac{\partial^2}{\partial \xi \partial \eta} (T_{\mathrm{B}} - T_{\mathrm{SB}})$$

$$- \eta_{\mathrm{B}}' \frac{\partial}{\partial \xi} (\tilde{T} - \tilde{T}_{\mathrm{S}}) = 0 \,, \tag{2.66}$$

where

$$\hat{S}(\xi) = \sqrt{\xi^2 + \eta_{\mathrm{B}}^2} \,. \tag{2.67}$$

4. The root condition: at $\xi = \xi_{\max} = \frac{C}{\varepsilon}$,

$$\{\tilde{T}, \tilde{T}_{\mathrm{S}}, \tilde{h}\} = 0 \,. \tag{2.68}$$

5. The tip smoothness condition: at $\xi = 0, \ \eta = \eta_{\mathrm{B}}(0, \varepsilon)$,

$$\frac{\partial}{\partial \xi} \{\tilde{T}, \tilde{T}_{\mathrm{S}}, \tilde{h}\} = 0 \,. \tag{2.69}$$

The above system (2.61)–(2.69) leads to a linear eigenvalue problem with two parameters ε, and η_0^2, when one looks for solutions of the type $\tilde{q} = \hat{q} e^{\sigma t}$. The eigenvalue σ must be a function of (η_0^2, ε).

This eigenvalue problem can be solved in two steps:

- Solve the system (2.61)–(2.68) for any given $(\sigma, \eta_0^2, \varepsilon)$. In doing so, we shall apply the multiple variables expansion (MVE) method (Kevorkian and Cole, 1996) to look for a uniformly valid asymptotic solution in the limit $\varepsilon \to 0$. The solution should satisfy all boundary conditions except the tip condition (2.69).
- Then apply the tip condition (2.69) to the asymptotic solution obtained. It shall lead to the quantization conditions, which determines the parameter σ as a function of η_0^2 and ε.

When finding the asymptotic solutions $\tilde{q}(\xi, \eta, \theta, t, \varepsilon)$, one first sees that there is a singularity at the tip $\xi = 0$. It will be also seen that the asymptotic expansion solutions have some other singular points in the complex plane. Hence, one must single out the tip region ($|\xi| \ll 1$) and the inner regions around the other singular points. We call the remaining region the outer region, and treat the outer solution first.

6. Outer Solution in the Outer Region away from the Singular Points

In the outer region, we shall look for the asymptotic expansion of the solutions $\tilde{q}(\xi, \eta, t, \varepsilon)$ in terms of the multiple variables expansion method. In doing so, we first introduce the fast variable η_+ along η direction as

$$\eta_+ = \frac{\eta - 1}{\varepsilon}. \tag{2.70}$$

The fast variable η_+ determines a boundary layer with the thickness of $O(\varepsilon)$ near the interface $\eta = 1$, which we call the *Interfacial Wave* (IW) layer. It will be seen that the perturbations of temperature field are only restricted inside of the IW layer. It vanishes exponentially away from this layer. Furthermore, we introduce the fast variable ξ_+ along ξ-direction. At the interface $\eta = 1$, this fast variable is defined as

$$\bar{\xi}_+ = \frac{1}{\varepsilon} \int_{\xi_0}^{\xi} k(\xi', \varepsilon) d\xi', \tag{2.71}$$

where the lower limit of the integral, ξ_0, will be specified later. Off the interface $\eta = 1$, we allow that the fast variable ξ_+ may slightly change. Namely, in the field of the liquid phase, we define

$$\xi_+(\xi, \eta, \varepsilon) = \bar{\xi}_+ + \varepsilon \int_0^{\eta_+} K(\xi, \eta'_+, \varepsilon) d\eta'_+, \tag{2.72}$$

while in the solid phase, the function $K(\xi, \eta_+, \varepsilon)$ is replaced by $K_S(\xi, \eta_+, \varepsilon)$. The second small correction term on the right-hand side of (2.72) is needed for deriving the higher-order solutions. This term may be dropped, if one is only interested in finding the leading-order solution.

The fast time variable t_+ can be defined as

$$t_+ = \frac{t}{\eta_0^2 \varepsilon}. \tag{2.73}$$

In terms of the above-defined multiple variables, $(\xi, \eta, \theta, \xi_+, \eta_+, t_+)$, we make the following multiple variables expansion (MVE) for general,

3D perturbed states:

$$\begin{aligned}
\tilde{T} &= \tilde{\mu}_0(\varepsilon)\left\{\tilde{T}_0(\xi,\eta,\xi_+,\eta_+) + \varepsilon\tilde{T}_1(\xi,\eta,\xi_+,\eta_+) + \cdots\right\} e^{im\theta+\sigma t_+},\\
\tilde{T}_S &= \tilde{\mu}_0(\varepsilon)\left\{\tilde{T}_{S0}(\xi,\eta,\xi_+,\eta_+) + \varepsilon\tilde{T}_{S1}(\xi,\eta,\xi_+,\eta_+) + \cdots\right\} e^{im\theta+\sigma t_+},\\
\tilde{h} &= \tilde{\mu}_0(\varepsilon)\left\{\tilde{h}_0(\xi,\theta,\bar{\xi}_+) + \varepsilon\tilde{h}_1(\xi,\bar{\xi}_+) + \cdots\right\} e^{im\theta+\sigma t_+},
\end{aligned} \tag{2.74}$$

where $m = 0, 1, 2, \ldots$, whereas

$$\begin{aligned}
k(\xi,\varepsilon) &= k_0(\xi) + \varepsilon k_1(\xi) + \cdots,\\
K(\xi,\eta_+,\varepsilon) &= K_0(\xi,\eta_+) + \varepsilon K_1(\xi,\eta_+) + \cdots,\\
K_S(\xi,\eta_+,\varepsilon) &= K_{S0}(\xi,\eta_+) + \varepsilon K_{S1}(\xi,\eta_+) + \cdots\\
\sigma &= \sigma_0 + \varepsilon\sigma_1 + \varepsilon^2\sigma_2 + \cdots .
\end{aligned} \tag{2.75}$$

Here, $\sigma = \sigma_R - i\omega$ $(\omega \geq 0)$ is, generally, a complex number. In the first step, we set σ_0 as an arbitrary constant. As before, the fast and slow variables $(\xi, \eta, \xi_+, \eta_+, t_+)$ are treated formally as independent variables. Hence, one must make the following replacements for the derivatives in the system (2.61)–(2.66):

$$\begin{aligned}
\tfrac{\partial}{\partial\xi} &\Longrightarrow \tfrac{k}{\varepsilon}\tfrac{\partial}{\partial\xi_+} + \tfrac{\partial}{\partial\xi} + \varepsilon\textstyle\int_0^{\eta_+} K_\xi(\xi,\eta'_+,\varepsilon)d\eta'_+,\\
\tfrac{\partial}{\partial\eta} &\Longrightarrow \tfrac{1}{\varepsilon}\tfrac{\partial}{\partial\eta_+} + \tfrac{\partial}{\partial\eta} + K\tfrac{\partial}{\partial\xi_+},\\
\tfrac{\partial^2}{\partial\xi^2} &\Longrightarrow \left\{\tfrac{k}{\varepsilon}\tfrac{\partial}{\partial\xi_+} + \tfrac{\partial}{\partial\xi} + \varepsilon\textstyle\int_0^{\eta_+} K_\xi(\xi,\eta'_+,\varepsilon)d\eta'_+\right\}^2,\\
\tfrac{\partial^2}{\partial\eta^2} &\Longrightarrow \left\{\tfrac{1}{\varepsilon}\tfrac{\partial}{\partial\eta_+} + \tfrac{\partial}{\partial\eta} + K\tfrac{\partial}{\partial\xi_+}\right\}^2.
\end{aligned} \tag{2.76}$$

The converted system with the multiple variables is as follows: The equation in the liquid phase is

$$\begin{aligned}
\left(k^2\frac{\partial^2}{\partial\xi_+^2} + \frac{\partial^2}{\partial\eta_+^2}\right)\tilde{T} &= \varepsilon\eta_0^2(\xi^2+\eta^2)\frac{\partial\tilde{T}}{\partial t_+} + \varepsilon\eta_0^2\left(\xi k\frac{\partial}{\partial\xi_+} - \eta\frac{\partial}{\partial\eta_+}\right)\tilde{T}\\
&- \varepsilon\left(\frac{k}{\xi}\frac{\partial}{\partial\xi_+} + \frac{1}{\eta}\frac{\partial}{\partial\eta_+}\right)\tilde{T} - \varepsilon\left(2k\frac{\partial^2}{\partial\xi\partial\xi_+} + k'\frac{\partial}{\partial\xi_+}\right)\tilde{T}\\
&- \varepsilon\left(2\frac{\partial^2}{\partial\eta\partial\eta_+} + 2K\frac{\partial^2}{\partial\xi_+\partial\eta_+} + \frac{\partial K}{\partial\eta_+}\frac{\partial}{\partial\xi_+}\right)\tilde{T}\\
&+ \varepsilon^2\left(\frac{1}{\xi^2} + \frac{1}{\eta^2}\right)\frac{\partial^2\tilde{T}}{\partial\theta^2} + O(\varepsilon^2).
\end{aligned} \tag{2.77}$$

The equation in the solid phase is

$$\Big(k^2\frac{\partial^2}{\partial\xi_+^2}+\frac{\partial^2}{\partial\eta_+^2}\Big)\tilde{T}_S=\varepsilon\eta_0^2(\xi^2+\eta^2)\frac{\partial\tilde{T}_S}{\partial t_+}+\varepsilon\eta_0^2\Big(\xi k\frac{\partial}{\partial\xi_+}-\eta\frac{\partial}{\partial\eta_+}\Big)\tilde{T}_S$$
$$-\varepsilon\Big(\frac{k}{\xi}\frac{\partial}{\partial\xi_+}+\frac{1}{\eta}\frac{\partial}{\partial\eta_+}\Big)\tilde{T}_S-\varepsilon\Big(2k\frac{\partial^2}{\partial\xi\partial\xi_+}+k'\frac{\partial}{\partial\xi_+}\Big)\tilde{T}_S$$
$$-\varepsilon\Big(2\frac{\partial^2}{\partial\eta\partial\eta_+}+2K_S\frac{\partial^2}{\partial\xi_+\partial\eta_+}+\frac{\partial K_S}{\partial\eta_+}\frac{\partial}{\partial\xi_+}\Big)\tilde{T}_S$$
$$+\varepsilon^2\Big(\frac{1}{\xi^2}+\frac{1}{\eta^2}\Big)\frac{\partial^2\tilde{T}_S}{\partial\theta^2}+O(\varepsilon^2). \tag{2.78}$$

The boundary conditions are:

1. As $\eta_+\to\infty$

$$\tilde{T}\to 0\,. \tag{2.79}$$

2. As $\eta_+\to-\infty$

$$\tilde{T}_S\to 0\,. \tag{2.80}$$

3. At the interface: $\eta_+=0,\eta=1$,

(i) the thermodynamic equilibrium condition

$$\tilde{T}=\tilde{T}_S+\tilde{h}-\varepsilon\eta_1\tfrac{\partial}{\partial\eta_+}(\tilde{T}-\tilde{T}_S)+O(\varepsilon^2)\,, \tag{2.81}$$

(ii) the Gibbs–Thomson condition

$$\tilde{T}_S+\varepsilon\eta_1\frac{\partial\tilde{T}_S}{\partial\eta_+}=\frac{1}{S(\xi)}\Big\{\Big(k^2\frac{\partial^2}{\partial\bar{\xi}_+^2}+2\varepsilon k\frac{\partial^2}{\partial\bar{\xi}_+\partial\xi}+\varepsilon\frac{\partial k}{\partial\xi}\frac{\partial}{\partial\bar{\xi}_+}$$
$$+\varepsilon^2\frac{\partial^2}{\partial\xi^2}\Big)+\varepsilon\Big(\frac{1}{\xi}+\frac{\xi}{S^2(\xi)}\Big)\Big(k\frac{\partial}{\partial\bar{\xi}_+}+\varepsilon\frac{\partial}{\partial\xi}\Big)-\frac{\varepsilon^2}{S^2(\xi)}\Big\}\tilde{h}$$
$$+\frac{\varepsilon^2}{\xi^2}\frac{\partial^2\tilde{h}}{\partial\theta^2}+O(\varepsilon^2), \tag{2.82}$$

(iii) the heat balance condition

$$\Big(\frac{\partial\tilde{T}}{\partial\eta_+}-\frac{\partial\tilde{T}_S}{\partial\eta_+}\Big)+\varepsilon\eta_1\frac{\partial^2}{\partial\eta_+^2}(\tilde{T}-\tilde{T}_S)+\varepsilon\Big(\frac{\partial\tilde{T}}{\partial\eta}-\frac{\partial\tilde{T}_S}{\partial\eta}\Big)$$

$$+\varepsilon\Big(\hat{K}\frac{\partial\tilde{T}}{\partial\bar{\xi}_+}-\hat{K}_S\frac{\partial\tilde{T}_S}{\partial\bar{\xi}_+}\Big)+S^2(\xi)\frac{\partial\tilde{h}}{\partial t_+}+\xi\Big(k\frac{\partial\tilde{h}}{\partial\bar{\xi}_+}+\varepsilon\frac{\partial\tilde{h}}{\partial\xi}\Big)$$

$$+\varepsilon\big(2+\eta_0^2\big)\tilde{h}+O(\varepsilon^2)=0\,, \tag{2.83}$$

where we have introduced the notation

$$\hat{K}_0=K_0(\xi,0),\quad \hat{K}_{S0}=K_{S0}(\xi,0). \tag{2.84}$$

4. The root condition: at $\xi\to\infty$ one has

$$\{\tilde{T};\ \tilde{T}_S;\ \tilde{h}\}=0\,. \tag{2.85}$$

By substituting (2.74) into the system (2.77)–(2.85), one can successively derive each order of approximation.

6.1 Zeroth-Order Approximation

As zeroth-order approximation, we derive the quasi-steady heat conduction equation

$$\begin{aligned}\Big(k_0^2\frac{\partial^2}{\partial\xi_+^2}+\frac{\partial^2}{\partial\eta_+^2}\Big)\tilde{T}_0&=0,\\ \Big(k_0^2\frac{\partial^2}{\partial\xi_+^2}+\frac{\partial^2}{\partial\eta_+^2}\Big)\tilde{T}_{S0}&=0,\end{aligned} \tag{2.86}$$

with the following boundary conditions:

1. In the liquid region $\eta>1$, as $\eta_+\to\infty$,

$$\tilde{T}_0\to 0\,. \tag{2.87}$$

2. In the solid region $\eta<1$, as $\eta_+\to-\infty$,

$$\tilde{T}_{S0}\to 0\,. \tag{2.88}$$

3. At the interface: $\eta_+=0$ or $\eta=\ 1$,

$$\tilde{T}_0=\tilde{T}_{S0}+\tilde{h}_0, \tag{2.89}$$

$$\tilde{T}_{S0}=\frac{k_0^2}{S(\xi)}\frac{\partial^2\tilde{h}_0}{\partial\bar{\xi}_+^2}, \tag{2.90}$$

$$\frac{\partial}{\partial\eta_+}(\tilde{T}_0-\tilde{T}_{S0})+\sigma_0S^2(\xi)\tilde{h}_0+k_0\xi\frac{\partial\tilde{h}_0}{\partial\bar{\xi}_+}=0\,. \tag{2.91}$$

4. As $\xi \to \infty$,

$$\tilde{h}_0 = 0\,. \tag{2.92}$$

The above system has the normal mode solutions

$$\begin{aligned} \tilde{T}_0 &= A_0(\xi,\eta)\exp\{\mathrm{i}\xi_+ - k_0\eta_+\}\,, \\ \tilde{T}_{S0} &= A_{S0}(\xi,\eta_+)\exp\{\mathrm{i}\xi_+ + k_0\eta_+\}\,, \\ \tilde{h}_0 &= \hat{D}_0\exp\{\mathrm{i}\bar{\xi}_+\}\,, \end{aligned} \tag{2.93}$$

where the coefficient $\hat{D}_0$ is set as a constant. Note that here, by introducing the form of the solutions (2.93), we have extended the solution analytically to the complex space (ξ_+, η_+). Setting

$$\begin{cases} \hat{A}_0(\xi) = A_0(\xi,1), \\ \hat{A}_{S0}(\xi) = A_{S0}(\xi,1)\,, \end{cases} \tag{2.94}$$

from (2.89)–(2.91) we derive the system of homogeneous equations

$$\begin{cases} \hat{A}_0 - \hat{A}_{S0} - \hat{D}_0 = 0, \\ \hat{A}_{S0} + \frac{k_0^2}{S}\hat{D}_0 = 0, \\ -k_0(\hat{A}_0 + \hat{A}_{S0}) + (\sigma_0 S^2 + \mathrm{i}\xi k_0)\hat{D}_0 = 0\,. \end{cases} \tag{2.95}$$

Obviously, for a nontrivial solution, one must have

$$\Delta = \det\begin{pmatrix} 1 & -1 & -1 \\ 0 & 1 & k_0^2/S \\ -k_0 & -k_0 & \sigma_0 S^2 + \mathrm{i}\xi k_0 \end{pmatrix} = 0\,, \tag{2.96}$$

which gives the local dispersion relation

$$\sigma_0 = \Sigma(\xi, k_0) = \frac{k_0}{S^2}\left(1 - \frac{2k_0^2}{S}\right) - \frac{\mathrm{i}\xi}{S^2}k_0. \tag{2.97}$$

Then, one solves

$$\begin{cases} \hat{A}_0 + \hat{A}_{S0} = \left(1 - \frac{2k_0^2}{S}\right)\hat{D}_0, \\ \hat{A}_0 - \hat{A}_{S0} = \hat{D}_0. \end{cases} \tag{2.98}$$

The amplitude functions $\hat{A}(\xi), \hat{A}_S(\xi)$ are fully determined up to an arbitrary constant $\hat{D}$.

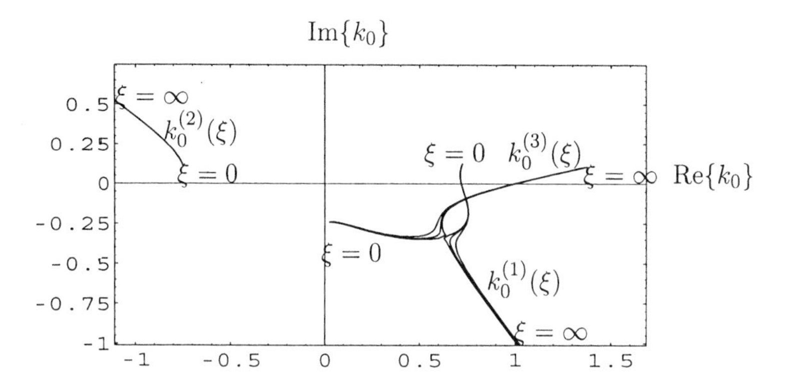

Figure 2.6. The variations of the wave number functions $\{k_0^{(1)}(\xi), k_0^{(2)}(\xi), k_0^{(3)}(\xi)\}$ in the complex k-plane with ξ for given $\sigma_0 = (0.035, -0.265)$; $(0.035, -0.269)$; $(0.035, -0.270)$

The local dispersion formula (2.97) obtained above is the generalization of the well-known Mullins–Sekerka formula in the unidirectional solidification (Mullins and Sekerka, 1963, 1964).

For any fixed parameter σ_0, one can solve for three wave numbers as functions of ξ (Fig. 2.6). Namely

$$\begin{cases} k_0^{(1)}(\xi) = M(\xi)\cos\left\{\frac{1}{3}\cos^{-1}\left(\frac{\sigma_0}{N(\xi)}\right)\right\} & \text{(short-wave branch)}, \\ k_0^{(2)}(\xi) = M(\xi)\cos\left\{\frac{1}{3}\cos^{-1}\left(\frac{\sigma_0}{N(\xi)}\right) + \frac{2\pi}{3}\right\}, & \\ k_0^{(3)}(\xi) = M(\xi)\cos\left\{\frac{1}{3}\cos^{-1}\left(\frac{\sigma_0}{N(\xi)}\right) + \frac{4\pi}{3}\right\} & \text{(long-wave branch)}, \end{cases} \tag{2.99}$$

where

$$\begin{cases} M(\xi) = \sqrt{\dfrac{2S(\xi)}{3}}(1 - \mathrm{i}\xi)^{\frac{1}{2}}, \\ N(\xi) = -\dfrac{M(\xi)}{3S^2(\xi)}(1 - \mathrm{i}\xi). \end{cases} \tag{2.100}$$

Given σ_0, the system allows three fundamental wave solutions corresponding to the wave number functions $\{k_0^{(1)}, k_0^{(2)}, k_0^{(3)}\}$:

H_1 wave (short wave branch) with larger $\mathrm{Re}\{k_0^{(1)}\} > 0$,

H_2 wave (no physical meaning) with $\mathrm{Re}\{k_0^{(2)}\} < 0$,

H_3 wave (long wave branch) with smaller $\mathrm{Re}\{k_0^{(3)}\} > 0$.

Among these solutions, the H_2 wave solution must be ruled out because the real part of its wave-number is negative. In the liquid region ($\eta > 1$), the H_2 wave solution yields $\mathrm{Re}\{\eta_+\} < 0$, hence, as $\eta_+ \to \infty$, its corresponding perturbed temperature field $\tilde{T}$ will grow exponentially. This would violate the boundary condition (2.87).

Consequently, in the zeroth-order approximation, the general H wave solution is

$$\tilde{h} \approx D_1 \mathrm{e}^{\sigma_0 t_+ + \mathrm{i}\left(\bar{\xi}_+^{(1)} + m\theta\right)} + D_3 \mathrm{e}^{\sigma_0 t_+ + \mathrm{i}\left(\bar{\xi}_+^{(1)} + m\theta\right)}, \tag{2.101}$$

where

$$\begin{aligned} \bar{\xi}_+^{(1)} &= \frac{1}{\varepsilon}\int_{\xi_0}^{\xi} k_0^{(1)} \mathrm{d}\xi_1, \\ \bar{\xi}_+^{(3)} &= \frac{1}{\varepsilon}\int_{\xi_0}^{\xi} k_0^{(3)} \mathrm{d}\xi_1, \end{aligned} \tag{2.102}$$

and the coefficients $\{D_1, D_3\}$ are arbitrary constants independent of ε which are to be determined.

6.2 First-Order Approximation

The first-order approximation solution will determine the amplitude functions $A_0(\xi,\eta)$, $A_{S0}(\xi,\eta)$, the functions $k_1(\xi)$, and σ_1. Our major concern is to determine σ_1. The equation for the first-order approximation can be obtained from (2.77) as the following. Noting that

$$\begin{aligned} \frac{\partial^2 \tilde{T}_0}{\partial\xi\partial\xi_+} &= \mathrm{i}\left(\frac{\partial A_0}{\partial\xi} - k_0'\eta_+ A_0\right)\mathrm{e}^{\mathrm{i}\xi_+ - k_0\eta_+}, \\ \frac{\partial^2 \tilde{T}_{S0}}{\partial\xi\partial\xi_+} &= \mathrm{i}\left(\frac{\partial A_{S0}}{\partial\xi} + k_0'\eta_+ A_{S0}\right)\mathrm{e}^{\mathrm{i}\xi_+ + k_0\eta_+}, \end{aligned} \tag{2.103}$$

we obtain

$$\begin{aligned} \left(k_0^2\frac{\partial^2}{\partial\xi_+^2} + \frac{\partial^2}{\partial\eta_+^2}\right)\tilde{T}_1 &= \Big[a_0(\xi,\eta) + \tilde{a}_0(\xi,\eta_+)\Big]\mathrm{e}^{\mathrm{i}\xi_+ - k_0\eta_+}, \\ \left(k_0^2\frac{\partial^2}{\partial\xi_+^2} + \frac{\partial^2}{\partial\eta_+^2}\right)\tilde{T}_{S1} &= \Big[a_{S0}(\xi,\eta) + \tilde{a}_{S0}(\xi,\eta_+)\Big]\mathrm{e}^{\mathrm{i}\xi_+ + k_0\eta_+}, \end{aligned} \tag{2.104}$$

where

$$\begin{cases} a_0 = 2k_0\Big(\dfrac{\partial A_0}{\partial \eta} - \mathrm{i}\dfrac{\partial A_0}{\partial \xi}\Big) + A_0\Big\{\sigma_0\eta_0^2(\xi^2+\eta^2) + k_0\eta_0^2(\mathrm{i}\xi+\eta) \\ \qquad + \dfrac{k_0}{\eta} - \mathrm{i}\dfrac{k_0}{\xi} - \mathrm{i}k_0' + 2k_0k_1\Big\} \\ a_{S0} = -2k_0\Big(\dfrac{\partial A_{S0}}{\partial \eta} + \mathrm{i}\dfrac{\partial A_{S0}}{\partial \xi}\Big) + A_{S0}\Big\{\sigma_0\eta_0^2(\xi^2+\eta^2) + k_0\eta_0^2(\mathrm{i}\xi-\eta) \\ \qquad - \dfrac{k_0}{\eta} - \mathrm{i}\dfrac{k_0}{\xi} - \mathrm{i}k_0' + 2k_0k_1\Big\}. \end{cases} \tag{2.105}$$

and

$$\begin{aligned} \tilde{a}_0(\xi,\eta_+) &= -\mathrm{i}A_0\left[\tfrac{\partial K_0}{\partial \eta_+} - 2k_0(K_0 + k_0'\eta_+)\right], \\ \tilde{a}_{S0}(\xi,\eta_+) &= -\mathrm{i}A_{S0}\left[\tfrac{\partial K_{S0}}{\partial \eta_+} + 2k_0(K_{S0} + k_0'\eta_+)\right]. \end{aligned} \tag{2.106}$$

To eliminate the secular terms on the right-hand side of (2.104), we set

$$a_0 = a_{S0} = \tilde{a}_0 = \tilde{a}_{S0} = 0\,. \tag{2.107}$$

Then, from (2.107) it follows that

$$\begin{aligned} \left(\frac{\partial}{\partial \eta} - \mathrm{i}\frac{\partial}{\partial \xi}\right)\ln\Phi_0(\xi,\eta) &= -\frac{\sigma_0\eta_0^2}{2k_0}(\xi^2+\eta^2) - \frac{\eta_0^2}{2}(\eta+\mathrm{i}\xi) - k_1, \\ \left(\frac{\partial}{\partial \eta} + \mathrm{i}\frac{\partial}{\partial \xi}\right)\ln\Phi_{S0}(\xi,\eta) &= \frac{\sigma_0\eta_0^2}{2k_0}(\xi^2+\eta^2) - \frac{\eta_0^2}{2}(\eta-\mathrm{i}\xi) + k_1\,, \end{aligned} \tag{2.108}$$

where

$$\begin{aligned} \Phi_0(\xi,\eta) &= A_0(\xi,\eta)k_0^{\frac{1}{2}}\xi^{\frac{1}{2}}\eta^{\frac{1}{2}}, \\ \Phi_{S0}(\xi,\eta) &= A_{S0}(\xi,\eta)k_0^{\frac{1}{2}}\xi^{\frac{1}{2}}\eta^{\frac{1}{2}}\,. \end{aligned} \tag{2.109}$$

We also obtain

$$\begin{aligned} K_0 &= -k_0'\eta_+ - \tfrac{k_0'}{2k_0}, \\ K_{S0} &= -k_0'\eta_+ + \tfrac{k_0'}{2k_0}. \end{aligned} \tag{2.110}$$

(2.108) are first-order hyperbolic equations. Once k_1 is found, the functions $A_0(\xi,\eta)$, $A_{S0}(\xi,\eta)$ can be solved in the ξ–η plane as an initial value problem given the initial values on the curve $\eta = 1$. Moreover, we derive

$$Q_0 = \frac{1}{\hat{D}_0}\frac{\partial}{\partial \eta}(A_0 - A_{S0})\bigg|_{\eta=1}$$

$$
\begin{aligned}
&= -\mathrm{i}2\frac{\partial}{\partial\xi}\left(\frac{k_0^2}{S}\right) - \left(1 - \frac{2}{S}k_0^2\right)\left[\frac{\eta_0^2}{2k_0}\sigma_0 S^2 + k_1\right.\\
&\left. + \frac{\mathrm{i}}{2}\left(\xi\eta_0^2 - \frac{1}{\xi} - \frac{k_0'}{k_0}\right)\right] - \left(\frac{1+\eta_0^2}{2}\right).
\end{aligned} \tag{2.111}
$$

These formulas will be needed later to solve for $k_1(\xi)$.

In terms of conditions (2.107), it follows that

$$
\begin{cases}
\left(k_0^2\dfrac{\partial^2}{\partial\xi_+^2} + \dfrac{\partial^2}{\partial\eta_+^2}\right)\tilde{T}_1 = 0,\\
\left(k_0^2\dfrac{\partial^2}{\partial\xi_+^2} + \dfrac{\partial^2}{\partial\eta_+^2}\right)\tilde{T}_{S1} = 0\,.
\end{cases} \tag{2.112}
$$

We obtain the first-order approximate solutions

$$
\begin{aligned}
\tilde{T}_1 &= A_1(\xi,\eta)\exp\{\mathrm{i}\xi_+ - k_0\eta_+\},\\
\tilde{T}_{S1} &= A_{S1}(\xi,\eta)\exp\{\mathrm{i}\xi_+ + k_0\eta_+\},\\
\tilde{h}_1 &= \hat{D}_1\exp\{\mathrm{i}\xi_+\}.
\end{aligned} \tag{2.113}
$$

By setting

$$
\begin{cases}
\hat{A}_1(\xi) = A_1(\xi,1),\\
\hat{A}_{S1}(\xi) = A_{S1}(\xi,1),
\end{cases} \tag{2.114}
$$

from (2.81)–(2.83) we find that

$$
\begin{aligned}
&\hat{A}_1 - \hat{A}_{S1} - \hat{D}_1 = I_1\hat{D}_0 = k_0\eta_1\left(1 - \tfrac{2k_0^2}{S}\right)\hat{D}_0,\\
&\hat{A}_{S1} + \tfrac{\hat{k}_0^2}{S}\hat{D}_1 = I_2\hat{D}_0,\\
&-\hat{k}_0(\hat{A}_1 + \hat{A}_{S1}) + (\sigma_0 S^2 + \mathrm{i}\xi\hat{k}_0)\hat{D}_1 = I_3\hat{D}_0,
\end{aligned} \tag{2.115}
$$

where we define

$$
I_1 = k_0\eta_1\left(1 - \frac{2k_0^2}{S}\right), \tag{2.116}
$$

$$
I_2 = \frac{1}{S}\left\{\mathrm{i}k_0' + \mathrm{i}k_0\left(\frac{1}{\xi} + \frac{\xi}{S^2}\right) - 2k_0k_1\right\} + \frac{k_0^3}{S}\eta_1, \tag{2.117}
$$

and

$$
\begin{aligned}
I_3 = &-(\sigma_1 S^2 + \mathrm{i}\xi k_1) - Q_0 - (2+\eta_0^2) + \mathrm{i}\frac{k_0'}{2k_0}\left(1 - \frac{2k_0^2}{S}\right)\\
&-\eta_1 k_0^2\,.
\end{aligned} \tag{2.118}
$$

The determinant Δ of the coefficient matrix of the above inhomogeneous system is zero. The necessary condition for the existence of a nontrivial solution for $\{\hat{A}_1, \hat{A}_{S1}, \hat{D}_1\}$ is:

$$\det\begin{pmatrix} 1 & -1 & I_1 \\ 0 & 1 & I_2 \\ -k_0 & -k_0 & I_3 \end{pmatrix} = 0, \tag{2.119}$$

which leads to the solvability condition:

$$I_3 + 2k_0 I_2 + I_1 k_0 = 0. \tag{2.120}$$

From (2.120), we obtain

$$\begin{aligned} k_1\left[1 - \mathrm{i}\xi - \frac{6k_0^2}{S}\right] &= \frac{\mathrm{i}}{2\xi}\left[1 - \mathrm{i}\xi - \frac{6k_0^2}{S}\right] + S^2\sigma_1 \\ &+\left(\frac{3}{2} + \eta_0^2\right) - \frac{\eta_0^2}{2}\left(1 - \frac{2}{S}k_0^2\right)^2 - \frac{1+\eta_0^2}{2} \\ &-\mathrm{i}\frac{6k_0k_0'}{S} \end{aligned} \tag{2.121}$$

or

$$k_1 = \frac{\mathrm{i}}{2\xi} + \frac{R_1(\xi)}{F(\xi)} - \mathrm{i}\frac{R_2(\xi)}{F(\xi)}\frac{k_0'}{k_0}, \tag{2.122}$$

where we denote

$$\begin{aligned} F(\xi) &= \left[1 - \mathrm{i}\xi - \frac{6k_0^2}{S}\right], \\ R_1(\xi) &= S^2\sigma_1 + 1 + \frac{\eta_0^2}{2} - \frac{\eta_0^2}{2}\left(1 - \frac{2k_0^2}{S}\right)^2, \\ R_2(\xi) &= \frac{6k_0^2}{S}. \end{aligned} \tag{2.123}$$

It is interesting to see that all the terms involving $\eta_1(\xi)$ are cancelled.

From (2.122), it is seen that the solution $k_1(\xi)$ has a singularity at the tip $\xi = 0$, and

$$k_1 \sim \frac{\mathrm{i}}{2\xi}, \tag{2.124}$$

as $\xi \to 0$. Moreover, from the dispersion formula (2.97), one finds that

$$F(\xi) = S^2\left(\frac{\partial \Sigma}{\partial k_0}\right). \tag{2.125}$$

Therefore, it is seen that at the root ξ_c of the equation

$$\left(\frac{\partial \Sigma}{\partial k_0}\right) = 0 \quad (\text{or} \quad \mathrm{F}(\xi) = 0), \tag{2.126}$$

the solution k_1, as well as the functions A_0 and A_{S0}, also has a singularity. Given σ_0, from the local dispersion formula (2.97) we have

$$\begin{aligned} \frac{k_0'(\xi)}{k_0(\xi)} &= \frac{R_0(\xi)}{F(\xi)}, \\ R_0(\xi) &= \mathrm{i} + \frac{2\xi(1-\mathrm{i}\xi)}{S^2} - \frac{6\xi k_0^2}{S^3}. \end{aligned} \tag{2.127}$$

Moreover, one can derive that as $\xi \to \xi_c$,

$$F(\xi) \sim 2^{\frac{1}{2}} m_0^{\frac{1}{2}} (\xi - \xi_c)^{\frac{1}{2}}, \tag{2.128}$$

and

$$k_1 \sim \frac{R_1(\xi_c)}{2^{\frac{1}{2}} m_0^{\frac{1}{2}} (\xi - \xi_c)^{\frac{1}{2}}} + \frac{\mathrm{i}}{4(\xi - \xi_c)} + O(1) \tag{2.129}$$

where

$$m_0 = -\frac{12k_0^2}{S} R_0(\xi_c). \tag{2.130}$$

The function $k_1(\xi)$ must be an analytic function in the complex ξ-plane. It can be verified by the matching condition with the inner solution near the singular point ξ_c that near the isolated singular point ξ_c, $k_1(\xi)$ must be expanded in the Laurent series, namely, from (2.129),

$$R_1(\xi_c) = 0 \tag{2.131}$$

must hold. Therefore, the singular point $\xi = \xi_c$, as well as $\xi = 0$, is a simple pole of the function $k_1(\xi)$ in the complex ξ-plane.

From condition (2.131), one obtains:

$$\sigma_1 = -\frac{1}{(1+\xi_c^2)}\left[1 + \frac{\eta_0^2}{2} - \frac{\eta_0^2}{18}\left(2 + \mathrm{i}\xi_c\right)^2\right]. \tag{2.132}$$

From (2.132), it is seen that up to the approximation of $O(\varepsilon)$, the eigenvalue σ, as well as the eigenfunctions are not affected by the interface shape correction, $\eta_1(\xi)$, caused by the isotropic surface tension. Furthermore, once σ_0 is known, then σ_1 is determined. From the above results, an important conclusion is drawn that up to the first approximation, one can use the Ivantsov's solution with zero surface tension as the basic state of dendritic growth for the calculations of perturbed states without producing additional error.

6.3 Singular Point ξ_c of the Outer Solution and Stokes Phenomenon

From (2.129) we see that the MVE solution (2.74) is not valid near the point ξ_c in the complex ξ-plane. The singularity at ξ_c was not seen in the zeroth-order approximation but it is now clearly seen in the first approximation. The existence of the singular point ξ_c plays a crucial role in pattern formation and selection in dendritic growth. This singularity was unknown by all previous researchers in the field until its discovery in 1989 (Xu, 1990b, 1991a).

The location of the singular point ξ_c in the complex ξ-plane depends on the value of σ_0. Combining (2.126) and (2.97), one finds that

$$\sigma_0 = \sqrt{\frac{2}{27}}\left(\frac{1-\mathrm{i}\xi_c}{S}\right)^{3/2}_{\xi=\xi_c} = \mathrm{e}^{-\mathrm{i}\frac{3\pi}{4}}\sqrt{\frac{2}{27}}\left(\frac{\xi_c+\mathrm{i}}{\xi_c-\mathrm{i}}\right)^{3/4}, \tag{2.133}$$

or

$$\sigma_0 = \sqrt{\frac{2}{27}}\left(\frac{r_2}{r_1}\right)^{3/4}\mathrm{e}^{\mathrm{i}\frac{3}{4}(\theta_2-\theta_1-\pi)}, \tag{2.134}$$

where

$$\xi_c+\mathrm{i} = r_2\mathrm{e}^{\mathrm{i}\theta_2}; \quad \xi_c-\mathrm{i} = r_1\mathrm{e}^{\mathrm{i}\theta_1}.$$

From Fig. 2.6, it is noted that for fixed $\varepsilon > 0$, as $\xi \gg 1$, we have $\mathrm{Im}\{k_0^{(1)}\} < 0$, while $\mathrm{Im}\{k_0^{(3)}\} > 0$. Hence, as $\xi \to \infty$, the solution H_1 increases exponentially whereas H_3 decreases exponentially. Thus in order to satisfy the root condition in the far field, the following asymptotic form must hold:

$$\tilde{h}_0 \sim D_3' H_3(\xi,\varepsilon) \quad \text{as } \xi \to \infty \quad (\varepsilon > 0). \tag{2.135}$$

As a consequence, in the far field one also has

$$\tilde{h}_0 \sim D_3' H_3(\xi,\varepsilon) \quad \text{as } \varepsilon \to 0. \tag{2.136}$$

This asymptotic condition shows that the solution (2.101), in the far-field, is approximated by a subdominant function $H_3(\xi,\varepsilon)$. As a consequence, for a fixed $\varepsilon > 0$, the asymptotic form (2.136) of the solution $\tilde{h}_0$ cannot be applied to the entire complex ξ-plane as a good approximation for the solution $\tilde{h}_0$, due to the so-called *Stokes phenomenon* (Bender and Orszag, 1978).

Let us first choose the singular point ξ_c as the lower limit, ξ_0, of the integral in the definition of ξ_+. Subsequently, the fundamental solutions are defined as

$$\begin{aligned} H_1(\xi,\varepsilon) &= \exp\left(\frac{\mathrm{i}}{\varepsilon}\int_{\xi_c}^{\xi} k_0^{(1)}\mathrm{d}\xi'\right), \\ H_3(\xi,\varepsilon) &= \exp\left(\frac{\mathrm{i}}{\varepsilon}\int_{\xi_c}^{\xi} k_0^{(3)}\mathrm{d}\xi'\right). \end{aligned} \tag{2.137}$$

The asymptotic solution (2.101) in the complex ξ-plane is an exponential function of $\frac{1}{\varepsilon}$. As $\varepsilon \to 0$, the two fundamental solutions (2.137) are of different orders of magnitude in the complex ξ-plane, except at some isolated lines known as the Stokes lines. The Stokes lines are defined by the integral

$$\mathrm{Im}\Big\{\int_{\xi_c}^{\xi}\left(k_0^{(1)} - k_0^{(3)}\right)\mathrm{d}\xi'\Big\} = 0\,. \tag{2.138}$$

On the other hand, the anti-Stokes lines are defined by

$$\mathrm{Re}\Big\{\int_{\xi_c}^{\xi}\left(k_0^{(1)} - k_0^{(3)}\right)\mathrm{d}\xi'\Big\} = 0\,. \tag{2.139}$$

A sketch of the Stokes lines (L_1), (L_2), (L_3) and anti-Stokes lines (A_1), (A_2), (A_3) of our system are shown in Fig. 2.7. The anti-Stokes line (A_2) divides the entire complex ξ-plane into the sector (S_1) and sector (S_2) . The far field $\xi = \infty$ belongs to sector (S_2), while the tip of the dendrite, $\xi = 0$, belongs to sector (S_1).

Note that when ξ is located at the right side of (L_2), $H_1(\xi,\varepsilon) \gg H_3(\xi,\varepsilon)$ exponentially as $\varepsilon \to 0$. As a result, the function $H_1(\xi,\varepsilon)$ is dominant while the function $H_3(\xi,\varepsilon)$ is subdominant. When ξ is located at the left side of (L_2), $H_3(\xi,\varepsilon) \gg H_1(\xi,\varepsilon)$ exponentially as $\varepsilon \to 0$. In this case, the function $H_3(\xi,\varepsilon)$ is dominant while the function $H_1(\xi,\varepsilon)$ is subdominant. Directly on the Stokes lines, $H_1(\xi,\varepsilon) = O(H_3(\xi,\varepsilon))$, the two functions have the same order of magnitudes as $\varepsilon \to 0$. When ξ moves across the Stokes line, the other of these two functions becomes dominant.

According to the turning point theory, the pair of coefficients of the solution (2.101) will be different in different sectors (S_1) and (S_2). Let us denote the coefficients of the solution (2.101) in (S_2) by $\{D_1', D_3'\}$ and by $\{D_1, D_3\}$ in (S_2). From the far field condition, we derive $D_1' = 0$, and we also have $D_3 = D_3'$. The connection condition between the coefficients $\{D_1, D_3 = D_3'\}$ in sector (S_1) and $\{D_1' = 0, D_3'\}$ in sector (S_2) is to be derived by matching the outer solution (2.101) with the inner solution near the singular point ξ_c.

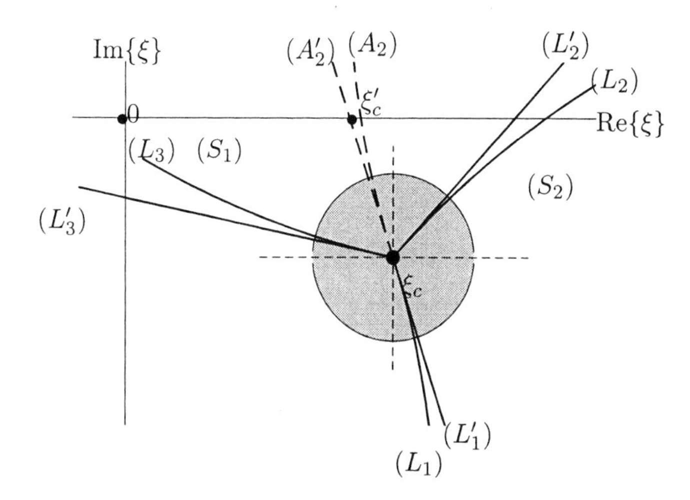

Figure 2.7. A sketch of the structure of Stokes lines for the system of dendritic growth

7. The Inner Solutions near the Singular Point ξ_c

As previously indicated, the MVE solution (2.101) is not valid at the singular point ξ_c. This implies that the solutions in the vicinity of ξ_c: $|\xi - \xi_c| \ll 1$; $|\eta - 1| \ll 1$, no longer have a multiple scale structure. To derive the inner solutions we must start with the perturbed system (2.61)–(2.66) and construct a different asymptotic expansion. For this purpose, we introduce the inner variables

$$\xi_* = \frac{\xi - \xi_c}{\varepsilon^\alpha}, \qquad \eta_* = \frac{\eta - 1}{\varepsilon^\alpha}, \tag{2.140}$$

where α is to be determined later. With the inner variables, the interface shape function is

$$\eta_s(\xi, t) = 1 + \frac{\tilde{h}}{\eta_0^2} = 1 + \varepsilon^\alpha \eta_{*s}. \tag{2.141}$$

Writing

$$\eta_{*s} = \frac{\hat{h}(\xi_*, t)}{\eta_0^2}, \tag{2.142}$$

we have

$$\tilde{h}(\xi, t) = \varepsilon^\alpha \hat{h}(\xi_*, t). \tag{2.143}$$

Accordingly, we put

$$\tilde{T}(\xi,\eta,t) = \varepsilon^{\alpha}\hat{T}(\xi_*,\eta_*,t); \quad \tilde{T}_{\rm S}(\xi,\eta,t) = \varepsilon^{\alpha}\hat{T}_{\rm S}(\xi_*,\eta_*,t). \qquad (2.144)$$

We seek the mode solutions and make the inner expansions

$$\begin{aligned}
\hat{T}(\xi_*,\eta_*,t) &= \Big[\hat{\nu}_0(\varepsilon)\hat{T}_0(\xi_*,\eta_*) + \nu_1(\varepsilon)\hat{T}_1(\xi_*,\eta_*) + \ldots\Big]e^{\frac{\sigma t}{\varepsilon\eta_0^2}}, \\
\hat{T}_{\rm S}(\xi_*,\eta_*,t) &= \Big[\hat{\nu}_0(\varepsilon)\hat{T}_{\rm S0}(\xi_*,\eta_*) + \nu_1(\varepsilon)\hat{T}_{\rm S1}(\xi_*,\eta_*) + \ldots\Big]e^{\frac{\sigma t}{\varepsilon\eta_0^2}}, \qquad (2.145) \\
\hat{h} &= \Big[\hat{\nu}_0(\varepsilon)\hat{h}_0 + \nu_1(\varepsilon)\hat{h}_1 + \cdots\Big]e^{\frac{\sigma t}{\varepsilon\eta_0^2}}.
\end{aligned}$$

To find the inner solution, we need further to apply the transformation first introduced in (Xu, 1989) and transform the solution $\tilde{h}_0$ into a new unknown function $W_0(\xi)$,

$$\tilde{h} = W(\xi)\exp\left[\frac{\rm i}{\varepsilon}\int_{\xi_c}^{\xi} k_c(\xi_1)\, {\rm d}\xi_1\right], \qquad (2.146)$$

where the reference wave number function $k_c(\xi)$ is to be chosen later. It will be seen that

$${\rm Re}\{k_0^{(3)}\} < {\rm Re}\{k_c\} < {\rm Re}\{k_0^{(1)}\}. \qquad (2.147)$$

In the outer region we have

$$H_1 = W_0^{(+)}(\xi)\exp\left[\frac{\rm i}{\varepsilon}\int_{\xi_c}^{\xi} k_c(\xi_1)\, {\rm d}\xi_1\right] \qquad (2.148)$$

and

$$H_3 = W_0^{(-)}(\xi)\exp\left[\frac{\rm i}{\varepsilon}\int_{\xi_c}^{\xi} k_c(\xi_1)\, {\rm d}\xi_1\right]. \qquad (2.149)$$

The general outer solutions for W_0 may be written in the form

$$W_0 = D_1 W_0^{(+)} + D_3 W_0^{(-)}. \qquad (2.150)$$

The wave number function for the $W_0^{(+)}$ wave is $k_0^{(+)} = k_0^{(1)} = k_0^{(1)} - k_c$. Since ${\rm Re}\{k_0^{(+)}\} > 0$, the phase velocity of the $W_0^{(+)}$ wave is positive and, as such. it is an outgoing wave. On the other hand, the wave number function for the $W^{(-)}$ wave is $k_0^{(-)} = k_0^{(3)} = k_0^{(3)} - k_c$, and ${\rm Re}\{k_0^{(-)}\} < 0$. Hence, the $W_0^{(-)}$ wave is an incoming wave with negative phase velocity.

One can draw the following diagram of the relation between the H waves and the W waves:

H_1 wave $\Longleftrightarrow$ $W^{(+)}$
(short-wave branch) (outgoing wave);

H_3 wave $\Longleftrightarrow$ $W^{(-)}$
(long-wave branch) (incoming wave).

In accordance with the above, in the inner region we can set

$$\hat{h} = \hat{W}(\xi_*) \exp\left\{ \frac{\mathrm{i}}{\varepsilon} \int_{\xi_c}^{\xi} k_c(\xi_1)\, \mathrm{d}\xi_1 \right\}. \tag{2.151}$$

We set

$$k_c(\xi) = \mathrm{e}^{-\mathrm{i}\frac{\pi}{4}} \sqrt{\frac{S}{6}\,(\xi + \mathrm{i})} \qquad (\,\mathrm{Re}\{k_c\} > 0) \tag{2.152}$$

and

$$\alpha = \frac{2}{3} \quad \text{and} \quad \xi_* = \frac{(\xi - \xi_c)}{\varepsilon^{\frac{2}{3}}}, \tag{2.153}$$

then the leading-order approximation of the inner equation is found to be the Airy equation

$$\frac{\mathrm{d}^2 \hat{W}_0}{\mathrm{d}\xi_*^2} + A^2 \xi_* \hat{W}_0 = 0, \tag{2.154}$$

where

$$A = \left(\frac{S^3}{6k_c} \frac{\partial \Sigma_c}{\partial \xi} \right)^{\frac{1}{2}}_{\xi=\xi_c} = \mathrm{i}\sqrt{\frac{1}{6}} \left(\frac{\xi_c + \mathrm{i}}{\xi_c - \mathrm{i}} \right)^{\frac{1}{4}} \tag{2.155}$$

$$\left(\frac{\pi}{2} < \arg\{A\} < \frac{3\pi}{4} \right).$$

This can be written as the standard Airy equation

$$\frac{\mathrm{d}^2 \hat{W}_0}{\mathrm{d}\hat{\xi}_*^2} + \hat{\xi}_* \hat{W}_0 = 0, \tag{2.156}$$

by introducing the new inner variable

$$\hat{\xi}_* = \frac{A^{\frac{2}{3}}}{\varepsilon^{\frac{2}{3}}} (\xi - \xi_c). \tag{2.157}$$

The general solution of the above Airy equation is

$$\hat{W}_0 = D_{*1}\hat{\xi}_*^{\frac{1}{2}} H^{(1)}_{\frac{1}{3}}(\zeta) + D_{*2}\hat{\xi}_*^{\frac{1}{2}} H^{(2)}_{\frac{1}{3}}(\zeta), \tag{2.158}$$

$$\left(\zeta = \frac{2}{3}\hat{\xi}_*^{\frac{3}{2}}\right),$$

where $H^{(1)}_\nu(z)$ is ν^{th}-order Hankel function of the first kind, while $H^{(2)}_\nu(z)$ is the Hankel function of the second kind. In order to match with the outer solution, which satisfies the downstream far-field condition (2.136) as $\hat{\xi}_* \to \infty$, the inner solution must be

$$\hat{W}_0 = D_*\hat{\xi}_*^{\frac{1}{2}} H^{(2)}_{\frac{1}{3}}(\zeta). \tag{2.159}$$

By matching the inner solution with the outer solution in the sector (S_1) and (S_2), we verify

$$k_1 \sim \frac{\mathrm{i}}{4(\xi - \xi_c)} + o(1), \quad R_1(\xi_c) = 0, \tag{2.160}$$

and obtain the connection condition

$$\frac{D_1}{D_3} = \mathrm{e}^{\frac{\mathrm{i}\pi}{2}}, \quad D_3 = D_3'. \tag{2.161}$$

8. Tip Inner Solution in the Tip Region

In the tip region, we define the tip inner variables

$$\begin{aligned} \hat{\xi} &= \frac{\hat{k}\xi}{\varepsilon}, \\ \hat{\eta} &= \frac{(\eta - 1)}{\varepsilon}, \\ \hat{t} &= \frac{t}{\eta_0^2 \varepsilon}, \end{aligned} \tag{2.162}$$

where $|\xi| \ll \varepsilon$, and $|\eta - 1| \ll 1$. The tip solution can be expressed as a function of these inner variables and expanded in the following asymptotic form as $\varepsilon \to 0$:

$$\begin{aligned} \hat{T}(\hat{\xi}, \hat{\eta}, \theta, \hat{t}) &= \left\{\hat{\mu}_0(\varepsilon)\hat{T}_0 + \hat{\mu}_1(\varepsilon)\hat{T}_1 + \cdots\right\}\mathrm{e}^{im\theta + \sigma\hat{t}}, \\ \hat{T}_S(\hat{\xi}, \hat{\eta}, \theta, \hat{t}) &= \left\{\hat{\mu}_0(\varepsilon)\hat{T}_{S0} + \hat{\mu}_1(\varepsilon)\hat{T}_{S1} + \cdots\right\}\mathrm{e}^{im\theta + \sigma\hat{t}}, \\ \hat{h}(\hat{\xi}, \theta, \hat{t}) &= \left\{\hat{\mu}_0(\varepsilon)\hat{h}_0 + \hat{\mu}_1(\varepsilon)\hat{h}_1 + \cdots\right\}\mathrm{e}^{im\theta + \sigma\hat{t}} \\ \hat{k} &= \hat{k}_0 + \varepsilon\hat{k}_1 + \cdots . \end{aligned} \tag{2.163}$$

At the zeroth-order in the tip region, the system (2.61)–(2.66) can be reduced to

$$\begin{aligned}\hat{k}_0^2\frac{\partial^2\hat{T}_0}{\partial\hat{\xi}^2}+\frac{\partial^2\hat{T}_0}{\partial\hat{\eta}^2}+\frac{\hat{k}_0^2}{\hat{\xi}}\frac{\partial\hat{T}_0}{\partial\hat{\xi}}-\frac{m^2\hat{k}_0^2}{\hat{\xi}^2}\hat{T}_0&=0\\ \hat{k}_0^2\frac{\partial^2\hat{T}_{S0}}{\partial\hat{\xi}^2}+\frac{\partial^2\hat{T}_{S0}}{\partial\hat{\eta}^2}+\frac{\hat{k}_0^2}{\hat{\xi}}\frac{\partial\hat{T}_{S0}}{\partial\hat{\xi}}-\frac{m^2\hat{k}_0^2}{\hat{\xi}^2}\hat{T}_{S0}&=0\end{aligned}\tag{2.164}$$

with the boundary conditions at $\hat{\eta}$=0

$$\hat{T}_0=\hat{T}_{S0}+\hat{h}_0,\tag{2.165}$$

$$\hat{T}_{S0}=\hat{k}_0^2\left(\frac{\partial^2\hat{h}_0}{\partial\hat{\xi}^2}+\frac{1}{\hat{\xi}}\frac{\partial\hat{h}_0}{\partial\hat{\xi}}-\frac{m^2}{\hat{\xi}^2}\hat{h}_0\right),\tag{2.166}$$

$$\frac{\partial}{\partial\hat{\eta}}(\hat{T}_0-\hat{T}_{S0})+\sigma_0\hat{h}_0=0.\tag{2.167}$$

This system admits the inner solutions

$$\begin{aligned}\hat{T}_0&=\hat{a}_0H_m^{(1)}(\hat{\xi})\mathrm{e}^{-\hat{k}_0\hat{\eta}},\\ \hat{T}_{S0}&=\hat{a}_{S0}H_m^{(1)}(\hat{\xi})\mathrm{e}^{\hat{k}_0\hat{\eta}},\\ \hat{h}_0&=\hat{d}_0H_m^{(1)}(\hat{\xi}),\end{aligned}\tag{2.168}$$

where $H_m^{(1)}(\hat{\xi})$ is the m-order Hankel function of the first kind.

From the boundary conditions (2.165)–(2.167), one obtains

$$\hat{a}_0=(1-\hat{k}_0^2)\hat{d}_0\,,\qquad \hat{a}_{S0}=-\hat{k}_0^2\hat{d}_0\,,\tag{2.169}$$

and the local dispersion relation in the tip region

$$\sigma_0=\hat{k}_0(1-2\hat{k}_0^2).\tag{2.170}$$

For fixed σ_0, (2.170) has three roots for $\hat{k}_0$. Comparing the local dispersion relation in the tip region (2.170) with the local dispersion relation in the outer region (2.97) one can evidently write

$$\begin{cases}\hat{k}_0^{(1)}=k_0^{(1)}(0),\\ \hat{k}_0^{(2)}=k_0^{(2)}(0),\\ \hat{k}_0^{(3)}=k_0^{(3)}(0).\end{cases}\tag{2.171}$$

The root $\hat{k}_0^{(2)}$ must be ruled out due to the fact that $\mathrm{Re}\{\hat{k}_0^{(2)}\} < 0$. Therefore, the general solution of $\hat{h}_0$ in the tip region is

$$\hat{h}_0 = \hat{d}_0^{(1)} H_m^{(1)}\left(\hat{k}_0^{(1)}\hat{\xi}\right) + \hat{d}_0^{(3)} H_m^{(1)}\left(\hat{k}_0^{(3)}\hat{\xi}\right). \tag{2.172}$$

The above tip inner solution must match with the outer solution. By setting

$$\hat{d}_0^{(1)} = d_0, \quad \hat{d}_0^{(3)} = -\hat{Q}_0 d_0, \tag{2.173}$$

we can express the tip inner solution as

$$\hat{h}_0 = d_0 \left\{ H_m^{(1)}\left(\hat{k}_0^{(1)}\hat{\xi}\right) - \hat{Q}_0 H_m^{(1)}\left(\hat{k}_0^{(3)}\hat{\xi}\right)\right\}. \tag{2.174}$$

As $\hat{\xi} \to \infty$, one has

$$\begin{aligned}\hat{h}_0 &\approx d_0\sqrt{\frac{2}{\pi}}\mathrm{e}^{-(m+\frac{1}{2})\frac{\pi}{2}} \\ &\times \left\{ \sqrt{\frac{\varepsilon}{k_0^{(1)}(0)\xi}}\,\mathrm{e}^{\frac{\mathrm{i}k_0^{(1)}(0)\xi}{\varepsilon}} - \hat{Q}_0\sqrt{\frac{\varepsilon}{k_0^{(3)}(0)\xi}}\,\mathrm{e}^{\frac{\mathrm{i}k_0^{(3)}(0)\xi}{\varepsilon}} \right\}. \end{aligned} \tag{2.175}$$

Now we need to match the tip inner solution (2.174) with the outer solution derived in the last section, namely,

$$\hat{\mu}_0(\varepsilon)\sqrt{\frac{2\varepsilon}{\pi}}d_0\left\{\sqrt{\frac{1}{\hat{k}_0^{(1)}\hat{\xi}}}\mathrm{e}^{\frac{\mathrm{i}\hat{k}_0^{(1)}\xi}{\varepsilon}} - \hat{Q}_0\sqrt{\frac{1}{\hat{k}_0^{(3)}\hat{\xi}}}\mathrm{e}^{\frac{\mathrm{i}\hat{k}_0^{(3)}\xi}{\varepsilon}}\right\}$$

$$\Updownarrow$$

$$\tilde{\mu}_0(\varepsilon)\Big(D_1 \exp\Big\{\frac{i}{\varepsilon}\int_{\xi_c}^{\xi} k_0^{(1)}\mathrm{d}\xi'\Big\} + D_3\exp\Big\{\frac{i}{\varepsilon}\int_{\xi_c}^{\xi} k_0^{(3)}\mathrm{d}\xi'\Big\}\Big).$$

To satisfy this matching condition, the parameter σ_0 must be a proper function of ε, such that as $\varepsilon \to 0$, the two functions, $\{\mathrm{e}^{\mathrm{i}\chi_1(\varepsilon)}, \mathrm{e}^{\mathrm{i}\chi_3(\varepsilon)}\}$ are of the same order of magnitude. Hereby, we have defined

$$\begin{aligned}\chi_1(\varepsilon) &= \frac{1}{\varepsilon}\int_0^{\xi_c} k_0^{(1)}(\xi')\mathrm{d}\xi', \\ \chi_3(\varepsilon) &= \frac{1}{\varepsilon}\int_0^{\xi_c} k_0^{(3)}(\xi')\mathrm{d}\xi'.\end{aligned} \tag{2.176}$$

In other words, the parameter σ_0 must be properly chosen, so that the turning point ξ_c and the structure of Stokes lines, as the functions of σ_0, are arranged in such a way, that the tip $\xi = 0$ is located on the Stokes line (L_3).

Thus it follows from the matching condition that

$$\hat{\mu}_0(\varepsilon)\varepsilon^{\frac{1}{2}} = \tilde{\mu}_0(\varepsilon), \tag{2.177}$$

and

$$\frac{D_3 \mathrm{e}^{-\mathrm{i}\chi_3}}{D_1 \mathrm{e}^{-\mathrm{i}\chi_1}} = -\mathrm{i}\mathrm{e}^{\mathrm{i}(\chi_1-\chi_3)} = -\hat{Q}_0 \left(\frac{k_0^{(1)}(0)}{k_0^{(3)}(0)}\right)^{\frac{1}{2}}, \tag{2.178}$$

or,

$$\hat{Q}_0 = \hat{Q}_0(\sigma_0, \varepsilon) = \mathrm{i}\mathrm{e}^{\mathrm{i}(\chi_1-\chi_3)} \left(\frac{k_0^{(1)}(0)}{k_0^{(3)}(0)}\right)^{\frac{1}{2}}. \tag{2.179}$$

Up to this point, the first order asymptotic solution contains a free parameter, σ_0.

GTW Modes and Quantization Condition]

9. Global Trapped-Wave (GTW) Modes and the Quantization Condition

We now turn to the second step: constructing the global eigenmodes and deriving the quantization condition for the eigenvalues by applying the tip smoothness condition (2.69) to the asymptotic solutions obtained above. Note that as $\hat{\xi} \to 0$,

$$H_m^{(1)}(\hat{k}_0\hat{\xi}) \sim \begin{cases} -\mathrm{i}\dfrac{(m-1)!}{\pi}\left(\dfrac{2}{\hat{k}_0\hat{\xi}}\right)^m + O\left(\dfrac{2}{\hat{k}_0\hat{\xi}}\right)^{m-2}, & (m>0), \\ \mathrm{i}\dfrac{2}{\pi}\ln(\hat{k}_0\hat{\xi}) + 1 + \mathrm{i}\dfrac{2}{\pi}(\gamma - \ln 2) + O\left(\hat{k}_0\hat{\xi}\right), & (m=0). \end{cases} \tag{2.180}$$

It is seen that in order for tip solution (2.174) to satisfy the tip condition, one must set

$$\hat{Q}_0 = \left(\frac{\hat{k}_0^{(3)}}{\hat{k}_0^{(1)}}\right)^m = \left(\frac{k_0^{(3)}(0)}{k_0^{(1)}(0)}\right)^m, \tag{2.181}$$

and only three modes $m = 0, 1, 2$ are allowable. The $m = 0$ mode gives an axi-symmetric interfacial travelling wave, while $m = 1$ and $m = 2$ mode, respectively, gives one-arm and two-arm spiral interfacial waves propagating along the Ivantsov paraboloid.

Formula (2.181) is the quantization condition for the eigenvalues. Combining with (2.179), one may re-write (2.181) in the form

$$\frac{1}{\varepsilon}\int_0^{\xi_c}\left(k_0^{(1)}-k_0^{(3)}\right)\mathrm{d}\xi=\left(2n+1+\frac{1}{2}+\frac{2m+1}{2}\theta_0\right)\pi$$
$$-\mathrm{i}\frac{2m+1}{2}\ln\alpha_0 \tag{2.182}$$
$$(m=0,1,2;\quad n=0,\pm1,\pm2,\ldots),$$

where

$$\alpha_0\,\mathrm{e}^{\mathrm{i}\theta_0\pi}=\frac{k_0^{(1)}(0)}{k_0^{(3)}(0)}\,. \tag{2.183}$$

This quantization condition gives a discrete set of complex eigenvalues

$$\sigma_0^{(n)}\quad(m=0,1,2;\quad n=0,\pm1,\pm2,\cdots),$$

as functions of the parameters ε. For the special axi-symmetric case (m=0), we have

$$\frac{1}{\varepsilon}\int_0^{\xi_c}\left(k_0^{(1)}-k_0^{(3)}\right)\mathrm{d}\xi=\left(2n+1+\frac{1}{2}+\frac{1}{2}\theta_0\right)\pi-\mathrm{i}\frac{1}{2}\ln\alpha_0. \tag{2.184}$$

This quantization condition gives a discrete set of complex eigenvalues

$$\sigma_0^{(n)}\quad(n=0,\pm1,\pm2,\cdots).$$

In Fig. 2.8, we show the variations of $\sigma_{0\mathrm{R}}$ with index $n=0$ for the modes $m=0,1,2$. It is seen that the axi-symmetric mode ($m=0$) is the most dangerous one. Hence, for stability analysis, one only needs to consider the axi-symmetric perturbations for three-dimensional dendritic growth.

In Figs. 2.10 and 2.11, we show the variations of $\sigma_{0\mathrm{R}}$ and ω_0 of global trapped-wave (GTW) modes $n=0,1,2$. The system under consideration has no real spectrum. It is seen that the system allows a unique neutral n mode ($\sigma_{\mathrm{R}}=0$) with the eigenvalue $\sigma=-\mathrm{i}\omega_{*n}$ when $\varepsilon=\varepsilon_{*n}$ (where $\varepsilon_{*0}=\varepsilon_*>\varepsilon_{*1}>\varepsilon_{*2}>\cdots$). Here, the critical number ε_* corresponds to the global neutrally stable mode with the index $n=0$. Obviously, when $\varepsilon>\varepsilon_*$ the system will be absolutely stable. When $\varepsilon_{*1}<\varepsilon<\varepsilon_{*0}$, the system has one growing mode and infinitely many decaying modes; when $\varepsilon_{*2}<\varepsilon<\varepsilon_{*1}$, the system has two growing modes and, in general, when $\varepsilon_{*k}<\varepsilon<\varepsilon_{*(k-1)}$, the system has k growing modes. As $\varepsilon\to0$, the eigenvalues of these growing modes apparently

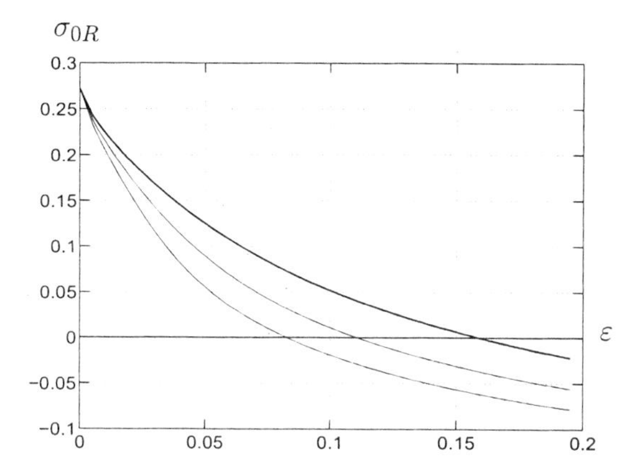

Figure 2.8. The variations of $\sigma_{0\mathrm{R}}$ with ε for the modes $m = 0, 1, 2$ with index $n = 0$, from top to bottom

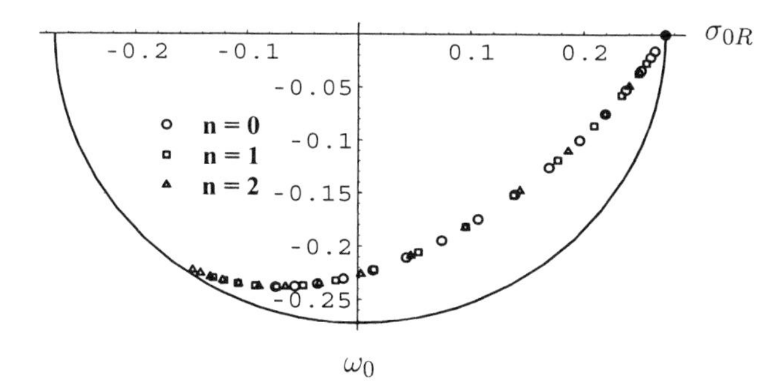

Figure 2.9. The variation of eigenvalues σ_0 with ε in the complex σ_0-plane. The real part of σ_0 decreases with increasing ε

tend to the limit $\sigma_0 = (0.2722, 0.0)$. We also show the variation of the eigenvalues on the complex σ_0-plane with ε in Fig. 2.9. It is very interesting to see that in the leading order approximation, all eigenvalues of the modes $n = 0, 1, 2, \ldots$ are on the same curve in the complex σ_0-plane.

In the leading order approximation, the eigenvalues $\sigma \approx \sigma_0$ are independent of the Peclet number η_0^2. We have calculated that the global neutrally stable mode with the index $n = 0$ has the eigenvalue $\sigma = -\mathrm{i}\omega_*^{(0)} = -0.2129\mathrm{i}$. It corresponds to the critical number neutrally stable mode

$$\varepsilon = \varepsilon_*^{(0)} = 0.1590. \tag{2.185}$$

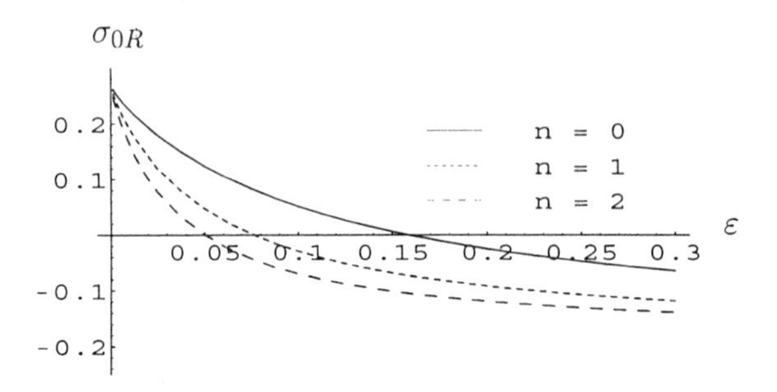

Figure 2.10. The variations of the real part of the zeroth-order approximation of eigenvalues, $\sigma_{0\mathrm{R}}$, of 3D, axially symmetrical GTW modes ($n = 0, 1, 2$) with ε

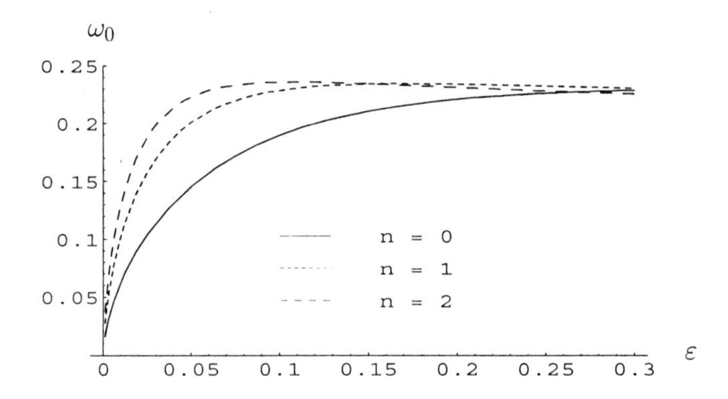

Figure 2.11. The variations of the imaginary part of the zeroth-order approximation of eigenvalues, ω_0, of 3D, axially symmetrical GTW modes ($n = 0, 1, 2$) with ε

In the first-order approximation, the eigenvalues $\sigma \approx \sigma_0 + \varepsilon\sigma_1$ will be a function of the Peclet number η_0^2. Consequently, the eigenvalue $\sigma = -\mathrm{i}\omega_*^{(1)}$ of the neutral mode ($n = 0$), as well as the corresponding critical number $\varepsilon = \varepsilon_*^{(1)}$, are functions of the Peclet number $\mathrm{Pe}_0 = \eta_0^2$. For small undercooling ($|T_\infty| \ll 1$), such a dependence is insensitive. We find that as $|T_\infty| = 0.002$,

$$\varepsilon_*^{(1)} \approx 0.1108, \quad \omega_*^{(1)} \approx 0.2183. \tag{2.186}$$

However, for large undercooling, say, $\mathrm{Pe}_0 > 1.0$, the situation is changed. The critical number $\varepsilon_*^{(1)}$ decreases significantly as the undercooling temperature increases (see Fig 2.12).

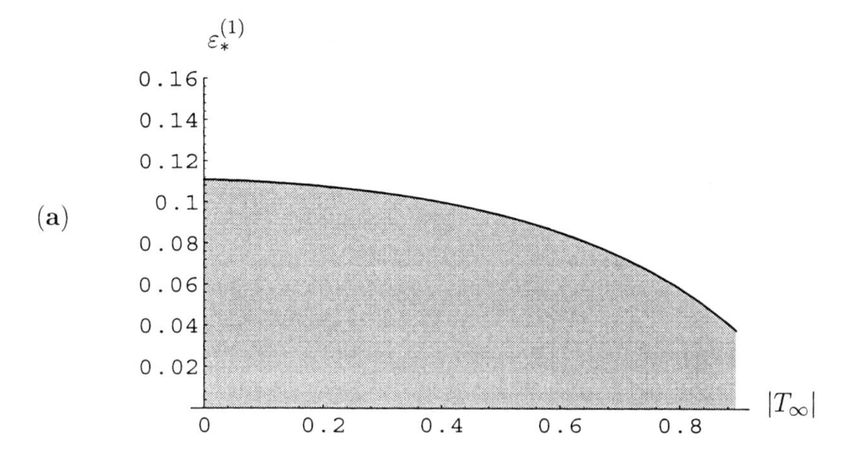

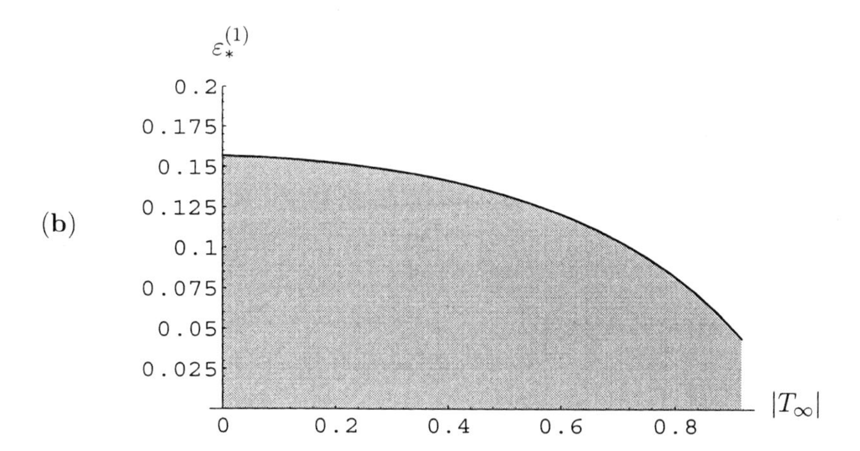

Figure 2.12. The variation of the critical number $\varepsilon_*^{(1)}$ for the cases: (a) symmetric model; (b) one-sided model. The shaded region is the linearly unstable region of steady solutions.

The global mode solutions obtained above have important physical implications. A wave diagram for these global modes is sketched in Fig. 2.13. It shows that an incident outgoing wave $W_0^{(+)}$ from the tip collides with an incoming wave $W_0^{(T)}$ from the far field at the point ξ_c' on the anti-Stokes line (A_2); the collision generates an incoming wave $W_0^{(-)}$ propagating towards the tip region. This incoming wave $W_0^{(-)}$ is then reflected at the tip region, and again becomes an outgoing wave $W_0^{(+)}$. The waves appear trapped in the sector (S_2) between the tip point and

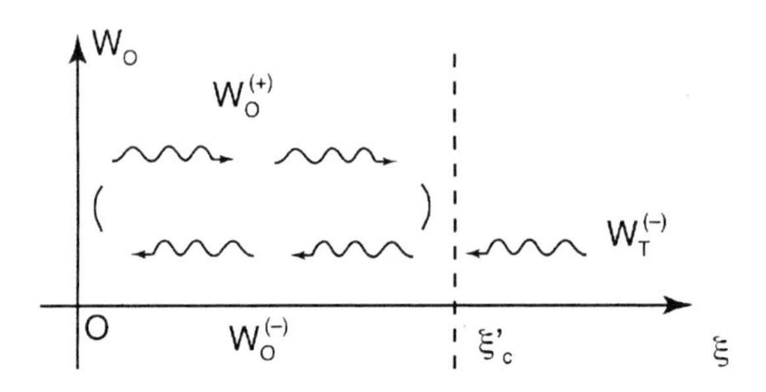

Figure 2.13. Wave diagram of the GTW mechanism

(a)

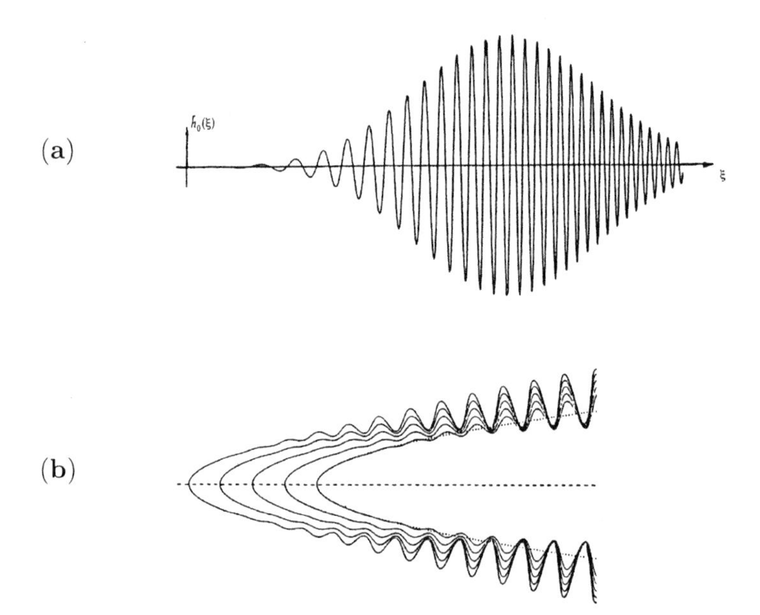

(b)

Figure 2.14. A typical GTW neutral mode: (**a**) the graphics of the eigenfunction; (**b**) the interface shape in a time sequence

the point ξ'_c. No wave escapes beyond the anti-Stokes line (A_2). This is the reason why we call these global modes the *Global Trapped-Wave* (GTW) modes. In the far field, the solution $\tilde{h}(\xi, t)$ describes a long outgoing H_3 wave.

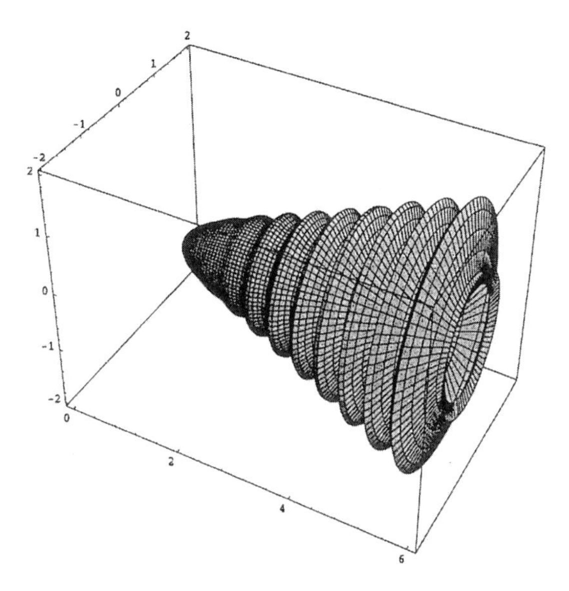

Figure 2.15. The 3D graphics of the interface shape of a typical GTW neutral mode

From the analytical form of the GTW modes, one can calculate the phase velocity of these travelling waves in the far field $\xi \to \infty$. The numerical computations show that for the GTW neutral modes the phase velocity $V_{\mathrm{p}} \approx 1.0$. This implies that the phase velocity of the GTW modes, in the moving frame fixed at the tip, is approximately equal to the tip velocity in the laboratory frame. This result is in agreement with experimental observations.

The existence of growing GTW modes explains the origin and essence of the dendritic structure in the solidifying system. Experimental observations show that dendritic growth always experiences such transient process that whose tip-velocity, U_{tip} accelerates from a smaller initial value and finally reaches to its limiting value. During this transient time, the corresponding parameter ε is smaller than the critical number ε_*. Hence, the system is unstable, any initial perturbation in the growth process may stimulate a spectrum of the above-described global modes. As $t \to \infty$, all decaying modes will vanish, while the amplitudes of the growing modes exponentially increase. Eventually, the GTW mode with the largest growth rate dominates the features of the microstructure of the dendrite. At the later stage of growth, the system will adjust itself, so that the dendrite-tip velocity is close to its limiting speed, while the corresponding ε is close to, but slightly smaller than the critical number ε_*. Thus, the system only remains a unique growing mode with a small growth rate, which is very close to the neutral GTW mode. With such a GTW mode, the head of the dendrite persistently emits a long, outgoing, interfacial wave-train propagating along the interface toward the far field with a phase velocity near unity. As a consequence, on the interface of the non-classic needle solution, the system exhibits a coherent side-branching structure described by the eigenfunction solution. Such structure self-sustained with no need of continuous presence of the perturbations. It is then deduced that as $t \to \infty$, the selection criterion of dendritic growth is $\varepsilon \to \varepsilon_*$.

The critical number ε_* is directly connected with the selected dendrite's tip velocity, tip radius, as well as the oscillation frequency of the dendrite. In fact, if one uses the capillary length ℓ_c as the length scale, one can write the dimensionless tip velocity, dimensionless tip radius and frequency of oscillation as follows:

$$U_{\text{tip}} = \frac{U\ell_c}{\kappa_T} = \frac{\ell_c}{\ell_T} = \varepsilon_*^2 \eta_0^2 , \tag{2.187}$$

$$R_{\text{tip}} = \frac{\ell_t}{\ell_c} = \frac{\ell_t}{\ell_T}\frac{\ell_T}{\ell_c} = \frac{\text{Pe}}{\varepsilon_*^2 \eta_0^4}, \tag{2.188}$$

$$\Omega_* = \frac{\omega_*}{\eta_0^2 \varepsilon_*} . \tag{2.189}$$

We recall that $\eta_0^2 \text{Pe}_0$ is the normalization parameter of the coordinate system or the Peclet number for the case of zero surface tension.

The eigenfunction of a typical selected global neutrally stable mode and its interface shape in a time sequence are shown in Fig. 2.14. The 3D graphics of dendritic growth is shown in Fig. 2.15.

10. The Comparison of Theoretical Predictions with Experimental Data

The growth theory developed above shows that dendritic growth is essentially a wave phenomenon involving the interaction of interfacial waves along the interface. This theory is therefore called the interfacial wave (IFW) theory.

The IFW theory states that when the surface tension is isotropic, at the later stage of growth, dendritic growth is not described by a steady state. Instead, it is described by a time periodic oscillatory state, the so-called global neutrally stable (GNS) state (Xu and Yu, 2001b, 2001a). Such a GNS state consists mathematically of three parts: (1) the Ivantsov solution, (2) the steady regular perturbation expansion (RPE) part due to the surface tension, and (3) the unsteady singular perturbation expansion (SPE) part. The interface shape of the dendrite can be expressed approximately in the form,

$$\eta_s(\xi, t) \approx 1 + \varepsilon^2 \eta_1(\xi) + \hat{h}_0(\xi) e^{\frac{\sigma t}{\eta_0^2 \varepsilon}} \quad (\text{as} \ \ \varepsilon \to 0). \tag{2.190}$$

In the above, $\eta_1(\xi)$ is the leading term in the RPE part; while $\hat{h}_0(\xi) e^{\frac{\sigma t}{\eta_0^2 \varepsilon}}$ is the leading term in the SPE part. The parameter $\sigma = \sigma_R - i\omega$ is the eigenvalue with the expansion form

$$\sigma = \sigma_0 + \varepsilon \sigma_1 + \cdots \quad (\text{as} \ \ \varepsilon \to 0). \tag{2.191}$$

For the selected solution, $\sigma_R = 0$.

In 1994, a series of careful experiments on free dendritic growth in pure succinonitrile (SCN) were conducted, for the first time, in the space shuttle Columbia by the research team headed by Glicksman (Glicksman et al, 1994, 1995). Under the micro-gravity environment, convective motion in the melt is greatly reduced. As expected, the new data for tip velocity and tip radius during the dendritic growth obtained by Glicksman et al. is more accurate.

Since the material SCN has a very small surface tension anisotropy, one can expect that the theoretical results for three-dimensional axially symmetric dendritic growth from a pure melt obtained in this chapter are comparable with these experimental results obtained by Glicksman et al.

According to the IFW theory, as we have seen, the selected tip growth velocity and tip radius are uniquely determined by the critical number ε_* via the formulas (2.187) and (2.188), respectively. Moreover, in the leading-order approximation, $\varepsilon_* = \varepsilon_*^{(0)} = 0.1590$, which is independent of the undercooling T_∞, while in the first-order approximation, $\varepsilon_* =$

Table 2.1. The thermodynamic properties of SCN

mol	0.092 g/mol
ΔH	11.051 cal/g
c_{pL}	0.4791 cal $\mathrm{g}^{-1}\mathrm{K}^{-1}$
c_{pS}	0.4468 cal $\mathrm{g}^{-1}\mathrm{K}^{-1}$
$c_{\mathrm{p}} = (c_{\mathrm{pL}} + c_{\mathrm{pS}})/2$	0.4630 cal $\mathrm{g}^{-1}\mathrm{K}^{-1}$
κ_{TL}	$1.127 \times 10^{-3}\ \mathrm{cm}^2\mathrm{s}^{-1}$
κ_{TS}	$1.155 \times 10^{-3}\ \mathrm{cm}^2\mathrm{s}^{-1}$
$\kappa_{\mathrm{T}} = (\kappa_{\mathrm{TL}} + \kappa_{\mathrm{TS}})/2$	$1.141 \times 10^{-3}\,\mathrm{cm}^2\mathrm{s}^{-1}$
T_{M0}	331.233 K
γ' (Gibbs–Thomson coefficient)	6.480×10^{-6}cm K
γ	$2.136 \times 10^{-7}\,\mathrm{cal\,cm}^{-2}$
β (thermal expansion coefficient)	$8.2 \times 10^{-4}\mathrm{K}^{-1}$

$\varepsilon_*^{(1)}(\eta_0^2)$, which rapidly decreases as the undercooling T_∞ approaches -1. But, in the range of small undercooling $|T_\infty| = 0.001 - 0.2$ under discussion, $\varepsilon_* = \varepsilon_*^{(1)} \approx 0.1108$.

The above theoretical prediction is free of adjustable parameters and can be directly tested in terms of the experimental data of Glicksman et al. It is therefore of great interest to compare quantitatively the predictions of this theory with these newest experimental data.

The thermodynamic data for SCN are listed in Table 2.1. Based on these data, one can calculate that the capillary length for SCN is $\ell_{\mathrm{c}} = 2.804 \times 10^{-7}$cm, the velocity unit is 4069.19 cm/s and the temperature unit is 23.067 K.

In Fig. 2.16, we show the tip velocity U_{tip} versus the undercooling temperature T_∞ and compare the experimental data with the theoretical curve determined by (2.187). In Fig. 2.17 we show the tip radius R_{tip} versus the undercooling T_∞. It is seen that the overall agreement between the theoretical curve and the experimental data is quite satisfactory, especially considering that the experimental data of the tip radius has an error of about 10%. Only in the regime of small undercooling, $|\Delta T| < 0.4$ K or $|T_\infty| < 0.01$, can one see a growing deviation between the theoretical curve and the experimental data.

To explore the deviations between the theoretical curve and the experimental data in more detail, we calculate $U_{\mathrm{tip}}^{\frac{1}{2}}/\mathrm{Pe}_0 = (\varepsilon_*)_{\mathrm{exp}}$ in terms

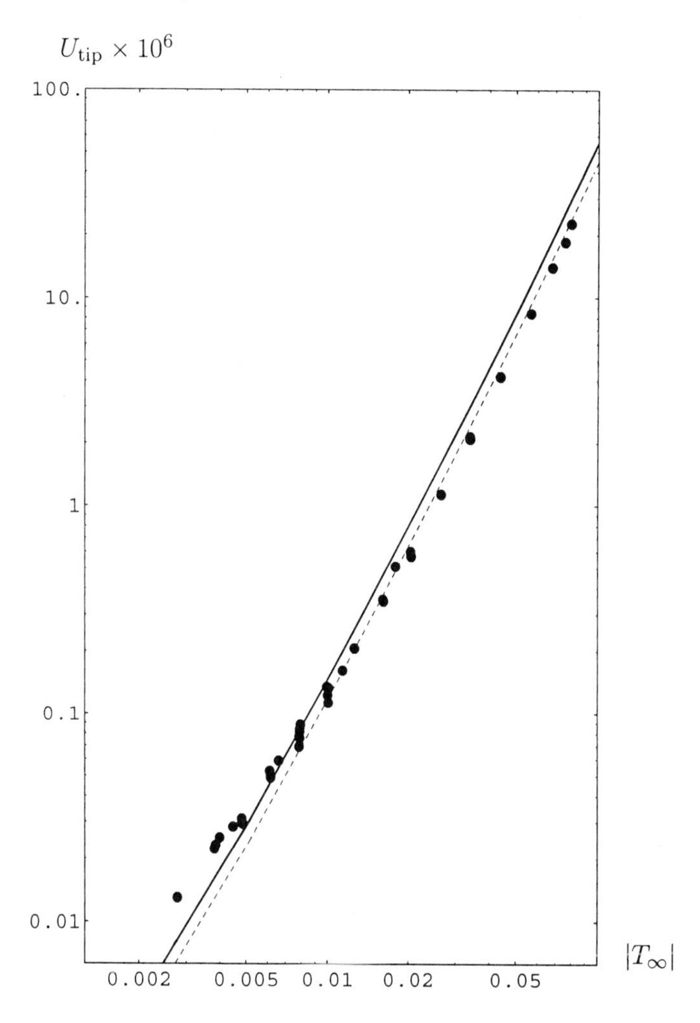

Figure 2.16. The variation of U_{tip} with $|T_\infty|$. The solid line is given by the IFW theory with zero anisotropy, $\alpha_4 = 0$. The dotted line is the modified IFW theory's results with the inclusion of the axial anisotropy of surface tension $\alpha_4 = 0.075$. The dots are the micro-gravity experimental data

of the data obtained from both the flight experiments and ground experiments shown in Fig. 2.18.

As a comparison, in Fig. 2.18 we also show the theoretical value of ε_* by a thin solid line. It is seen that in the entire undercooling regime under discussion, the theoretical curve, which remains flat with $\varepsilon_* \approx 0.1108$, agrees with the experimental data in the high T_∞ regime reasonably well.

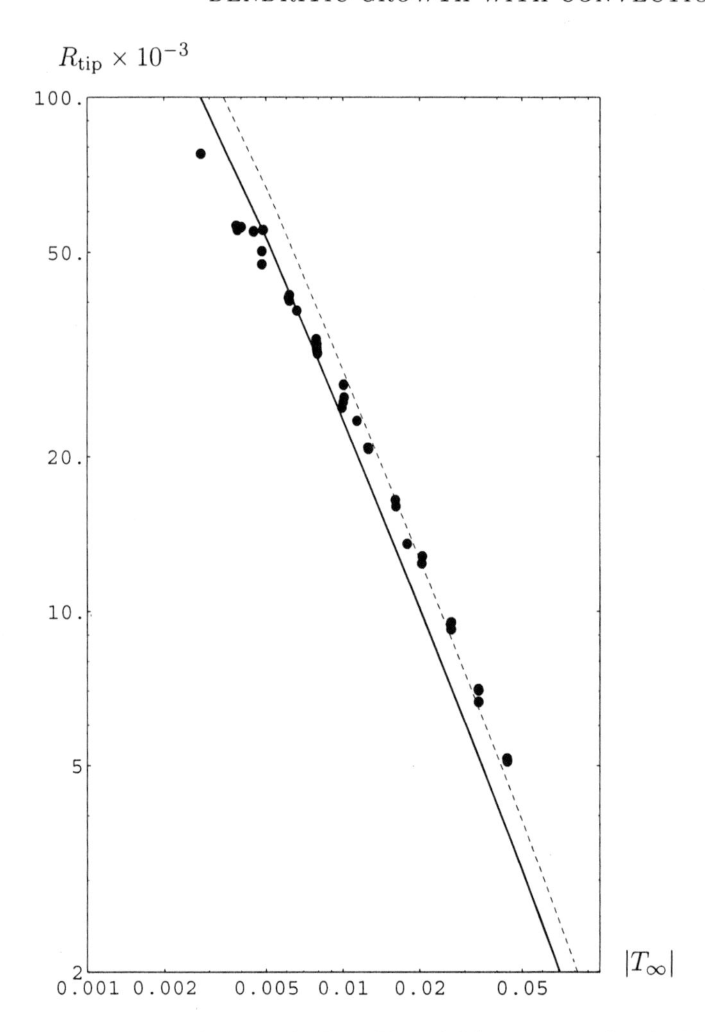

Figure 2.17. The variation of R_{tip} with T_∞. The solid line is given by the IFW theory with zero anisotropy, $\alpha_4 = 0$. The dotted line is the modified IFW theory's results with the inclusion of the axial anisotropy of surface tension $\alpha_4 = 0.075$. The dots are the micro-gravity experimental data

Some notes should be made here.

1. The theory given above neglects the effects of many physical parameters, in particular, as the anisotropy of surface tension α_4 at the interface and the convection in the melt. The effect of anisotropy of surface tension plays a significant role in the early stage of growth, by selecting the orientation of dendrite, initiating dendritic growth

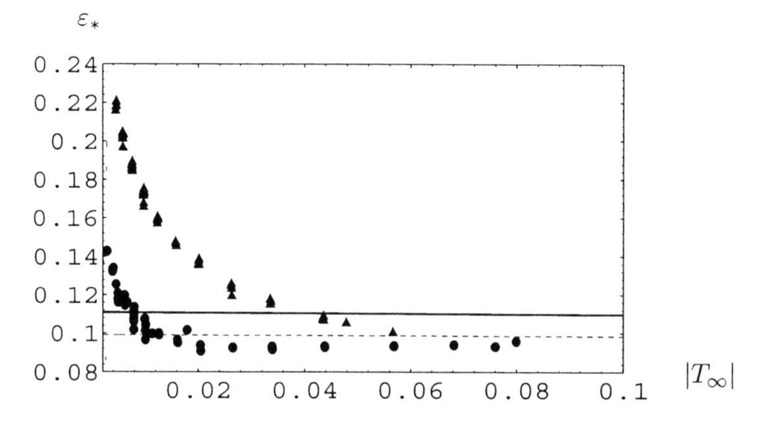

Figure 2.18. Comparison of ε_* with $(\varepsilon_*)_{\rm exp}$ within the region of undercooling, $0.002 < |T_\infty| < 0.1$. The dots are the micro-gravity experimental data, while the triangles are the ground experimental data. The solid line is the stability criterion, $\varepsilon_* = 0.01108$, predicted by IFW theory, with zero anisotropy, $\alpha_4 = 0$. The dashed line is the stability criterion with the anisotropy correction, $\varepsilon_* = 0.0991$, assuming SCN has an axial anisotropy $\alpha_4 = 0.075$

and affecting the morphology of the interface profoundly. The presence of anisotropy of surface tension may also invoke a new instability mechanism, the so-called low frequency (LF) instability, which may dominate the GTW instability, when anisotropy is sufficiently large. However, it has been shown in (Xu, 1997) that when the anisotropy of surface tension is smaller than a critical number, once the dendritic growth is initiated, the effect of the anisotropy of surface tension on the selection of dendrite-tip velocity **at the later stage of growth** is not important. It only changes the numerical value of the selection criterion ε_* slightly. For instance, the anisotropy for SCN is $\alpha_4 \approx 0.075$. Taking this effect into account, the value of ε_* is modified to $\varepsilon_* \approx 0.0991$. This number is shown by the dashed line in Fig. 2.18, which displays an excellent quantitative agreement with experimental data in the regime, where the undercooling is not too small.

2. It is seen from Fig. 2.18 that two sets of data are well separated. The data $(\varepsilon_*)_{\rm exp}$ derived from the experiments in micro-gravity remain approximately constant at 0.094 within the regime of large undercooling. On the other hand, in the regime of small T_∞, starting from $|T_\infty| = 0.01$, the data $(\varepsilon_*)_{\rm exp}$ increase as $|T_\infty|$ decreases. The data derived from the ground experiments show a similar tendency, but the increase starts at a higher undercooling temperature $|T_\infty| = 0.06$. This fact clearly shows that the increase of $(\varepsilon_*)_{\rm exp}$ in the low un-

dercooling regime is due to the convection caused by the buoyancy effect, as gravity is greatly reduced in the flight experiment but not completely eliminated. The effect of buoyancy on the selection of dendritic growth is the topic discussed in Chap. 7–9.

Chapter 3

STEADY DENDRITIC GROWTH FROM MELT WITH CONVECTIVE FLOW

1. Mathematical Formulation of Problem with Navier–Stokes Model

We now consider a more general system of dendritic growth, in which convective flow is generated by all kinds of sources, such as density change during phase transition, enforced uniform flow in the far field, and buoyancy effect in gravitational field. We still adopt the moving paraboloidal coordinate system (ξ, η, θ) fixed at the dendrite tip as defined in the last chapter (See Fig. 3.1). We let $\mathbf{u}(\xi, \eta, \theta, t) = (u, v, w)$ represent the relative velocity field in the liquid state. Here, (u, v) are the components of the relative velocity along ξ- and η-direction, respectively, in the moving frame at the instant t. Furthermore, let $\boldsymbol{\Omega} = \nabla \times \mathbf{u}$ denote the vorticity; $\eta_s(\xi, \theta, t)$ denote the interface shape function; T and T_S denote the temperature field in the melt and in the solid state, respectively. The subscript 'S' refers to the solid state. The governing equations for the dendritic growth process now consist of the fluid dynamical equations and the heat conduction equation.

As in the last chapter, we only consider the case with axi-symmetric basic state. One has seen that the most dangerous modes for the perturbed states were axi-symmetrical. Therefore, for the system under study, it is sensible to restrict ourselves to the axi-symmetrical case for both basic and perturbed states. In this case, the system allows the stream function $\Psi = \Psi(\xi, \eta, t)$, $\mathbf{u} = (u, v, 0)$, and the vorticity vector $\boldsymbol{\Omega} = (0, 0, \omega_3)$ has only one nonzero component. We shall use the stream function $\Psi(\xi, \eta, t)$ and the vorticity function $\zeta(\xi, \eta, t) = \eta_0^2 \xi \eta \omega_3$ as the basic hydrodynamical quantities.

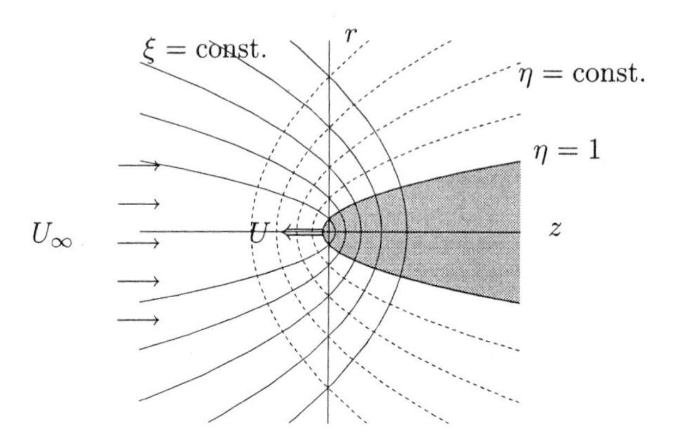

Figure 3.1. A sketch of dendritic growth in external flow

In the paraboloidal coordinate system, we can write

$$\begin{cases} u = \dfrac{1}{\eta_0^4 \xi \eta \sqrt{\xi^2+\eta^2}} \dfrac{\partial \Psi}{\partial \eta}, \\ v = -\dfrac{1}{\eta_0^4 \xi \eta \sqrt{\xi^2+\eta^2}} \dfrac{\partial \Psi}{\partial \xi}. \end{cases} \tag{3.1}$$

and the non-dimensional governing equations in the following forms:

1 Kinematic equation:

$$D^2 \Psi = -\eta_0^4 (\xi^2 + \eta^2) \zeta. \tag{3.2}$$

2 Vorticity equation:

$$\begin{aligned} \mathrm{Pr} D^2 \zeta = {} & \eta_0^2 (\xi^2 + \eta^2) \frac{\partial \zeta}{\partial t} + \frac{2\zeta}{\eta_0^4 \xi^2 \eta^2} \frac{\partial(\Psi, \eta_0^2 \xi \eta)}{\partial(\xi, \eta)} - \frac{1}{\eta_0^2 \xi \eta} \frac{\partial(\Psi, \zeta)}{\partial(\xi, \eta)} \\ & + \frac{\eta_0^4 \mathrm{Gr}}{\mathrm{Pr} |T_\infty|} \xi \eta \left(\eta \frac{\partial T}{\partial \xi} + \xi \frac{\partial T}{\partial \eta} \right). \end{aligned} \tag{3.3}$$

3 Heat conduction equation in liquid phase:

$$\nabla^2 T = \eta_0^4 (\xi^2 + \eta^2) \frac{\partial T}{\partial t} + \frac{1}{\eta_0^2 \xi \eta} \left(\frac{\partial \Psi}{\partial \eta} \frac{\partial T}{\partial \xi} - \frac{\partial \Psi}{\partial \xi} \frac{\partial T}{\partial \eta} \right), \tag{3.4}$$

and heat conduction equation in solid phase:

$$\nabla^2 T_{\mathrm{S}} = \eta_0^4 (\xi^2 + \eta^2) \frac{\partial T_{\mathrm{S}}}{\partial t} + \left(\xi \frac{\partial T_{\mathrm{S}}}{\partial \xi} - \eta \frac{\partial T_{\mathrm{S}}}{\partial \eta} \right). \tag{3.5}$$

In the above, the differentiation operators ∇^2 and D^2 are defined as

$$\nabla^2 = \left\{ \frac{\partial^2}{\partial \xi^2} + \frac{\partial^2}{\partial \eta^2} + \frac{1}{\xi}\frac{\partial}{\partial \xi} + \frac{1}{\eta}\frac{\partial}{\partial \eta} \right\}, \tag{3.6}$$

$$D^2 = \left\{ \frac{\partial}{\partial \xi^2} + \frac{\partial^2}{\partial \eta^2} - \frac{1}{\xi}\frac{\partial}{\partial \xi} - \frac{1}{\eta}\frac{\partial}{\partial \eta} \right\}. \tag{3.7}$$

As we have defined in Chap. 1, in the above

$$\mathrm{Pr} = \frac{\nu}{\kappa_{\mathrm{T}}}, \tag{3.8}$$

is the Prandtl number,

$$\mathrm{Gr} = \frac{g\beta\nu}{U^3}\Big|T_{\mathrm{M0}} - (T_\infty)_{\mathrm{D}}\Big|, \tag{3.9}$$

is the Grashof number, which measures the strength of buoyancy effect.
The boundary conditions are:

1. The up-stream far-field conditions: We assumed that a uniform external flow is imposed in the far field. In practice, it implies that the flux of the uniform flow in the far field is given. Hence, one has

$$\Psi \sim \frac{1}{2}\lambda_0 \eta_0^4 \xi^2 \eta^2 + o(1); \quad \zeta \to 0, \tag{3.10}$$

as $\eta \to \infty$, where

$$\lambda_0 = (1 + \overline{U}_\infty) \quad \text{and} \quad \overline{U}_\infty = \frac{U_\infty}{U}. \tag{3.11}$$

Furthermore,

$$T \to T_\infty, \tag{3.12}$$

where the undercooling parameter T_∞ is defined in the last chapter as

$$T_\infty = \frac{(T_\infty)_{\mathrm{D}} - T_{\mathrm{M0}}}{\Delta H/(c_{\mathrm{p}}\rho)} < 0\,. \tag{3.13}$$

2. Axi-symmetrical condition: at the symmetrical axis $\xi = 0, \eta > 1$,

$$\Psi = \zeta = \frac{\partial \Psi}{\partial \xi} = \frac{\partial \zeta}{\partial \xi} = \frac{\partial \Psi}{\partial \eta} = \frac{\partial \zeta}{\partial \eta} = \frac{\partial T}{\partial \xi} = 0. \tag{3.14}$$

3. The interface condition: at $\eta = \eta_s(\xi, t)$,

(i) Thermo-dynamical equilibrium condition:

$$T = T_S \tag{3.15}$$

(ii) Gibbs–Thomson condition:

$$T_S = -\varepsilon^2 \eta_0^2 \mathcal{K}\left\{\frac{d^2}{d\xi^2}, \frac{d}{d\xi}\right\}\eta_s(\xi, t) \tag{3.16}$$

where the curvature operator

$$\mathcal{K}\left\{\frac{d^2}{d\xi^2}, \frac{d}{d\xi}\right\}\eta_s = -\frac{1}{\sqrt{\xi^2+\eta_s^2}}\left\{\frac{\eta_s''}{(1+\eta_s'^2)^{3/2}} - \frac{1}{\eta_s(1+\eta_s'^2)^{1/2}} + \frac{\eta_s'(\eta_s^2+2\xi^2) - \xi\eta_s}{\xi(\xi^2+\eta_s^2)(1+\eta_s'^2)^{1/2}}\right\}; \tag{3.17}$$

(iii) Enthalpy conservation condition:

$$\left(\frac{\partial T}{\partial \eta} - \eta_s' \frac{\partial T}{\partial \xi}\right) - (1+\alpha)\left(\frac{\partial T_S}{\partial \eta} - \eta_s' \frac{\partial T_S}{\partial \xi}\right) + \eta_0^2(\xi\eta_s)' + \eta_0^4\left(\xi^2+\eta_s^2\right)\frac{\partial \eta_s}{\partial t} = 0, \tag{3.18}$$

where density change parameter α is defined by

$$\alpha = \frac{\rho_S - \rho}{\rho}, \tag{3.19}$$

and measures the density change.

(iv) Mass conservation condition:

$$\left(\frac{\partial \Psi}{\partial \xi} + \eta_s' \frac{\partial \Psi}{\partial \eta}\right) = (1+\alpha)\xi\eta_s(\xi\eta_s)' + \alpha\eta_0^2\xi\eta_s\left(\xi^2+\eta_s^2\right)\frac{\partial \eta_s}{\partial t}, \tag{3.20}$$

(v) Continuity condition of the tangential component of velocity:

$$\left(\frac{\partial \Psi}{\partial \eta} - \eta_s' \frac{\partial \Psi}{\partial \xi}\right) + \xi\eta_s(\eta_s\eta_s' - \xi) = 0. \tag{3.21}$$

As before the notation prime represents the derivative with respect to ξ and the interfacial stability parameter ε is defined as

$$\varepsilon = \frac{1}{\eta_0^2}\sqrt{\frac{\ell_c}{\ell_T}}. \tag{3.22}$$

The system derived contains three parameters: the density change parameter α, Grashof number Gr and the enforced flow parameter $\overline{U}_\infty$, which constitutes the driving forces of convection in the system. We shall examine these effects separately.

Chapter 4

STEADY VISCOUS FLOW PAST A PARABOLOID OF REVOLUTION

In the next few chapters, we shall turn to study dendritic growth with convection in the framework of the IFW theory. For this purpose, we need, first of all, to derive the basic steady state with different types of convective flows for the case of zero surface tension. This solution will play the same role as the Ivantsov solution plays for the system with no convection.

We start with the special case, a pure fluid dynamic problem: enforced, uniform viscous flow past a paraboloidal needle as sketched in Fig 4.1. This is the case of dendritic growth in uniform flow with vanished growth velocity. In this case, the temperature field is to be uniform, the system does not have the thermal diffusion length ℓ_{T}. So that, instead of ℓ_{T}, we shall use the viscous diffusion length $\ell_{\mathrm{d}} = \frac{\nu}{U_\infty}$ as the length scale.

1. Mathematical Formulation of the Problem

The non-dimensional governing equations for this problem are:

$$D^2\Psi = -\eta_0^4(\xi^2 + \eta^2)\zeta, \tag{4.1}$$

$$D^2\zeta = \frac{2\zeta}{\eta_0^4\xi^2\eta^2}\frac{\partial(\Psi, \eta_0^2\xi\eta)}{\partial(\xi, \eta)} - \frac{1}{\eta_0^2\xi\eta}\frac{\partial(\Psi, \zeta)}{\partial(\xi, \eta)}. \tag{4.2}$$

The boundary conditions are:

1. Up-stream far-field conditions: Assume that the flux of the imposed uniform flow in the far field is prescribed. Therefore, as $\eta \to \infty$, one

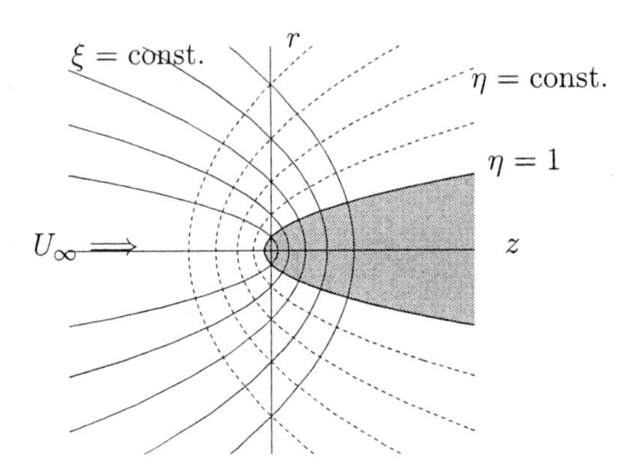

Figure 4.1. A sketch of viscous flow past a paraboloid of revolution

has

$$\Psi \sim \frac{1}{2}\eta_0^4\xi^2\eta^2 + o(1), \quad \zeta \to 0. \tag{4.3}$$

2. Axi-symmetrical condition: at the symmetrical axis $\xi = 0$,

$$\Psi = \zeta = \frac{\partial \Psi}{\partial \xi} = \frac{\partial \Psi}{\partial \eta} = 0. \tag{4.4}$$

3. The interface condition: at $\eta = 1$,

$$\frac{\partial \Psi}{\partial \eta} = 0, \tag{4.5}$$

$$\Psi = 0. \tag{4.6}$$

The above system contains a single parameter: $\eta_0^2 = \frac{\ell_t}{\ell_d}$. The Reynolds number is defined based on the tip radius, $\mathrm{Re} = \frac{U_\infty \ell_t}{\nu}$. Hence, we have $\eta_0^2 = \mathrm{Re}$.

In the far field, the velocity field is nearly uniform and has the unit velocity, hence,

$$\begin{cases} \bar{\zeta}_* = 0, \\ \bar{\Psi}_* = \frac{1}{2}\eta_0^4\xi^2\eta^2 . \end{cases} \tag{4.7}$$

We set

$$\begin{aligned} \Psi(\xi,\eta) &= \bar{\Psi}_*(\xi,\eta) + \bar{\Psi}(\xi,\eta), \\ \zeta(\xi,\eta) &= \bar{\zeta}(\xi,\eta). \end{aligned} \tag{4.8}$$

Thus, from system (4.1)–(4.6), one can derive the system of perturbed fields, $\{\bar{\Psi}, \bar{\zeta}\}$, as follows:

$$D^2\bar{\Psi} = -\eta_0^4(\xi^2 + \eta^2)\bar{\zeta}\ , \tag{4.9}$$

$$D^2\bar{\zeta} = \eta_0^2\left(\xi\frac{\partial\bar{\zeta}}{\partial\xi} - \eta\frac{\partial\bar{\zeta}}{\partial\eta}\right) + \frac{2\bar{\zeta}}{\eta_0^2\xi^2\eta^2}\left(\eta\frac{\partial\bar{\Psi}}{\partial\eta} - \xi\frac{\partial\bar{\Psi}}{\partial\xi}\right)$$
$$-\frac{1}{\eta_0^2\xi\eta}\frac{\partial(\bar{\Psi},\bar{\zeta})}{\partial(\xi,\eta)}\ . \tag{4.10}$$

In what follows, for more convenience, we shall utilize the new variables

$$\sigma = \tfrac{\eta_0^2}{2}\xi^2, \qquad \tau = \tfrac{\eta_0^2}{2}\eta^2\ . \tag{4.11}$$

With the variables $(\sigma;\tau)$, we have

$$u = \frac{1}{2\sqrt{\sigma(\sigma+\tau)}}\frac{\partial\Psi}{\partial\tau}, \qquad w = -\frac{1}{2\sqrt{\tau(\sigma+\tau)}}\frac{\partial\Psi}{\partial\sigma}. \tag{4.12}$$

Furthermore, the operator D^2 becomes

$$D^2 = 2\eta_0^2 L_2, \tag{4.13}$$

where

$$L_2 = \left(\sigma\frac{\partial^2}{\partial\sigma^2} + \tau\frac{\partial^2}{\partial\tau^2}\right). \tag{4.14}$$

The system (4.9)– (4.10) is then changed to

$$L_2[\bar{\Psi}] = -(\sigma+\tau)\bar{\zeta}, \tag{4.15}$$

$$L_2[\bar{\zeta}] = \left(\sigma\frac{\partial}{\partial\sigma} - \tau\frac{\partial}{\partial\tau}\right)\bar{\zeta} - \frac{\bar{\zeta}}{2\tau\sigma}\left(\tau\frac{\partial\bar{\Psi}}{\partial\tau} - \sigma\frac{\partial\bar{\Psi}}{\partial\sigma}\right)$$
$$-\frac{1}{2}\frac{\partial(\bar{\Psi},\bar{\zeta})}{\partial(\sigma,\tau)}\ . \tag{4.16}$$

Moreover, with the new variables, the interface $\eta = 1$, can be described as $\tau = \tau_0 = \frac{\eta_0^2}{2} = \frac{1}{2}\mathrm{Re}$. The surface conditions become: at $\tau = \frac{1}{2}\mathrm{Re}$,

$$\frac{\partial\bar{\Psi}}{\partial\sigma} + \mathrm{Re} = 0, \tag{4.17}$$

$$\frac{\partial \bar{\Psi}}{\partial \tau} + 2\sigma = 0. \tag{4.18}$$

2. The Oseen Model Problem

Viscous flow past a small or slender, finite or semi-infinite body have been widely studied classic subjects in fluid dynamics (Van Dyke, 1975). The simplest case is viscous flow past a fine sphere, which was first studied by Stokes in 1851 (Stokes, 1851).

Stokes considered the case of small Reynolds number Re, and assumed that when the Reynolds number is sufficiently small, the nonlinear inertia terms can be neglected in the whole flow region. The assumption adopted by Stokes is called the Stokes model, and his solution is called the Stokes solution, which satisfies all the interface conditions, as well as the far-field conditions. It was soon recognized that the Stokes solution failed in the asymptotic sense. When $0 < \mathrm{Re} \ll 1$, the Stokes solution may be considered as the leading term of the Stokes expansion solution for the Navier–Stokes equation. However, this Stokes expansion solution is only valid in the region near the sphere, it is invalid in the far field away from the interface. Approaching to the far field, all the higher-order terms in the Stokes expansion solution have the same order of magnitude as its leading term — the Stokes solution. This is the so-called Whitehead paradox, which reveals the nature of singular perturbation of the problem in the far field. To overcome the difficulty, Oseen took into account the inertia terms in the Navier–Stokes equations (Oseen, 1910). Rather than neglecting the inertia terms altogether, he approximated them by their linearized forms that were valid far from the body. The assumption adopted by Oseen is called the Oseen model, and the solution obtained by Oseen is called the Oseen solution. The Oseen solution satisfies all the far-field conditions and the interface conditions. It, however, in the asymptotic sense fails in the near field close to the sphere. The Oseen solution may be considered as the leading term of the so-called, Oseen asymptotic expansion solution, which is only valid in the far field, but invalid in the region close to the sphere. Approaching the sphere, all the higher-order terms in the Oseen expansion solution have the same order of magnitude as its leading term — the Oseen solution. The proper, uniformly valid solution then was later obtained by the matching asymptotic expansion method, in which the Stokes expansion as the inner expansion matches the Oseen expansion as the outer expansion in the intermediate region (Van Dyke, 1975).

The Oseen model solution, Stokes model solution and the uniformly valid asymptotic expansion solution for viscous flow past a sphere are well-known classic results in fluid dynamics.

For the case of viscous flow past a semi-infinite body, the situation is much more complicated.

Dennis and Warsh (1971), Davis (1972), Davis and Werle (1972) and Veldman (1973) carried out numerical studies for the cases of paraboloid cylinder and paraboloid revolution.

The analytical study with the Oseen model for the problem of viscous flow past a paraboloid with elliptic section was first obtained by Wilkinson (Wilkinson, 1955). Much later on, when Ananth and Gill (Ananth and Gill, 1989, 1991) studied dendritic growth in external flow, they considered the same problem and gained the same solution, but in a different form, as obtained by Wilkinson. For the special case of viscous flow past a paraboloid of revolution, the solution obtained by Ananth and Gill may be expressed in some analytical form in the paraboloid coordinate system (ξ, η), defined as (2.1). As $\eta \to \infty$, in the up-stream far field their solution yields a velocity field approaching to the velocity of the imposed external uniform flow, as required by their boundary condition in the far field. On the other hand, the stream function given by their solution has the following asymptotic expression:

$$\Psi(\xi, \eta) \sim \Psi_\infty + A\xi^2 + o(1), \quad \text{as} \quad \eta \to \infty, \tag{4.19}$$

where Ψ_∞ represents the stream function of the external uniform flow, A is some constant. The second term $A\xi^2$ in this asymptotic expression describes a non-vanishing perturbed flow flux, which is induced by the presence of the paraboloid body. However, in practice, the imposed external flow is always given with a fixed flow flux. For instance, in wind tunnel experiments, the uniform external flow in the far field is generated by pumping a prescribed flow flux into the tunnel, whereas in water channel experiments, the uniform external flow in the far field is generated by moving the experimental object along the closed channel. A free dendritic growth experiment is performed in a large growth chamber. In practice, the uniform external flow in the far field is generated by moving the chamber with constant velocity against the growth direction. Hence, in the moving coordinate system fixed at the dendrite-tip, the flow flux in the far field is prescribed independently from the growth, and cannot be changed by the presence of the paraboloid body. For these practical situations, it appears that the solution obtained by Ananth and Gill, and Wilkinson is not the right solution for the Oseen model problem under study.

In this section, we shall first present the solution of the Oseen model problem of viscous flow past a paraboloid of revolution. Then in the next section, we shall derive the uniformly valid asymptotic solution for the problem with Navier–Stokes model in the limit Re $\to 0$. The approach developed in this section will be an important basis for the further studies of dendritic growth problems to be conducted in the later chapters.

With the Oseen model, neglecting the non-linear terms, the differential operator for the velocity field is then reduced to

$$L_2[\bar{\Psi}_0] = -(\sigma+\tau)\bar{\zeta}_0, \tag{4.20}$$

$$L_2[\bar{\zeta}_0] = \left(\sigma\frac{\partial}{\partial\sigma} - \tau\frac{\partial}{\partial\tau}\right)\bar{\zeta}_0. \tag{4.21}$$

The equation (4.21) allows the form of the solutions to be

$$\bar{\zeta}_0 = \mathcal{F}(\tau)\mathcal{G}(\sigma). \tag{4.22}$$

We derive that

$$\frac{\sigma\mathcal{G}''}{\mathcal{G}} + \frac{\tau\mathcal{F}''}{\mathcal{F}} = \frac{\sigma\mathcal{G}'}{\mathcal{G}} - \frac{\tau\mathcal{F}'}{\mathcal{F}}, \tag{4.23}$$

so that,

$$\mathcal{G}'' - \mathcal{G}' + \frac{k}{\sigma}\mathcal{G} = 0, \tag{4.24}$$

$$\mathcal{F}'' + \mathcal{F}' - \frac{k}{\tau}\mathcal{F} = 0. \tag{4.25}$$

To find $\mathcal{G}(\eta)$, we make the variables transformation

$$\mathcal{G} = R(\sigma)\mathrm{e}^{\frac{\sigma}{2}}. \tag{4.26}$$

Then (4.24) is transformed to the Whittaker equation

$$\frac{\mathrm{d}^2R}{\mathrm{d}\sigma^2} + \left(-\frac{1}{4} + \frac{k}{\sigma}\right)R = 0, \tag{4.27}$$

whose solutions are the Whittaker functions with index $\left(k, \mu = \frac{1}{2}\right)$:

$$W_{k,\mu}(\sigma) = \begin{cases} \sigma\mathrm{e}^{-\frac{1}{2}\sigma}M(1-k, 2, \sigma), \\ \sigma\mathrm{e}^{-\frac{1}{2}\sigma}U(1-k, 2, \sigma). \end{cases} \tag{4.28}$$

Note that as $\sigma \to 0$, $U(1-k, 2, \sigma)$ has a logarithmic singularity. So, it should be ruled out from the solution. On the other hand, when $1-k = -n, \quad n = 0, 1, \ldots$, $M(1-k, 2, \sigma) = M(-n, 2, \sigma) = \frac{n!}{(2)_n} L_n^{(1)}(\sigma) \to \infty$ algebraically, where $L_n^{(\alpha)}(\sigma)$ is the general Laguerre function,

$$L_n^{(1)}(x) = \sum_{m=0}^{n} (-1)^m \frac{(n+1)!}{(n-m)!(1+m)!m!} x^m. \qquad (4.29)$$

For the other index k, $M(1-k, 2, \sigma) \to \infty$ exponentially. Since one cannot allow $\zeta_0 \to \infty$ exponentially, we derive that

$$k = 1 + n, \quad (n = 0, 1, \ldots), \qquad (4.30)$$

and

$$\mathcal{G}(\sigma) = \mathcal{G}_n(\sigma) = \sigma M(-n, 2, \sigma) = \frac{\sigma}{n+1} L_n^{(1)}(\sigma). \qquad (4.31)$$

Similarly, to find $\mathcal{F}(\tau)$, we make the variables transformation

$$\mathcal{F} = W(\tau) \mathrm{e}^{-\frac{\tau}{2}}. \qquad (4.32)$$

Then (4.25) is transformed to the Whittaker equation

$$\frac{\mathrm{d}^2 W}{\mathrm{d}\tau^2} + \left(-\frac{1}{4} - \frac{k}{\tau}\right) W = 0. \qquad (4.33)$$

With the far-field condition,

$$W(\tau) \to 0, \qquad (4.34)$$

as $\tau \to \infty$, we obtain

$$W = W_{-k, \frac{1}{2}}(\tau) = \tau \mathrm{e}^{-\frac{\tau}{2}} U(1+k, 2, \tau), \qquad (4.35)$$

so that,

$$\begin{aligned} \mathcal{F}(\tau) = \mathcal{F}_n(\tau) &= \tau \mathrm{e}^{-\tau} U(n+2, 2, \tau) \\ &= \sum_{k=0}^{n} \frac{(-1)^k}{k!(n-k)!} E_{k+2}(\tau), \end{aligned} \qquad (4.36)$$

where

$$E_k(z) = \int_1^\infty \frac{\mathrm{e}^{-zt}}{t^k} \mathrm{d}t \qquad (4.37)$$

is the exponential function (Abramovitz and Stegun, 1964). Particularly,

$$\begin{aligned} \mathcal{F}_0(\tau) &= E_2(\tau), \\ \mathcal{F}_1(\tau) &= E_2(\tau) - E_3(\tau), \\ \mathcal{F}_2(\tau) &= \tfrac{1}{2!}\Big[E_2(\tau) - 2E_3(\tau) + E_4(\tau)\Big], \\ &\vdots \end{aligned} \tag{4.38}$$

With the above results, we can write

$$\bar{\zeta}_0 = \sigma \sum_{n=0}^{\infty} a_n \mathcal{F}_n(\tau) L_n^{(1)}(\sigma). \tag{4.39}$$

Furthermore, with the conditions:

$$\begin{aligned} &\bar{\Psi}_0 \to 0 && (\text{as} \quad \tau \to \infty), \\ &\bar{\Psi}_0 \quad \text{algebraically increases or decreases} && (\text{as} \quad \sigma \to \infty), \\ &\bar{\Psi}_0 < \infty && (\text{at} \quad \sigma = 0), \end{aligned} \tag{4.40}$$

one may derive the general solutions for the associated homogeneous equation of the stream function as

$$\bar{\Psi}_{\mathrm{H},0} = \mathcal{C}_0(\lambda)\sigma^{\frac{1}{2}}\tau^{\frac{1}{2}} K_1(2\lambda\tau^{\frac{1}{2}}) J_1(2\lambda\sigma^{\frac{1}{2}}), \tag{4.41}$$

where λ and $\mathcal{C}_0$ are arbitrary real constants. For the special case $\lambda \to 0$,

$$\bar{\Psi}_{\mathrm{H},0} = \bar{d}_* \sigma. \tag{4.42}$$

The solution (4.39) gives us an extremely important clue that in order to obtain the solutions of the problem, one may use the Laguerre series representation.

2.1 Laguerre Series Representation of Solutions

Note that the Laguerre functions $\{L_n^{(1)}(\sigma)\}$, $(n = 0, 1, 2, \ldots)$ form a set of complete orthogonal functions; furthermore, the solutions that we are concerned with must satisfy the axi-symmetric condition, $u(0,\tau) = \bar{\Psi}_0(0,\tau) = 0$. We, therefore, may expand our solutions in the Laguerre series

$$\begin{cases} \bar{\zeta}_0 = \sigma \sum\limits_{n=0}^{\infty} A_n(\tau, \mathrm{Re}) L_n^{(1)}(\sigma), \\ \bar{\Psi}_0 = \sigma \sum\limits_{n=0}^{\infty} B_n(\tau, \mathrm{Re}) L_n^{(1)}(\sigma). \end{cases} \tag{4.43}$$

The Oseen equations (4.20)–(4.21) can then be transformed into the following system of difference-differential equations:

$$\begin{aligned}\frac{\mathrm{d}^2\bar{B}_n}{\mathrm{d}\tau^2} &= \frac{n+2}{\tau}\bar{B}_{n+1} - \left[1 + \frac{2(n+1)}{\tau}\right]\bar{A}_n \\ &\quad + \frac{1}{\tau}\Big[(n+2)\bar{A}_{n+1} + n\bar{A}_{n-1}\Big], \\ \frac{\mathrm{d}^2\bar{A}_n}{\mathrm{d}\tau^2} &+ \frac{\mathrm{d}\bar{A}_n}{\mathrm{d}\tau} - \frac{n+1}{\tau}\bar{A}_n = 0 \\ &(n = 0, 1, 2, \ldots).\end{aligned} \tag{4.44}$$

The general solution for the homogeneous equation of $\bar{A}_n$ is found to be

$$\bar{A}_n(\tau) = \bar{a}_n \mathcal{F}_n(\tau) \tag{4.45}$$

where $\bar{a}_n$ is an arbitrary constant.

The particular solution for $\bar{B}_n(\tau)$ is found to be (see Appendix A for the derivation)

$$\bar{B}_n(\tau) = \bar{b}_n \mathcal{F}_n(\tau), \tag{4.46}$$

where

$$\bar{b}_n = -\bar{a}_n + n\bar{a}_{n-1}. \tag{4.47}$$

To write the general solution for $\bar{B}_n(\tau)$, $(n = 0, 1, \ldots)$, and finally for the stream function $\bar{\Psi}_0(\sigma, \tau)$, we need the solution of the stream function for the system of the associated homogeneous equations, $\bar{\Psi}_{\mathrm{H},0}$. Such solutions have already been found in the last section. Its Laguerre series representation can be directly derived from (4.41) by using the formula,

$$J_1(2\lambda\sigma^{\frac{1}{2}}) = \sigma^{\frac{1}{2}} \mathrm{e}^{-\lambda^2} \sum_{n=0}^{\infty} (-1)^n \frac{\lambda^{2n+1}}{(n+1)!} L_n^{(1)}(\sigma), \tag{4.48}$$

as shown in (Gradshteyn and Ryzhik, 1980, page 1038, [8.975-3]). As a matter of fact, one can transform (4.41) in the form

$$\begin{aligned}&\bar{\Psi}_{\mathrm{H},0}(\sigma, \tau) = \psi_0(\sigma, \tau) \\ &= \sigma\tau^{\frac{1}{2}} \sum_{n=0}^{\infty} \left\{ \int_0^{\infty} (-1)^n \mathcal{C}_0(\lambda) \mathrm{e}^{-\lambda^2} \frac{\lambda^{2n+1}}{(n+1)!} K_1(2\lambda\tau^{\frac{1}{2}}) \mathrm{d}\lambda \right\} L_n^{(1)}(\sigma),\end{aligned} \tag{4.49}$$

where $\mathcal{C}_0(\lambda)$ is a real function of λ. Noted that as $\mathcal{C}_0(\lambda)$ is the Dirac Delta function,

$$\mathcal{C}_0(\lambda) = \delta_0(\lambda), \tag{4.50}$$

the solution $\psi_0(\sigma,\tau)$ is reduced to

$$\psi_0(\sigma,\tau) = \bar{d}_*\sigma. \tag{4.51}$$

For continuous function $\mathcal{C}_0(\lambda)$, we may expand it into the Laguerre series

$$\mathcal{C}_0(\lambda) = \sum_{k=0}^{\infty} \mathcal{C}_{0,k} L_k^{(1)}(\lambda). \tag{4.52}$$

With the above results, we may write the general solution for the associated homogeneous equation of the stream function in the form

$$\bar{\Psi}_{\mathrm{H},0}(\sigma,\tau) = \bar{d}_*\sigma + \psi_0(\sigma,\tau). \tag{4.53}$$

The function ψ_0 and its derivative $\dfrac{\partial\psi_0}{\partial\tau}$ can be expressed explicitly with the coefficients $\mathcal{C}_{0,k}$ as follows:

$$\psi_0 = \sigma \sum_{n=0}^{\infty} L_n^{(1)}(\sigma) \sum_{k=0}^{\infty} \mathcal{C}_{0,k}\mathcal{A}_{n,k}(\tau), \tag{4.54}$$

and

$$\frac{\partial\psi_0}{\partial\tau} = \sigma \sum_{n=0}^{\infty} L_n^{(1)}(\sigma) \sum_{k=0}^{\infty} \mathcal{C}_{0,k}\hat{\mathcal{B}}_{n,k}(\tau). \tag{4.55}$$

Here, we define

$$\mathcal{A}_{n,k}(\tau) = \tau^{\frac{1}{2}} \int_0^{\infty} (-1)^n \mathrm{e}^{-\lambda^2} \frac{\lambda^{2n+1}}{(n+1)!} K_1(2\lambda\tau^{\frac{1}{2}}) L_k^{(1)}(\lambda)\,\mathrm{d}\lambda, \tag{4.56}$$

$$\hat{\mathcal{A}}_{n,k}(\tau) = \tau^{\frac{1}{2}} \int_0^{\infty} (-1)^n \mathrm{e}^{-\lambda^2} \frac{\lambda^{2n+2}}{(n+1)!} K_1(2\lambda\tau^{\frac{1}{2}}) L_k^{(1)}(\lambda)\,\mathrm{d}\lambda, \tag{4.57}$$

and

$$\mathcal{B}_{n,k}(\tau) = \int_0^{\infty} (-1)^n \mathrm{e}^{-\lambda^2} \frac{\lambda^{2n+1}}{(n+1)!} K_0(2\lambda\tau^{\frac{1}{2}}) L_k^{(1)}(\lambda)\,\mathrm{d}\lambda, \tag{4.58}$$

$$\hat{\mathcal{B}}_{n,k}(\tau) = \int_0^{\infty} (-1)^n \mathrm{e}^{-\lambda^2} \frac{\lambda^{2n+2}}{(n+1)!} K_0(2\lambda\tau^{\frac{1}{2}}) L_k^{(1)}(\lambda)\,\mathrm{d}\lambda. \tag{4.59}$$

The four series of the functions $\mathcal{A}_{n,k}(\tau)$, $\hat{\mathcal{A}}_{n,k}(\tau)$, $\mathcal{B}_{n,k}(\tau)$ and $\hat{\mathcal{B}}_{n,k}(\tau)$ $(n,k=0,1,\ldots)$ can be calculated through the recurrence formulas given in the Appendix (A). It can be seen that as $\tau\to\infty$, all functions $\hat{\mathcal{A}}_{n,k}(\tau)$

tend to zero algebraically, and the speed approaching zero rapidly increases, as the index n increases.

Formulas (4.53)–(4.54) give a large class of solutions of the associated homogeneous equation of the stream function, $\Psi_{\mathrm{H},0}(\sigma, \tau)$. Particularly, with any finite number of nonzero constants $\mathcal{C}_{0,k}$, $(k = 0, 1, \ldots, N)$, a function defined by (4.54) is a solution of the associated homogeneous equation. This part of the solutions was missed by all researchers who had previously worked on this problem.

The general solution of the whole stream function is obtained as

$$\bar{\Psi}_0 = \bar{d}_* \sigma + \psi_0(\sigma, \tau) + \sigma \sum_{n=0}^{\infty} \bar{b}_n \mathcal{F}_n(\tau) L_n^{(1)}(\sigma), \tag{4.60}$$

while the general solution of the vorticity function is

$$\bar{\zeta}_0 = \sigma \sum_{n=0}^{\infty} \bar{a}_n \mathcal{F}_n(\tau) L_n^{(1)}(\sigma). \tag{4.61}$$

The solution (4.60) contains arbitrary constants $\bar{d}_*$, $\bar{b}_n$ $(n = 0, 1, \ldots)$ and $\mathcal{C}_{0,k}$ $(k = 0, 1, 2, \ldots)$, which are to be determined by the surface conditions: at $\tau = \tau_0 = \frac{1}{2}\mathrm{Re}$,

$$\frac{\partial \bar{\Psi}_0}{\partial \sigma} + \mathrm{Re} = 0, \tag{4.62}$$

$$\frac{\partial \bar{\Psi}_0}{\partial \tau} + 2\sigma = 0. \tag{4.63}$$

2.2 Solution of the Oseen Model and the Paradox

In this section, we shall derive the Oseen model solution. To find the solution, we need to determine two sequences of real unknowns: $\{\bar{b}_n\}$, $\{\mathcal{C}_{0,k}\}$, $(k = 0, 1, 2, \ldots, N, \ldots)$ with another real unknown number $\bar{d}_*$. These unknowns are to be determined by the surface conditions at $\tau = \tau_0$:

$$\bar{\Psi}_0 = \bar{d}_* \sigma + \psi_0 + \sigma \sum_{n=0}^{\infty} \bar{b}_n \mathcal{F}_n(\tau_0) L_n^{(1)}(\sigma) = -\sigma \mathrm{Re}, \tag{4.64}$$

and

$$\frac{\partial \bar{\Psi}_0}{\partial \tau} = \frac{\partial \psi_0}{\partial \tau} + \sigma \sum_{n=0}^{\infty} \bar{b}_n \mathcal{F}'_n(\tau_0) L_n^{(1)}(\sigma) = -2\sigma. \tag{4.65}$$

For $n = 0$, we have

$$\begin{aligned}
\sum_{k=0}^{\infty} \mathcal{C}_{0,k}\mathcal{A}_{0,k}(\tau_0) &= -\bar{d}_* - \mathrm{Re} - \bar{b}_0\mathcal{F}_0(\tau_0), \\
\sum_{k=0}^{\infty} \mathcal{C}_{0,k}\hat{\mathcal{B}}_{0,k}(\tau_0) &= -\bar{b}_0\mathcal{F}_0'(\tau_0) - 2,
\end{aligned} \tag{4.66}$$

while for $(n = 1, 2, \ldots)$,

$$\begin{aligned}
\sum_{k=1}^{\infty} \mathcal{C}_{0,k}\mathcal{A}_{n,k}(\tau_0) + \bar{b}_n\mathcal{F}_n(\tau_0) &= -\mathcal{C}_{0,0}\mathcal{A}_{n,0}(\tau_0), \\
\sum_{k=1}^{\infty} \mathcal{C}_{0,k}\hat{\mathcal{B}}_{n,k}(\tau_0) + \bar{b}_n\mathcal{F}_n'(\tau_0) &= -\mathcal{C}_{0,0}\hat{\mathcal{B}}_{n,0}(\tau_0).
\end{aligned} \tag{4.67}$$

System (4.67) determines the parameters $\bar{b}_n$, and $\mathcal{C}_{0,k}$ $(n, k = 1, 2, \ldots, N, \ldots)$ as the functions of the parameter $\mathcal{C}_{0,0}$. Letting

$$\mathcal{C}_{0,k} = \widehat{\mathcal{C}}_{0,k}\mathcal{C}_{0,0}, \qquad \widehat{\mathcal{C}}_{0,0} = 1, \qquad \bar{b}_n = \widehat{b}_n\mathcal{C}_{0,0}, \tag{4.68}$$

we derive that

$$\left(\mathcal{Q}_{n,k}\right) \cdot \begin{pmatrix} \widehat{\mathcal{C}}_{0,1} \\ \widehat{\mathcal{C}}_{0,2} \\ \vdots \\ \widehat{\mathcal{C}}_{0,N} \\ \vdots \end{pmatrix} = \begin{pmatrix} \mathcal{H}_1 \\ \mathcal{H}_2 \\ \vdots \\ \mathcal{H}_N \\ \vdots \end{pmatrix} \tag{4.69}$$

and

$$\widehat{b}_n = -\frac{1}{\mathcal{F}_n(\tau_0)} \sum_{k=0}^{\infty} \widehat{\mathcal{C}}_{0,k}\mathcal{A}_{n,k}(\tau_0), \quad (n = 1, 2, \ldots, N, \ldots) \tag{4.70}$$

where the matrix $\left(\mathcal{Q}_{n,k}\right)$ is defined as

$$\mathcal{Q}_{n,k} = \mathcal{F}_n'(\tau_0)\mathcal{A}_{n,k}(\tau_0) - \mathcal{F}_n(\tau_0)\hat{\mathcal{B}}_{n,k}(\tau_0) \tag{4.71}$$

and

$$\mathcal{H}_n = \left[\mathcal{F}_n(\tau_0)\hat{\mathcal{B}}_{n,0}(\tau_0) - \mathcal{F}_n'(\tau_0)\mathcal{A}_{n,0}(\tau_0)\right]. \tag{4.72}$$

In Fig. 4.2, we show the curves of $\mathcal{H}_n(\tau)$ for $(n = 1, 2, 3, 4)$, while in Fig. 4.3, we show the curves of $\mathcal{Q}_{n,k}(\tau)$ with index $k = 1$ for $(n = 1, 2, 3, 4)$. It is seen that all the elements $\mathcal{H}_n(\tau)$ and $\mathcal{Q}_{n,k}(\tau)$ decrease with τ exponentially. On the other hand, for a given $\tau = \tau_0$, these

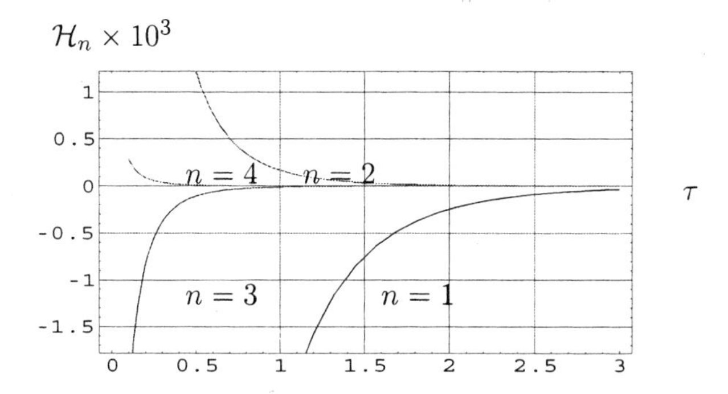

Figure 4.2. The figures of functions $\mathcal{H}_n(\tau), n = 1, 2, 3, 4$

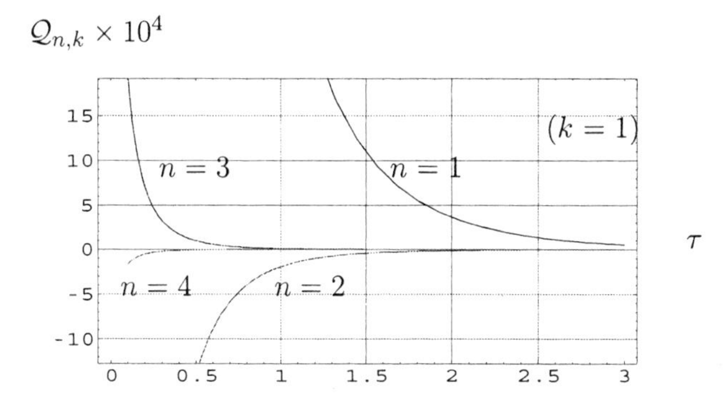

Figure 4.3. The figures of functions $\mathcal{Q}_{n,k}(\tau), n = 1, 2, 3, 4$

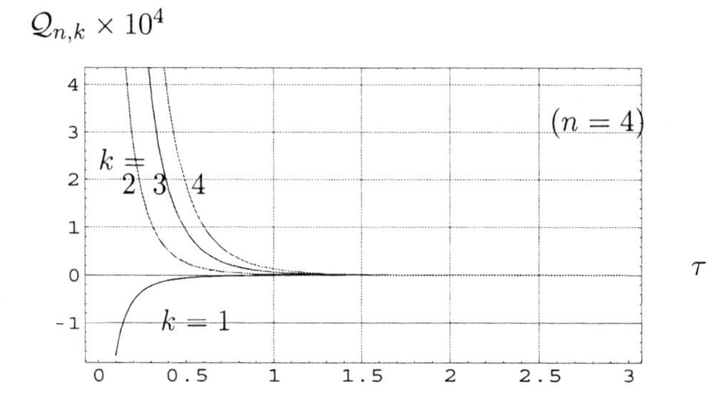

Figure 4.4. The figures of functions $\mathcal{Q}_{4,k}(\tau), k = 1, 2, 3, 4$

functions also decrease with index n, rapidly. Furthermore, in Fig. 4.4, we show the curves of $\mathcal{Q}_{n,k}(\tau)$ with index $n = 4$ for $(k = 1, 2, 3, 4)$. It is seen that given $\tau = \tau_0$, the elements of matrix $\mathcal{Q}_{n,k}(\tau_0)$ increase with the index k. Thus, for given $\tau = \tau_0$, one can find an integer $N(\tau_0)$, such that as the index $n > N$, all the elements $\mathcal{H}_n(\tau)$ as well as the elements $\mathcal{Q}_{n,k}(\tau)$ with the index $k = 1, 2, \ldots, N$ will be negligibly small. The numerical calculations show that the larger τ_0, the smaller the integer N. For instance, as $\tau_0 = 0.5$, we have

$$\mathcal{Q}_{n,k} = \begin{pmatrix} 1.9885 \times 10^{-2} & 1.1512 \times 10^{-2} & -1.6394 \times 10^{-2} & -6.6070 \times 10^{-2} & \cdots \\ -1.3951 \times 10^{-3} & 5.4131 \times 10^{-4} & 5.2799 \times 10^{-3} & 1.2895 \times 10^{-2} & \cdots \\ 9.5328 \times 10^{-5} & -1.8037 \times 10^{-4} & -7.9609 \times 10^{-4} & -1.7281 \times 10^{-3} & \cdots \\ -5.7928 \times 10^{-6} & 2.5796 \times 10^{-5} & 9.3070 \times 10^{-5} & 1.8986 \times 10^{-4} & \cdots \\ 2.9184 \times 10^{-7} & -2.8125 \times 10^{-6} & -9.2176 \times 10^{-6} & -1.7992 \times 10^{-5} & \cdots \\ \cdots & \cdots & \cdots & \cdots & \cdots \end{pmatrix}$$

and

$$\mathcal{H}_n = \Big(-1.3767 \times 10^{-2}, 1.2179 \times 10^{-3}, -1.0884 \times 10^{-4}, 9.1848 \times 10^{-6}, -7.1960 \times 10^{-7}, \cdots\Big)^{\mathrm{T}}.$$

By neglecting the small terms of $O(10^{-7})$, we may choose $N = 4$.

In view of the above, in general, we may simplify equation (4.69) as the following:

$$\left(\begin{array}{ccc|ccc} \mathcal{Q}_{1,1} & \cdots & \mathcal{Q}_{1,N} & * & * & \cdots \\ \vdots & \vdots & \vdots & \vdots & \vdots & \vdots \\ \mathcal{Q}_{N,1} & \cdots & \mathcal{Q}_{1,N} & * & * & \cdots \\ \hline 0 & \cdots & 0 & * & * & \cdots \\ 0 & \cdots & 0 & * & * & \cdots \\ \vdots & \vdots & \vdots & \vdots & \vdots & \vdots \end{array}\right) \cdot \left(\begin{array}{c} \widehat{C}_{0,1} \\ \vdots \\ \widehat{C}_{0,N} \\ \hline \widehat{C}_{0,N+1} \\ \vdots \\ \vdots \end{array}\right) = \left(\begin{array}{c} \mathcal{H}_1 \\ \vdots \\ \mathcal{H}N \\ \hline 0 \\ \vdots \\ \vdots \end{array}\right). \tag{4.73}$$

The solution of (4.73) can be obtained as $\widehat{C}_{0,n} = 0, (n > N)$, and

$$\begin{pmatrix} \widehat{C}_{0,1} \\ \vdots \\ \widehat{C}_{0,N} \end{pmatrix} = \begin{pmatrix} \mathcal{Q}_{1,1} & \cdots & \mathcal{Q}_{1,N} \\ & \vdots & \\ \mathcal{Q}_{N,1} & \cdots & \mathcal{Q}_{1,N} \end{pmatrix}^{-1} \cdot \begin{pmatrix} \mathcal{H}_1 \\ \vdots \\ \mathcal{H}_N \end{pmatrix}, \tag{4.74}$$

Then, it follows that

$$\widehat{b}_n = -\frac{1}{\mathcal{F}_n(\tau_0)}\left[\mathcal{A}_{n,0}(\tau_0) + \sum_{k=1}^{N} \widehat{\mathcal{C}}_{0,k}\mathcal{A}_{n,k}(\tau_0)\right], \quad (n = 1, 2, \cdots, N, \cdots) \tag{4.75}$$

One can solve (4.74) and (4.75) first, then substitute the solutions of (4.74) into (4.66). This results in two equations for three constants $\bar{d}_*$, $\bar{b}_0$ and $\mathcal{C}_{0,0}$.

$$\begin{aligned} \mathcal{C}_{0,0} \sum_{k=0}^{N} \widehat{\mathcal{C}}_{0,k}\mathcal{A}_{0,k}(\tau_0) + \bar{b}_0\mathcal{F}_0(\tau_0) &= -2\tau_0 - \bar{d}_*, \\ \mathcal{C}_{0,0} \sum_{k=0}^{N} \widehat{\mathcal{C}}_{0,k}\hat{\mathcal{B}}_{0,k}(\tau_0) + \bar{b}_0\mathcal{F}_0'(\tau_0) &= -2. \end{aligned} \tag{4.76}$$

We derive the solution:

$$\begin{cases} \mathcal{C}_{0,0} = \dfrac{2\mathcal{F}_0(\tau_0) - 2\tau_0\mathcal{F}_0' - \bar{d}_*}{\mathcal{F}_0'(\tau_0) \sum_{k=0}^{N} \widehat{\mathcal{C}}_{0,k}\mathcal{A}_{0,k}(\tau_0) - \mathcal{F}_0(\tau_0) \sum_{k=0}^{N} \widehat{\mathcal{C}}_{0,k}\hat{\mathcal{B}}_{0,k}(\tau_0)}, \\ \bar{b}_0 = \dfrac{-2\tau_0 - \mathcal{C}_{0,0} \sum_{k=0}^{N} \widehat{\mathcal{C}}_{0,k}\mathcal{A}_{0,k}(\tau_0)}{\mathcal{F}_0(\tau_0)}. \end{cases} \tag{4.77}$$

It is, therefore, seen that the solution of the Oseen model equations may be not unique. In order to show the numerical differences of the different solutions, in what follows, we shall examine two special solutions:

- The solution of type (I): $\mathcal{C}_{n,k,} = 0$, for all $(n, k = 0, 1, \ldots)$, but $\bar{d}_* \neq 0$;
- The solution of type (II): $\bar{d}_* = 0$.

2.3 The Solution of Type (I)

By setting $\mathcal{C}_{n,k} = 0$ for all $(n, k = 0, 1, 2, \ldots)$, (4.76) is reduced to

$$\begin{cases} \bar{b}_0\mathcal{F}_0(\tau_0) = -\bar{d}_* - \mathrm{Re}, \\ \bar{b}_0\mathcal{F}_0'(\tau_0) = -2. \end{cases} \tag{4.78}$$

So, we have

$$\begin{aligned} \bar{b}_0 &= \frac{2}{E_1(\tau_0)}, \\ \bar{d}_* &= -\mathrm{Re} - 2\frac{E_2(\tau_0)}{E_1(\tau_0)}. \end{aligned} \tag{4.79}$$

The stream function is obtained as

$$\bar{\Psi}_0 = \sigma S_{0*}(\tau), \tag{4.80}$$

where

$$S_{0*}(\tau) = \left\{ \frac{2}{E_1(\tau_0)}[E_2(\tau) - E_2(\tau_0)] - \mathrm{Re} \right\}. \tag{4.81}$$

Furthermore, due to

$$\begin{aligned} \bar{a}_0 &= -\bar{b}_0, \\ \bar{a}_n &= n!\bar{a}_0 = -\frac{2n!}{E_1(\tau_0)}, \end{aligned} \tag{4.82}$$

we derive the vorticity function:

$$\bar{\zeta}_0 = -\frac{2}{E_1(\tau_0)}\sigma \sum_{n=0}^{\infty} n!\,\mathcal{F}_n(\tau) L_n^{(1)}(\sigma). \tag{4.83}$$

With the solution $\bar{\Psi}_0$ of form (4.80), one can also derive the solution $\bar{\zeta}_0$ directly from (4.20) as

$$\bar{\zeta}_0 = -\frac{2}{E_1(\tau_0)} \frac{\sigma e^{-\tau}}{\sigma + \tau}. \tag{4.84}$$

It is easy to verify that the solution (4.84) is equivalent to (4.83) . The special solution (4.80)–(4.84), which contains a source flow term, $\psi_{*0} = \bar{d}_* \sigma$, is just the one previously obtained by Ananth and Gill (1989, 1991).

2.4 The Solution of Type (II)

By setting $\bar{d}_* = 0$, from (4.74), (4.75) and (4.77) one can uniquely solve the coefficients $\{\widehat{C}_{n,k}\}$ and subsequently determine a unique solution for the stream function $\Psi_0(\sigma, \tau)$. The cases: $\tau_0 = 0.1 - 10$, which corresponds to $\mathrm{Re} = 0.2 - 20$, have been calculated. The typical numerical results are to be described below. In Fig. 4.5, we show the perturbed stream function $\bar{\Psi}_0(\sigma, \tau)/\sigma$ versus τ as $\sigma = 0.1, 1.0, 3.0$. It is seen that, unlike the solution of type (I), the perturbed stream function type (II) is not a linear function of the variable σ. Its perturbed stream function contains higher-order components of Laguerre expansion. However, these higher-order components in the solution of type (II) are relatively small, hence, the function $\Psi_0(\sigma, \tau)/\sigma$ is not very sensitive to the variation of σ. Furthermore, the perturbed stream function with all σ vanish

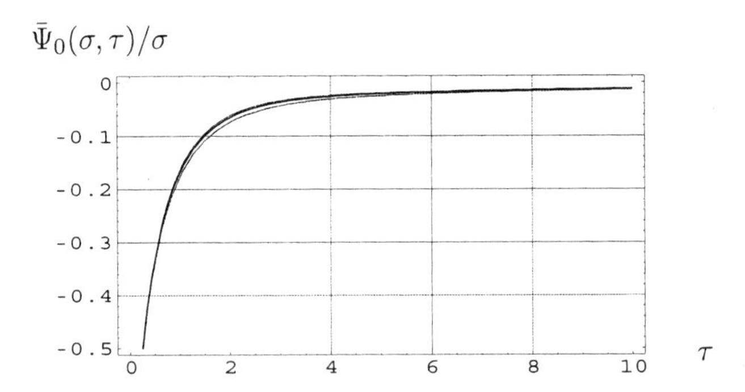

Figure 4.5. The perturbed stream functions of solution type (II) at $\sigma = 0.1, 1, 3$ for the case $\tau_0 = 0.25$ versus τ

in the far field as $\tau \to \infty$. In Fig. 4.6, we show the total stream function $\Psi_0(\sigma, \tau)$ for the case $\tau_0 = 0.25$ versus τ as $\sigma = 1.0$. For comparison, in the same figure, we include the stream function of uniform external flow and the total stream function of type (I) solution. It is seen that the solutions of type (I) and (II) are qualitatively different with each other, not only in the far field, but also in the field near the surface of the body. In Fig. 4.7, Fig. 4.8 and Fig. 4.9 we show the results of the perturbed stream functions $\bar{\Psi}_0(\sigma, \tau)$ versus τ as $\sigma = 1.0$ for the cases $\tau_0 = 0.1, 1.0, 10$, and compared with the corresponding solutions of type (I).

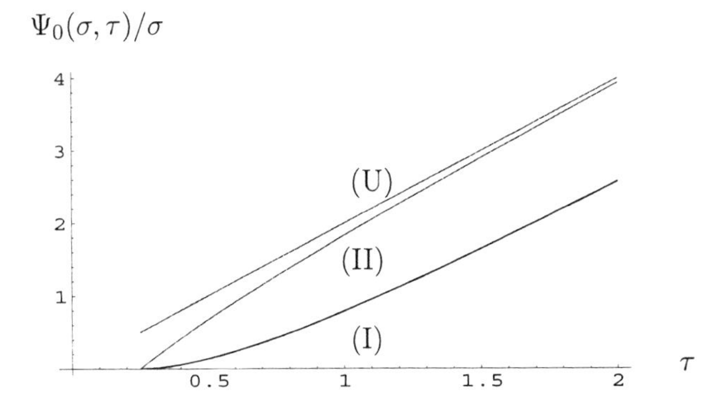

Figure 4.6. The distribution of total stream functions $\Psi_0(\sigma, \tau)/\sigma$ at $\sigma = 1$ for the case $\tau_0 = 0.25$ along τ-axis, where (I) represents the solution of type (I); (II) represents the solution of type (II); (U) represents the uniform external flow

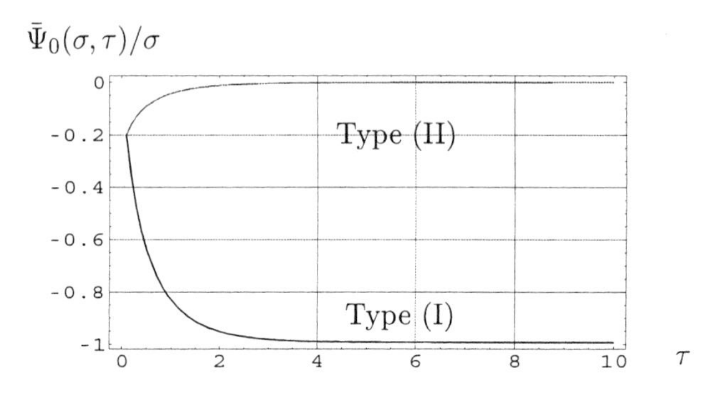

Figure 4.7. The comparison of the perturbed stream functions of type (I) and (II) at $\sigma = 1$ for the case $\tau_0 = 0.1$, or Re $= 0.2$

2.5 The Paradox of Oseen Model Solutions and Its Resolution

The different types of solutions of Oseen model equations explored in the last section leads to the paradox that if one accepts the solution of type (I), obtained by Gill and others, then the whole family of solutions with a free parameter $\bar{d}_*$ must be acceptable for the Oseen model problem as well. These solutions all satisfy the far-field condition of the velocity field: as $\eta \to \infty$,

$$u \to 0, \quad w \to -1. \tag{4.85}$$

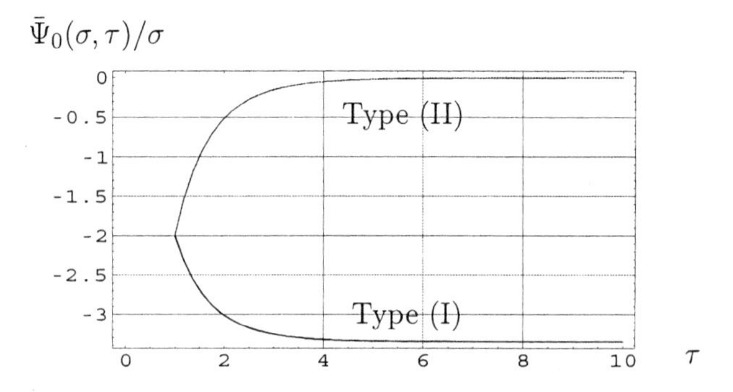

Figure 4.8. The comparison of the perturbed stream functions of type (I) and (II) at $\sigma = 1$ for the case $\tau_0 = 1$, or Re $= 2$

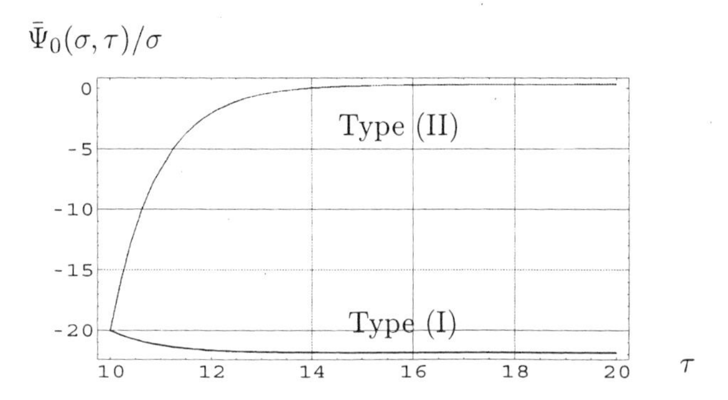

Figure 4.9. The comparison of the perturbed stream functions of type (I) and (II) at $\sigma = 1$ as $\tau_0 = 10$, or Re = 20

The far-field velocity condition (4.85) is equivalent to the condition of the stream function

$$\Psi \sim \frac{1}{2}\eta_0^4\xi^2\eta^2 + O(1) = 2\sigma\tau + O(1). \tag{4.86}$$

This implies that the uniform flow in the far-field condition is not imposed with a fixed flow flux, as the system allows undetermined, perturbed flow flux in the far field described by $\psi_{*,0} = \bar{d}_*\sigma$.

In order to screen the multiple mathematical solutions and guarantee the uniqueness of the solution, one might have different choices. A quite natural choice is to strengthen the up-stream far-field condition, as we have given in (3.10): namely, as $\eta \to \infty$,

$$\Psi \sim \frac{1}{2}\eta_0^4\xi^2\eta^2 + o(1) = 2\sigma\tau + o(1). \tag{4.87}$$

It implies that the uniform flow in the far field is imposed with a given flow flux, which is not changed by the presence of the paraboloid. In practice, in realistic experiments, this appears to be the case. With the far-field condition (4.87), one may conclude that the unique physically acceptable solution for the Oseen model problem is the solution of case (II):

$$\Psi_0 = 2\sigma\tau + \psi_0(\sigma,\tau) + \sigma\sum_{n=0}^{\infty}\bar{b}_n\mathcal{F}_n(\tau)L_n^{(1)}(\sigma). \tag{4.88}$$

and the solutions (4.80)–(4.84) obtained by Wilkinson (1955), and Ananth and Gill (1989,1991), which has been widely accepted in the fluid-dynamic literature for quite a long time, should be abandoned.

2.6 Appendix (A)

2.6.1 The Properties of Laguerre Functions

$$\int_0^\infty x\mathrm{e}^{-x}L_n^{(1)}(x)L_m^{(1)}(x)\mathrm{d}x = \begin{cases} n+1, & m=n, \\ 0, & m \neq n; \end{cases} \tag{4.89}$$

$$\begin{cases} L_0^{(1)}(x) = 1, \\ xL_n^{(1)''}(x) + (2-x)L_n^{(1)'}(x) + nL_n^{(1)}(x) = 0, \\ (n+1)L_{n+1}^{(1)}(x) = (2n+2-x)L_n^{(1)}(x) - (n+1)L_{n-1}^{(1)}(x), \\ xL_n^{(1)'}(x) = nL_n^{(1)}(x) - (n+1)L_{n-1}^{(1)}(x). \end{cases} \tag{4.90}$$

The following integrals are derived:

$$\frac{1}{n+1}\int_0^\infty x^2\mathrm{e}^{-x}L_m^{(1)''}(x)L_n^{(1)}(x)\mathrm{d}x = \begin{cases} 2, & m \geq n+2, \\ -n, & m = n+1, \\ 0, & m \leq n; \end{cases} \tag{4.91}$$

$$\frac{1}{n+1}\int_0^\infty x^2\mathrm{e}^{-x}L_m^{(1)}(x)L_n^{(1)}(x)\mathrm{d}x = \begin{cases} -(n+2), & m = n+1, \\ 2(n+1), & m = n, \\ -n, & m = n-1 \\ 0, & \text{otherwise}; \end{cases} \tag{4.92}$$

$$\frac{1}{n+1}\int_0^\infty x^2\mathrm{e}^{-x}L_m^{(1)'}(x)L_n^{(1)}(x)\mathrm{d}x = \begin{cases} -(n+2), & m = n+1, \\ n, & m = n, \\ 0, & \text{otherwise}; \end{cases} \tag{4.93}$$

$$\frac{1}{n+1}\int_0^\infty x\mathrm{e}^{-x}L_m^{(1)}(x)L_n^{(1)}(x)\mathrm{d}x = \begin{cases} -1, & m \geq n+1, \\ 0, & m \leq n; \end{cases} \tag{4.94}$$

$$\frac{1}{n+1}\int_0^\infty x\mathrm{e}^{-x}\left[xL_m^{(1)}(x)\right]'' L_n^{(1)}(x)\mathrm{d}x = \begin{cases} -(n+2), & m = n+1, \\ 0, & m \neq n+1; \end{cases} \tag{4.95}$$

$$\frac{1}{n+1}\int_0^\infty x\mathrm{e}^{-x}\left[xL_m^{(1)}(x)\right]' L_n^{(1)}(x)\mathrm{d}x = \begin{cases} -(n+2), & m = n+1, \\ n+1, & m = n, \\ 0, & \text{otherwise}; \end{cases} \tag{4.96}$$

2.6.2 Important Formulas

1. (see (Gradshteyn and Ryzhik, 1980)-[**13.4.17**])

$$U(a, b-1, z) = U(a, b, z) + aU(a+1, b, z), \tag{4.97}$$

2. For $n \neq 0, 1$ and $a \neq n$, (see (Gradshteyn and Ryzhik, 1980)-[**13.1.6**])

$$\begin{aligned} U(a, n+1, z) = & \frac{(-1)^{n+1}}{n!\Gamma(a-n)}\Big[M(a, n+1, z)\ln z \\ & + \sum_{k=0}^{\infty}\frac{(a)_k z^k}{(n+1)_k k!}\{\psi(a+k) - \psi(1+k) \\ & -\psi(1+n+k)\}\Big] \\ & + \frac{(n-1)!}{\Gamma(a)} z^{-n} M(a-n, 1-n, z)_n, \end{aligned} \tag{4.98}$$

where

$$\begin{aligned} & M(a, b, z)_n = 1 + \frac{az}{b} + \frac{(a)_2 z^2}{(b)_2 2!} + \cdots + \frac{(a)_{n-1} z^{n-1}}{(b)_{n-1}(n-1)!}, \\ & (a)_0 = 1, \quad 0! = 1, \quad (a)_n = a(a+1)(a+2)\cdots(a+n), \\ & M(a, b, z)_0 = 0, \qquad M(a, b, z)_1 = 1. \end{aligned} \tag{4.99}$$

3.

$$E_n(x) = \int_1^{\infty} \frac{\mathrm{e}^{-xt}}{t^n} \mathrm{d}t \quad (n = 0, 1, 2, \ldots.), \tag{4.100}$$

$$E_{n+1}(x) = \frac{1}{n}\Big[\mathrm{e}^{-x} - xE_n(x)\Big] \quad (n = 1, 2, 3, \ldots). \tag{4.101}$$

4. Derivatives: (see (Gradshteyn and Ryzhik, 1980)-[**5.1.26**])

$$\frac{\mathrm{d}E_n(x)}{\mathrm{d}x} = -E_{n-1}(x) \quad (n = 1, 2, 3, \ldots). \tag{4.102}$$

2.6.3 The derivation of the solution $\{A_n, B_n\}$ for (4.44)

We attempt to solve following difference-differential equations:

$$\frac{\mathrm{d}^2 \mathrm{B}_{\mathrm{n},0}}{\mathrm{d}\tau^2} = \frac{n+2}{\tau} B_{n+1,0} - \left(1 + \frac{2n+2}{\tau}\right) A_{n,0} + \frac{n+2}{\tau} A_{n+1,0} + \frac{n}{\tau} A_{n-1,0} \tag{4.103}$$

$$\frac{\mathrm{d}^2 \mathrm{A}_{\mathrm{n},0}}{\mathrm{d}\tau^2} + \frac{\mathrm{d}\mathrm{A}_{\mathrm{n},0}}{\mathrm{d}\tau} - \frac{n+1}{\tau} A_{n,0} = 0 \tag{4.104}$$

for $n = 0, 1, 2, \cdots$. It is easily seen that the solution of $A_{n,0}$ can be written in the form:

$$A_{n,0}(\tau) = a_{n,0} \tau \mathrm{e}^{-\tau} U(n+2, 2, \tau) \tag{4.105}$$

We derive that the equation (4.103) has the particular solution:

$$\begin{cases} B_{n,0}(\tau) = b_{n,0} \tau \mathrm{e}^{-\tau} U(n+2, 2, \tau) \\ b_{n,0} = -a_{n,0} + n a_{n-1,0} \end{cases} \tag{4.106}$$

To prove this, we substitute (4.106) into (4.103) and observe that

$$\begin{aligned} \text{L.H.S} &= \frac{\mathrm{d}^2 \mathrm{B}_{\mathrm{n},0}}{\mathrm{d}\tau^2} = \frac{\mathrm{d}^2}{\mathrm{d}\tau^2} \left[b_{n,0} \tau \mathrm{e}^{-\tau} U(n+2, 2, \tau)\right] \\ &= (-a_{n,0} + n a_{n-1,0}) \mathrm{e}^{-\tau} U(n+1, 2, \tau) \end{aligned} \tag{4.107}$$

and

$$\begin{aligned} \text{R.H.S} &= \frac{n+2}{\tau} B_{n+1,0} - \left(1 + \frac{2n+2}{\tau}\right) A_{n,0} + \frac{n+2}{\tau} A_{n+1,0} + \frac{n}{\tau} A_{n-1,0} \\ &= a_{n,0} \mathrm{e}^{-\tau} \Big[(n+1)(n+2) U(n+3, 2, \tau) \\ &- (2n+2+\tau) U(n+2, 2, \tau)\Big] + n a_{n-1,0} \mathrm{e}^{-\tau} U(n+1, 2, \tau) \end{aligned} \tag{4.108}$$

By applying the following recurrence relation with $a = n+1$ and $b = 2$ (see (Gradshteyn and Ryzhik, 1980)-[**3.4.15**]):

$$U(a-1, b, z) + (b - 2a - z) U(a, b, z) + a(1 + a - b) U(a+1, b, z) = 0$$

one obtains

$$U(n, 2, x) = (2n + x) U(n+1, 2, x) - n(n+1) U(n+2, 2, x).$$

This proves that L.H.S = R.H.S.

2.6.4 The Determination of the Functions: $\{\mathcal{A}_{n,k}(\tau), \hat{\mathcal{A}}_{n,k}(\tau), \mathcal{B}_{n,k}(\tau), \hat{\mathcal{B}}_{n,k}(\tau)\}$

We define

$$\mathcal{A}_{n,k}(\tau) = \tau^{\frac{1}{2}} \int_0^\infty (-1)^n \mathrm{e}^{-\lambda^2} \frac{\lambda^{2n+1}}{(n+1)!} K_1(2\lambda\tau^{\frac{1}{2}}) L_k^{(1)}(\lambda)\, \mathrm{d}\lambda, \quad (4.109)$$

$$\hat{\mathcal{A}}_{n,k}(\tau) = \tau^{\frac{1}{2}} \int_0^\infty (-1)^n \mathrm{e}^{-\lambda^2} \frac{\lambda^{2n+2}}{(n+1)!} K_1(2\lambda\tau^{\frac{1}{2}}) L_k^{(1)}(\lambda)\, \mathrm{d}\lambda; \quad (4.110)$$

and

$$\mathcal{B}_{n,k}(\tau) = \int_0^\infty (-1)^n \mathrm{e}^{-\lambda^2} \frac{\lambda^{2n+1}}{(n+1)!} K_0(2\lambda\tau^{\frac{1}{2}}) L_k^{(1)}(\lambda)\, \mathrm{d}\lambda, \quad (4.111)$$

$$\hat{\mathcal{B}}_{n,k}(\tau) = \int_0^\infty (-1)^n \mathrm{e}^{-\lambda^2} \frac{\lambda^{2n+2}}{(n+1)!} K_0(2\lambda\tau^{\frac{1}{2}}) L_k^{(1)}(\lambda)\, \mathrm{d}\lambda. \quad (4.112)$$

In terms of the recurrence formula

$$\begin{cases} L_0^{(1)}(x) = 1, \\ (n+1)L_{n+1}^{(1)}(x) = (2n+2-x)L_n^{(1)}(x) - (n+1)L_{n-1}^{(1)}(x), \end{cases} \quad (4.113)$$

and the formula

$$\begin{aligned} \int_0^\infty \lambda^m \mathrm{e}^{-\lambda^2} K_\nu(a\lambda)\mathrm{d}\lambda &= \tfrac{1}{2a}\Gamma\left(\tfrac{m+1+\nu}{2}\right)\Gamma\left(\tfrac{m+1-\nu}{2}\right)\exp\left(\tfrac{a^2}{8}\right) \\ &\quad \times W_{-\frac{1}{2}m,\frac{1}{2}\nu}\left(\tfrac{a^2}{4}\right) \qquad (m > \nu - 1) \end{aligned} \quad (4.114)$$

(see (Gradshteyn and Ryzhik, 1980), page 717, [**6.6311 – 3**]), where

$$W_{-\frac{1}{2}m,\frac{1}{2}\nu}(z) = \mathrm{e}^{-\frac{1}{2}z} z^{\frac{1+\nu}{2}} U\Big(\frac{1+m+\nu}{2},\ 1+\nu,\ z\Big), \quad (4.115)$$

we obtain

$$\begin{aligned} \mathcal{A}_{n,0}(\tau) &= (-1)^n \tfrac{\tau}{4} \tfrac{n+\frac{1}{2}}{(n+1)!} \left[\Gamma(n+\tfrac{1}{2})\right]^2 U\left(n+\tfrac{3}{2}, 2, \tau\right), \\ \hat{\mathcal{A}}_{n,0}(\tau) &= (-1)^n \tfrac{\tau}{4} n! U\left(n+2, 2, \tau\right), \end{aligned} \quad (4.116)$$

$$\begin{aligned} \mathcal{B}_{n,0}(\tau) &= (-1)^n \tfrac{1}{4(n+1)!} \left[\Gamma\left(n+\tfrac{1}{2}\right)\right]^2 U\left(n+1, 1, \tau\right), \\ \hat{\mathcal{B}}_{1,0}(\tau) &= (-1)^n \tfrac{n!}{4(n+1)} U\left(n+\tfrac{3}{2}, 1, \tau\right). \end{aligned} \quad (4.117)$$

Moreover, we derive the following recurrence formulas for the functions $\mathcal{A}_{n,k}(\tau)$, $\hat{\mathcal{A}}_{n,k}(\tau)$, $\mathcal{B}_{n,k}(\tau)$ and $\hat{\mathcal{B}}_{n,k}(\tau)$, respectively:

$$\begin{aligned} \mathcal{A}_{n,k+1} &= 2\mathcal{A}_{n,k} - \mathcal{A}_{n,k-1} - \tfrac{1}{k+1}\hat{\mathcal{A}}_{n,k}\,, \\ \hat{\mathcal{A}}_{n,k+1} &= 2\hat{\mathcal{A}}_{n,k} - \hat{\mathcal{A}}_{n,k-1} - \tfrac{n+2}{k+1}\mathcal{A}_{n+1,k}, \end{aligned} \tag{4.118}$$

$$\begin{aligned} \mathcal{A}_{n,1} &= 2\mathcal{A}_{n,0} - \hat{\mathcal{A}}_{n,0}\,, \\ \hat{\mathcal{A}}_{n,1} &= 2\hat{\mathcal{A}}_{n,0} - (n+2)\mathcal{A}_{n+1,0}\,, \end{aligned} \tag{4.119}$$

and

$$\begin{aligned} \mathcal{B}_{n,k+1} &= 2\mathcal{B}_{n,k} - \mathcal{B}_{n,k-1} - \tfrac{1}{k+1}\hat{\mathcal{B}}_{n,k}\,, \\ \hat{\mathcal{B}}_{n,k+1} &= 2\hat{\mathcal{B}}_{n,k} - \hat{\mathcal{B}}_{n,k-1} - \tfrac{n+2}{k+1}\mathcal{B}_{n+1,k}\,, \end{aligned} \tag{4.120}$$

$$\begin{aligned} \mathcal{B}_{n,1} &= 2\mathcal{B}_{n,0} - \hat{\mathcal{B}}_{n,0}\,, \\ \hat{\mathcal{B}}_{n,1} &= 2\hat{\mathcal{B}}_{n,0} - (n+2)\mathcal{B}_{n+1,0}. \end{aligned} \tag{4.121}$$

3. Uniformly Valid Asymptotic Solution for Steady Viscous Flow past a Slender Paraboloid of Revolution

The Oseen model solution derived in the last section, in principle, is valid for viscous flow past any semi-infinite paraboloid of revolution under any Reynolds number $\mathrm{Re} = O(1)$. It, however, cannot be used as the leading term of the uniformly valid asymptotic expansion solution for the problem with the Navier–Stokes model. On the other hand, one can expect that a continuous solution passing the paraboloid may be unstable, when Re number is sufficiently large. Therefore, for practically observable continuous solutions, it would be more sensible to seek for the solutions in the regime of small Re number. In order to derive a uniformly valid asymptotic solution with the Navier–Stokes model, in this section we shall consider viscous flow past a slender paraboloid body. The slenderness of the paraboloid body can be measured by the ratio of the tip radius of the paraboloid, ℓ_{t} and the viscous diffusion length $\ell_{\mathrm{d}} = \nu/U_\infty$, namely by the small parameter, $\epsilon_0 = \frac{\ell_{\mathrm{t}}}{\ell_{\mathrm{d}}}$. On the other hand, the Reynolds number of flow based on the tip radius, $\mathrm{Re} = \frac{\ell_{\mathrm{t}} U_\infty}{\nu} = \epsilon_0$. Hence, our slender body assumption is equivalent to the small flow Reynolds number assumption. The problem under study is formulated as the so-called *singular boundary problem.* By utilizing the Laguerre series representation and the matched asymptotic expansion technique, we shall derive a uniformly valid asymptotic expansion solution for the problem in the limit $\epsilon_0 \to 0$.

3.1 Mathematical Formulation of the Problem

The complete mathematical formulation of the problem with the Navier–Stokes model has been given in section 1 of this chapter already. Now with the slenderness assumption: $\eta_0^2 = \text{Re} = \epsilon_0$, the surface conditions become: at $\tau = \tau_0 = \frac{1}{2}\epsilon_0$,

$$\frac{\partial \bar{\Psi}}{\partial \sigma} + \epsilon_0 = 0, \tag{4.122}$$

$$\frac{\partial \bar{\Psi}}{\partial \tau} + 2\sigma = 0. \tag{4.123}$$

It is seen from the above that the surface shape function $\tau = \tau_0 \to 0$, as $\epsilon_0 \to 0$. The problem under study, therefore, is a singular boundary problem. We are going to solve this problem with the matched asymptotic expansion method. In doing so, we divide the whole flow field into two regions:

- the outer region away from the surface, $\tau = O(1)$,
- the inner region near the surface, $\tau \ll 1$.

We shall separately find the asymptotic expansion solution in each region, then match these solutions in the intermediate region.

3.2 Laguerre Series Representation of Solutions

To proceed, we use the Laguerre series representation method developed in section 2 of this chapter. Noting the axi-symmetric condition, $u(0,\tau) = \bar{\Psi}_0(0,\tau) = 0$, we first expand the solution in the Laguerre series

$$\begin{cases} \bar{\zeta} = \sigma \sum\limits_{n=0}^{\infty} A_n(\tau,\epsilon_0) L_n^{(1)}(\sigma) \\ \bar{\Psi} = \sigma \sum\limits_{n=0}^{\infty} B_n(\tau,\epsilon_0) L_n^{(1)}(\sigma). \end{cases} \tag{4.124}$$

By substituting the above expansion in the equations (4.15)–(4.16) we derive the following two sets of difference–differential equations:

$$\frac{\partial^2 B_n}{\partial \tau^2} + A_n = \frac{n+2}{\tau} B_{n+1} - \frac{1}{\tau}\Big[2(n+1)A_n - (n+2)A_{n+1} - nA_{n-1}\Big],$$
$$\frac{\partial^2 A_n}{\partial \tau^2} + \frac{\partial A_n}{\partial \tau} - \frac{n+1}{\tau} A_n = -\mathcal{N}\Big\{A_n; B_n\Big\}, \tag{4.125}$$
$$(n = 0, 1, 2, \ldots,)$$

where

$$\mathcal{N}\Big\{A_n; B_n\Big\} = \frac{\partial(\bar{\zeta}, \bar{\Psi})}{\partial(\sigma, \tau)} \tag{4.126}$$

is the nonlinear part of the differential operator. The problem now is to find the solution $\{A_n, B_n\}$ $(n = 0, 1, \ldots)$ for (4.125).

3.3 Outer Asymptotic Expansion Solution in the Limit Re $\to 0$

In the limit $\epsilon_0 \to 0$, we may consider the following general asymptotic expansion forms for the outer solution $\{A_n = \bar{A}_n,\ B_n = \bar{B}_n\}$ $(n = 0, 1, \ldots)$:

$$\begin{aligned}\bar{A}_n(\tau, \epsilon_0) = \ & \bar{\nu}_0(\epsilon_0)\Big\{\bar{A}_{0,n,0}(\tau) + \epsilon_0 \bar{A}_{0,n,1}(\tau) + \cdots\Big\} \\ & + \bar{\nu}_1(\epsilon_0)\Big\{\bar{A}_{1,n,0}(\tau) + \epsilon_0 \bar{A}_{1,n,1}(\tau) + \cdots\Big\} \\ & + \bar{\nu}_2(\epsilon_0)\Big\{\bar{A}_{2,n,0}(\tau) + \epsilon_0 \bar{A}_{2,n,1}(\tau) + \cdots\Big\} + \cdots,\end{aligned} \tag{4.127}$$

$$\begin{aligned}\bar{B}_n(\tau, \epsilon_0) = \ & \bar{\nu}_0(\epsilon_0)\Big\{\bar{B}_{0,n,0}(\tau) + \epsilon_0 \bar{B}_{0,n,1}(\tau) + \cdots\Big\} \\ & + \bar{\nu}_1(\epsilon_0)\Big\{\bar{B}_{1,n,0}(\tau) + \epsilon_0 \bar{B}_{1,n,1}(\tau) + \cdots\Big\} \\ & + \bar{\nu}_2(\epsilon_0)\Big\{\bar{B}_{2,n,0}(\tau) + \epsilon_0 \bar{B}_{2,n,1}(\tau) + \cdots\Big\} + \cdots\end{aligned} \tag{4.128}$$

where the asymptotic factors,

$$\bar{\nu}_0(\epsilon_0) \gg \bar{\nu}_1(\epsilon_0) \gg \bar{\nu}_2(\epsilon_0) \gg \cdots,$$
$$\bar{\nu}_i(\epsilon_0) \neq \bar{\nu}_j(\epsilon_0)\epsilon_0^k, \quad (i, j, k = 0, 1, \ldots)$$

are to be determined later. In accordance with the above, the solution $\bar{\zeta}$ and $\bar{\Psi}$ may have the general asymptotic structure

$$\begin{aligned}\bar{\zeta} &= \bar{\nu}_0(\epsilon_0)\bar{\zeta}_0 + \bar{\nu}_1(\epsilon_0)\bar{\zeta}_1 + \bar{\nu}_2(\epsilon_0)\bar{\zeta}_2 + \cdots, \\ \bar{\Psi} &= \bar{\nu}_0(\epsilon_0)\bar{\Psi}_0 + \bar{\nu}_1(\epsilon_0)\bar{\Psi}_1 + \bar{\nu}_2(\epsilon_0)\bar{\Psi}_2 + \cdots.\end{aligned} \tag{4.129}$$

With the factor $\bar{\nu}_m(\epsilon_0)$, the solution $\bar{\zeta}_m$ and $\bar{\Psi}_m$ have the following asymptotic structure:

$$\begin{aligned}\bar{\zeta}_m(\sigma,\tau,\epsilon_0) &= \Big[\bar{A}_{m,0,0}(\tau)+\epsilon_0\bar{A}_{m,0,1}(\tau)+\cdots\Big]L_0^{(1)}(\sigma)\\ &+\Big[\bar{A}_{m,1,0}(\tau)+\epsilon_0\bar{A}_{m,1,1}(\tau)+\cdots\Big]L_1^{(1)}(\sigma)\\ &+\Big[\bar{A}_{m,2,0}(\tau)+\epsilon_0\bar{A}_{m,2,1}(\tau)+\cdots\Big]L_2^{(1)}(\sigma)+\cdots\end{aligned} \tag{4.130}$$

and

$$\begin{aligned}\bar{\Psi}_m(\sigma,\tau,\epsilon_0) &= \Big[\bar{B}_{m,0,0}(\tau)+\epsilon_0\bar{B}_{m,0,1}(\tau)+\cdots\Big]L_0^{(1)}(\sigma)\\ &+\Big[\bar{B}_{m,1,0}(\tau)+\epsilon_0\bar{B}_{m,1,1}(\tau)+\cdots\Big]L_1^{(1)}(\sigma)\\ &+\Big[\bar{B}_{m,2,0}(\tau)+\epsilon_0\bar{B}_{m,2,1}(\tau)+\cdots\Big]L_2^{(1)}(\sigma)+\cdots.\end{aligned} \tag{4.131}$$

3.3.1 Zeroth-Order Solution of Velocity Field $O(\bar{\nu}_0(\epsilon_0))$

We assume that the leading factor $\bar{\nu}_0(\epsilon_0)\to 0$, as $\epsilon_0\to 0$, which will be verified later. Then, in the leading-order approximation, the system is reduced to

$$\begin{aligned}&\frac{\mathrm{d}^2\bar{B}_{0,n,0}}{\mathrm{d}\tau^2}=\frac{n+2}{\tau}\bar{B}_{0,n+1,0}-\left[1+\frac{2(n+1)}{\tau}\right]\bar{A}_{0,n,0}\\ &\quad+\tfrac{1}{\tau}\Big[(n+2)\bar{A}_{0,n+1,0}+n\bar{A}_{0,n-1,0}\Big],\\ &\frac{\mathrm{d}^2\bar{A}_{0,n,0}}{\mathrm{d}\tau^2}+\frac{\mathrm{d}\bar{A}_{0,n,0}}{\mathrm{d}\tau}-\frac{n+1}{\tau}\bar{A}_{0,n,0}=0\\ &\quad(n=0,1,2,\cdots).\end{aligned} \tag{4.132}$$

Evidently, this system is just the Oseen model problem discussed in the last subsection, where the general solution of the stream function has been given in the analytical form (4.60) and the related general solution for the associated homogeneous equation of the stream function has been written in the form

$$\psi_0=\sigma\sum_{n=0}^{\infty}L_n^{(1)}(\sigma)\sum_{k=0}^{\infty}C_{0,k}\mathcal{A}_{n,k}(\tau), \tag{4.133}$$

$$\frac{\partial\psi_0}{\partial\tau}=\sigma\sum_{n=0}^{\infty}L_n^{(1)}(\sigma)\sum_{k=0}^{\infty}C_{0,k}\hat{\mathcal{B}}_{n,k}(\tau). \tag{4.134}$$

With these results, we can write the following two outer solutions:

$$
\begin{aligned}
\bar{\nu}_{*0}(\epsilon_0)\bar{\Psi}_{*0} &= \bar{\nu}_{*0}(\epsilon_1)\sigma \sum_{n=0}^{N} L_n^{(1)}(\sigma)\left[\sum_{k=0}^{N} C_{*,k}\mathcal{A}_{n,k} + \bar{b}_{*,n}\mathcal{F}_n(\tau)\right], \\
\bar{\nu}_{*0}(\epsilon_0)\bar{\zeta}_{*0} &= -\bar{\nu}_{*0}(\epsilon_0)\ \sigma \sum_{n=0}^{\infty} \bar{a}_{*,0}\mathcal{F}_n(\tau)L_n^{(1)}(\sigma)
\end{aligned}
\tag{4.135}
$$

and

$$
\begin{aligned}
\bar{\nu}_0(\epsilon_0)\bar{\Psi}_0 &= \bar{\nu}_0(\epsilon_1)\sigma \sum_{n=0}^{\infty} \bar{b}_{0,n,0}\mathcal{F}_n(\tau)L_n^{(1)}(\sigma), \\
\bar{\nu}_0(\epsilon_0)\bar{\zeta}_0 &= -\bar{\nu}_0(\epsilon_0)\ \sigma \sum_{n=0}^{\infty} \bar{a}_{0,n,0}\mathcal{F}_n(\tau)L_n^{(1)}(\sigma),
\end{aligned}
\tag{4.136}
$$

where

$$
\begin{aligned}
\bar{b}_{*,n} &= -\bar{a}_{*,n} + n\bar{a}_{*,n-1}, \\
\bar{b}_{0,n,0} &= -\bar{a}_{0,n,0} + n\bar{a}_{0,n-1,0}.
\end{aligned}
\tag{4.137}
$$

The above outer solutions obviously satisfy the far-field condition (3.10), since all the functions $\mathcal{A}_{n,k}(\tau)$, $\hat{\mathcal{A}}_{n,k}(\tau)$, $\mathcal{B}_{n,k}(\tau)$ and $\hat{\mathcal{B}}_{n,k}(\tau)$ tend to zero as $\tau \to \infty$. The outer solution (4.135) involves two sequences of unknowns: $\{\bar{b}_{*,n}\ C_{*,k}\ (n,k = 0,1,2,\ldots)\}$, while (4.136) involves two sequences of unknowns: $\{\bar{b}_{0,n,0}\ C_{0,k}(n,k = 0,1,2,\ldots)\}$, which are to be determined by matching conditions with the inner solution.

3.4 Inner Asymptotic Expansion of the Solution

In the inner region, we introduce the inner variables $(\sigma, \hat{\tau})$, where

$$
\hat{\tau} = \frac{\tau}{\epsilon_0}, \tag{4.138}
$$

and accordingly let

$$
\begin{cases}
\hat{\Psi}(\sigma,\hat{\tau}) = \bar{\Psi}(\sigma,\epsilon_0\hat{\tau}), \\
\hat{\zeta}(\sigma,\hat{\tau}) = \bar{\zeta}(\sigma,\epsilon_0\hat{\tau}),
\end{cases}
\tag{4.139}
$$

We make the Laguerre expansion for the inner solutions:

$$
\begin{cases}
\hat{\zeta}(\sigma,\hat{\tau},\epsilon_0) = \sigma \sum_{n=0}^{\infty} \hat{A}_n(\hat{\tau},\epsilon_0)L_n^{(1)}(\sigma), \\
\hat{\Psi}(\sigma,\hat{\tau},\epsilon_0) = \sigma \sum_{n=0}^{\infty} \hat{B}_n(\hat{\tau},\epsilon_0)L_n^{(1)}(\sigma),
\end{cases}
\tag{4.140}
$$

where we have

$$\begin{cases} \hat{A}_n(\sigma, \hat{\tau}) = A(\sigma, \epsilon_0 \hat{\tau}), \\ \hat{B}_n(\sigma, \hat{\tau}) = B(\sigma, \epsilon_0 \hat{\tau}). \end{cases} \tag{4.141}$$

Equations of $\{\hat{A}_n, \hat{B}_n\}$ can be derived from (4.125) by setting $\tau = \epsilon_0 \hat{\tau}$:

$$\begin{aligned} \frac{\mathrm{d}^2 \hat{\mathrm{B}}_\mathrm{n}}{\mathrm{d}\hat{\tau}^2} = \epsilon_0 \frac{n+2}{\hat{\tau}} \hat{B}_{n+1} - \epsilon_0 \left(\epsilon_0 + \frac{2n+2}{\hat{\tau}} \right) \hat{A}_n \\ + \epsilon_0 \frac{n+2}{\hat{\tau}} \hat{A}_{n+1} + \epsilon_0 \frac{n}{\hat{\tau}} \bar{A}_{n-1}, \end{aligned} \tag{4.142}$$

$$\frac{\mathrm{d}^2 \hat{\mathrm{A}}_\mathrm{n}}{\mathrm{d}\hat{\tau}^2} = -\epsilon_0 \frac{\mathrm{d}\hat{\mathrm{A}}_\mathrm{n}}{\mathrm{d}\hat{\tau}} + \epsilon_0 \frac{n+1}{\hat{\tau}} \hat{A}_n + \epsilon_0 \widehat{\mathcal{N}}_n \Big\{ \hat{A}_k, \hat{B}_k \Big\}, \tag{4.143}$$

where

$$\widehat{\mathcal{N}}_n \Big\{ \hat{A}_k, \hat{B}_k \Big\} = \frac{\partial(\hat{\zeta}, \hat{\Psi})}{\partial(\sigma, \hat{\tau})}. \tag{4.144}$$

On the other hand, with the inner variable $(\sigma, \hat{\tau})$, the body shape function changes to

$$\hat{\tau} = \hat{\tau}_0 = \frac{1}{2} \tag{4.145}$$

and the surface conditions change to

$$\hat{\Psi} + \epsilon_0 \sigma = 0, \tag{4.146}$$

and

$$\frac{\partial \hat{\Psi}}{\partial \hat{\tau}} + 2\epsilon_0 \sigma = 0. \tag{4.147}$$

Similar to (4.127)–(4.128), the following asymptotic expansion can be made for $\{\hat{A}_n, \hat{B}_n, (n = 0, 1, 2, \ldots)\}$:

$$\begin{aligned} \hat{A}_n(\hat{\tau}, \epsilon_0) = & \ \hat{\nu}_0(\epsilon_0) \Big\{ \hat{A}_{0,n,0}(\hat{\tau}) + \epsilon_0 \hat{A}_{0,n,1}(\hat{\tau}) + \cdots \Big\} \\ & + \hat{\nu}_1(\epsilon_0) \Big\{ \hat{A}_{1,n,0}(\hat{\tau}) + \epsilon_0 \hat{A}_{1,n,1}(\hat{\tau}) + \cdots \Big\} \\ & + \hat{\nu}_2(\epsilon_0) \Big\{ \hat{A}_{2,n,0}(\hat{\tau}) + \epsilon_0 \hat{A}_{2,n,1}(\hat{\tau}) + \cdots \Big\} \\ & + \cdots, \end{aligned} \tag{4.148}$$

$$
\begin{aligned}
\hat{B}_n(\hat{\tau},\epsilon_0) = \; & \epsilon_0\hat{\nu}_0(\epsilon_0)\Big\{\hat{B}_{0,n,0}(\hat{\tau}) + \epsilon_0\hat{B}_{0,n,1}(\hat{\tau}) + \cdots\Big\} \\
& + \epsilon_0\hat{\nu}_1(\epsilon_0)\Big\{\hat{B}_{1,n,0}(\hat{\tau}) + \epsilon_0\hat{B}_{1,n,1}(\hat{\tau}) + \cdots\Big\} \\
& + \epsilon_0\hat{\nu}_2(\epsilon_0)\Big\{\hat{B}_{2,n,0}(\hat{\tau}) + \epsilon_0\hat{B}_{2,n,1}(\hat{\tau}) + \cdots\Big\} \\
& + \cdots,
\end{aligned}
\tag{4.149}
$$

where the leading asymptotic factors

$$
\hat{\nu}_0(\epsilon_0) \gg \hat{\nu}_1(\epsilon_0) \gg \hat{\nu}_2(\epsilon_0) \gg \cdots
\tag{4.150}
$$

are to be determined by applying interface conditions and matching conditions with the outer solution. The solution $\hat{\zeta}$ and $\hat{\Psi}$ then have the following general asymptotic expansion:

$$
\begin{aligned}
\hat{\zeta} &= \hat{\nu}_0(\epsilon_0)\hat{\zeta}_0 + \hat{\nu}_1(\epsilon_0)\hat{\zeta}_1 + \hat{\nu}_2(\epsilon_0)\hat{\zeta}_2 + \cdots, \\
\hat{\Psi} &= \hat{\Psi}_{*,0}(\sigma,\hat{\tau}) + \epsilon_0\Big[\hat{\nu}_0(\epsilon_0)\hat{\Psi}_0 + \hat{\nu}_1(\epsilon_0)\hat{\Psi}_1 + \hat{\nu}_2(\epsilon_0)\hat{\Psi}_2\Big],
\end{aligned}
\tag{4.151}
$$

where the function

$$
\hat{\Psi}_{*,0}(\sigma,\hat{\tau}) = -2\epsilon_0\sigma\hat{\tau}
\tag{4.152}
$$

is introduced to satisfy the inhomogeneous boundary conditions (4.146)–(4.146).

With the factor $\hat{\nu}_m(\epsilon_0)$, the solution $\hat{\zeta}_m$ will have the following asymptotic structure:

$$
\begin{aligned}
\hat{\zeta}_m(\sigma,\hat{\tau},\epsilon_0) = \; & \Big[\hat{A}_{m,0,0}(\hat{\tau}) + \epsilon_0\hat{A}_{m,0,1}(\hat{\tau}) + \cdots\Big]L_0^{(1)}(\sigma) \\
& + \Big[\bar{A}_{m,1,0}(\hat{\tau}) + \epsilon_0\hat{A}_{m,1,1}(\hat{\tau}) + \cdots\Big]L_1^{(1)}(\sigma) \\
& + \Big[\hat{A}_{m,2,0}(\hat{\tau}) + \epsilon_0\hat{A}_{m,2,1}(\hat{\tau}) + \cdots\Big]L_2^{(1)}(\sigma) \\
& + \cdots
\end{aligned}
\tag{4.153}
$$

and

$$
\begin{aligned}
\hat{\Psi}_m(\sigma,\hat{\tau},\epsilon_0) = \; & \Big[\hat{B}_{m,0,0}(\hat{\tau}) + \epsilon_0\hat{B}_{m,0,1}(\hat{\tau}) + \cdots\Big]L_0^{(1)}(\sigma) \\
& + \Big[\bar{B}_{m,1,0}(\hat{\tau}) + \epsilon_0\hat{B}_{m,1,1}(\hat{\tau}) + \cdots\Big]L_1^{(1)}(\sigma) \\
& + \Big[\hat{B}_{m,2,0}(\hat{\tau}) + \epsilon_0\hat{B}_{m,2,1}(\hat{\tau}) + \cdots\Big]L_2^{(1)}(\sigma) \\
& + \cdots.
\end{aligned}
\tag{4.154}
$$

3.4.1 The Zeroth-Order Inner Solution

With the leading factor $\hat{\nu}_0(\epsilon_0)$, the equations of $\{\hat{A}_{0,n,0}, \hat{B}_{0,n,0}\}$ have the forms

$$\frac{\mathrm{d}^2\hat{\mathrm{B}}_{0,\mathrm{n},0}}{\mathrm{d}\hat{\tau}^2} = -\frac{2n+2}{\hat{\tau}}\hat{A}_{0,n,0} + \frac{n+2}{\hat{\tau}}\hat{A}_{0,n+1,0} + \frac{n}{\hat{\tau}}\bar{A}_{0,n-1,0}, \quad (4.155)$$

$$\frac{\mathrm{d}^2\hat{\mathrm{A}}_{0,\mathrm{n},0}}{\mathrm{d}\hat{\tau}^2} = 0. \quad (4.156)$$

In the zeroth-order approximation, we have the surface conditions

$$\begin{aligned} &\sum_{n=0}^{\infty} \hat{B}_{0,n,0}(\hat{\tau}_0) L_n^{(1)}(\sigma) = 0, \\ &\sum_{n=0}^{\infty} \hat{B}'_{0,n,0}(\hat{\tau}_0) L_n^{(1)}(\sigma) = 0. \end{aligned} \quad (4.157)$$

It follows that

$$\begin{aligned} &\hat{B}_{0,n,0}(\hat{\tau}_0) = 0 \quad (n \geq 0), \\ &\hat{B}'_{0,n,0}(\hat{\tau}_0) = 0 \quad (n \geq 0). \end{aligned} \quad (4.158)$$

From equation (4.156), we find

$$\hat{A}_{0,n,0} = \hat{a}_{0,n,0}\hat{\tau} + \hat{b}_{0,n,0}. \quad (4.159)$$

From equation (4.155) with surface conditions (4.158), we obtain

$$\hat{B}_{0,n,0} = \tilde{b}_{0,n,0}\Big[\hat{\tau}\ln\hat{\tau} - \hat{\tau}(1+\ln\hat{\tau}_0) + \hat{\tau}_0\Big] + \tilde{a}_{0,n,0}\left(\hat{\tau} - \hat{\tau}_0\right)^2, \quad (4.160)$$

where

$$\begin{aligned} \tilde{a}_{0,n,0} &= \tfrac{1}{2}\Big[-2(n+1)\,\hat{a}_{0,n,0} + (n+2)\,\hat{a}_{0,n+1,0} + n\,\hat{a}_{0,n-1,0}\Big], \\ \tilde{b}_{0,n,0} &= -2(n+1)\,\hat{b}_{0,n,0} + (n+2)\,\hat{b}_{0,n+1,0} + n\,\hat{b}_{0,n-1,0}, \end{aligned} \quad (4.161)$$

for $n = 1, 2, \cdots$.

The final form of inner expansion solution is

$$\left\{\begin{aligned} &\hat{\Psi}_{*,0}(\sigma,\hat{\tau},\epsilon_0) + \epsilon_0\hat{\nu}_0(\epsilon_0)\hat{\Psi}_0(\sigma,\hat{\tau},\epsilon_0) + \cdots \\ &= -2\epsilon_0\sigma\hat{\tau} + \epsilon_0\,\hat{\nu}_0(\epsilon_0)\sigma\sum_{n=0}^{\infty}\Big\{\tilde{b}_{0,n,0}\Big[\hat{\tau}\ln\hat{\tau} - \hat{\tau}(1+\ln\hat{\tau}_0) + \hat{\tau}_0\Big] \\ &\quad + \tilde{a}_{0,n,0}(\hat{\tau} - \hat{\tau}_0)^2\Big\}L_n^{(1)}(\sigma) + O(\epsilon_0^2\hat{\nu}_0(\epsilon_0)), \\ &\hat{\nu}_0(\epsilon_0)\ \hat{\zeta}_0(\sigma,\hat{\tau},\epsilon_0) + \cdots \\ &= \hat{\nu}_0(\epsilon_0)\sigma\Big\{\sum_{n=0}^{\infty}\Big[\hat{a}_{0,n,0}\hat{\tau} + \hat{b}_{0,n,0}\Big]L_n^{(1)}(\sigma)\Big\} + O(\epsilon_0\hat{\nu}_0(\epsilon_0)). \end{aligned}\right. \quad (4.162)$$

3.5 Matching Conditions of the Solutions

To match the outer solutions with the inner solutions, we first rewrite the outer solution (4.135) and (4.136) with inner variables $\{\bar{\sigma}, \bar{\tau}\} = \{\bar{\sigma},\ \epsilon_1\hat{\tau}\}$. and apply the following asymptotic formula: as $x \to 0$,

$$\begin{aligned} E_2(x) &= 1 + x\ln x + (\gamma_0 - 1)x + O(x^2), \\ \mathcal{F}_n(x) &= \tfrac{1}{(n+1)!} + O(x\ln x), \end{aligned} \tag{4.163}$$

where $\gamma_0 = 0.5772\cdots$ is the Euler constant. Before matching, we point out that the system allows a special type of outer solutions $\Psi_{*,0}(\sigma, \tau)$, which as $\epsilon_0 \to 0$ have the asymptotic behaviors

$$\begin{aligned} &\bar{\Psi}_{*,0}(\sigma, \epsilon_0\hat{\tau}) \sim \sigma L_0^{(1)}(\sigma),\ \tfrac{\partial\bar{\Psi}_{*,0}}{\partial\tau}(\sigma, \epsilon_0\hat{\tau}) \sim O(1), \\ &\bar{\zeta}_{*,0}(\sigma, \epsilon_0\hat{\tau}) \sim 0. \end{aligned} \tag{4.164}$$

The proof of this statement and derivation of this special type of outer solutions is given in Appendix (B). We now write the outer solution:

$$\begin{aligned} \bar{\Psi}(\sigma, \epsilon_0\hat{\tau}, \epsilon_0) &= \bar{\nu}_{*,0}(\epsilon_0)\gamma_{0,0}\bar{\Psi}_{*0} + \nu_0(\epsilon_0)\bar{\Psi}_0 + \cdots \\ &= \bar{\nu}_{*,0}(\epsilon_0)\sigma\gamma_{0,0}L_0^{(1)}(\sigma) \\ &\quad + \bar{\nu}_0(\epsilon_0)\sigma\Big\{\bar{b}_{0,0,0}\Big[1 + \epsilon_0\ln\epsilon_0\hat{\tau} + \epsilon_0(\gamma_0 - 1)\hat{\tau} + \epsilon_0\hat{\tau}\ln\hat{\tau}\Big] \\ &\quad + \sum_{n=1}^{\infty}\frac{\bar{b}_{0,n,0}}{(n+1)!}L_n^{(1)}(\sigma)\Big\} + O(\epsilon_0\nu_0(\epsilon_0))\cdots, \end{aligned} \tag{4.165}$$

$$\begin{aligned} \bar{\zeta}(\sigma, \epsilon_0\hat{\tau}, \epsilon_0) &= \bar{\nu}_{*,0}(\epsilon_0)\zeta_{*0} + \nu_0(\epsilon_0)\zeta_0 + \cdots \\ &= -\bar{\nu}_0(\epsilon_0)\ \sigma\sum_{n=0}^{\infty}\frac{\bar{a}_{0,n,0}}{(n+1)!}L_n^{(1)}(\sigma) \\ &\quad + O(\epsilon_0\nu_0(\epsilon_0)). \end{aligned} \tag{4.166}$$

On the other hand, we have the inner solution:

$$\begin{aligned} \hat{\Psi}(\sigma, \hat{\tau}, \epsilon_0) &= -2\epsilon_0\sigma\hat{\tau} + \epsilon_0\hat{\nu}_0(\epsilon_0)\hat{\Psi}_0 + \cdots \\ &= -2\epsilon_0\sigma\hat{\tau} + \epsilon_0\ \hat{\nu}_0(\epsilon_0)\sigma\sum_{n=0}^{\infty}\Big\{\tilde{b}_{0,n,0}\Big[\hat{\tau}\ln\hat{\tau} - \hat{\tau}(1 + \ln\hat{\tau}_0) + \hat{\tau}_0\Big] \\ &\quad + \tilde{a}_{0,n,0}(\hat{\tau} - \hat{\tau}_0)^2\Big\}L_n^{(1)}(\sigma) +\ O(\epsilon_0^2\hat{\nu}_0(\epsilon_0)), \end{aligned} \tag{4.167}$$

$$\hat{\zeta}(\sigma, \hat{\tau}, \epsilon_0) = \hat{\nu}_0(\epsilon_0)\hat{\zeta}_0 + \cdots$$

$$= \hat{\nu}_0(\epsilon_0)\,\sigma \sum_{n=0}^{\infty} (\hat{a}_{0,n,0}\hat{\tau} + \hat{b}_{0,n,0}) L_n^{(1)}(\sigma)\Big\}$$
$$+ O(\epsilon_0 \hat{\nu}_0(\epsilon_0)). \tag{4.168}$$

We first match the solutions of stream functions (4.165) and (4.167).
(1) By matching the terms of $\{\sigma\hat{\tau}\}$, we obtain

$$\begin{aligned} \bar{\nu}_0(\epsilon_0) &= \frac{1}{\ln \epsilon_0}, \\ \bar{b}_{0,0,0} &= -2. \end{aligned} \tag{4.169}$$

(2) By matching the terms of $\{\sigma\}$, we obtain

$$\begin{aligned} &\bar{\nu}_{*,0}(\epsilon_0) = \bar{\nu}_0(\epsilon_0) = \frac{1}{\ln \epsilon_0}, \\ &\gamma_{0,0} + \bar{b}_{0,0,0} = 0, \\ &\bar{b}_{0,n,0} = 0 \qquad (n = 1, 2, \ldots). \end{aligned} \tag{4.170}$$

From the above, we obtain

$$\gamma_{0,0} = 2. \tag{4.171}$$

Furthermore, from (4.137), we derive that

$$\begin{aligned} \bar{a}_{0,0,0} &= -\bar{b}_{0,0,0} = 2, \\ \bar{a}_{0,n,0} &= n!\bar{a}_{0,0,0} = 2(n!). \end{aligned} \tag{4.172}$$

(3) By matching the solutions of vorticity functions (4.166) and (4.168), it is found that

$$\begin{aligned} &\hat{\nu}_0(\epsilon_0) = \bar{\nu}_0(\epsilon_0) = \frac{1}{\ln \epsilon_0}, \\ &\hat{a}_{0,n,0} = 0, \\ &\hat{b}_{0,n,0} = -\frac{\bar{a}_{0,n,0}}{(n+1)!} = \frac{2}{n+1}. \end{aligned} \tag{4.173}$$

From (4.161), we derive

$$\begin{aligned} &\tilde{b}_{0,0,0} = -2, \\ &\tilde{b}_{0,n,0} = 0, \qquad (n = 1, 2, \ldots) \\ &\tilde{a}_{0,n,0} = 0, \qquad (n = 0, 1, 2, \ldots). \end{aligned} \tag{4.174}$$

Now, we match the higher-order solutions of stream functions.

(4) Note that the term,

$$\tilde{b}_{0,0,0}\frac{\epsilon_0}{\ln\epsilon_0}\{\sigma\hat{\tau}\ln\hat{\tau}\}$$

in the inner solution automatically matches with the term

$$\bar{b}_{0,0,0}\frac{\epsilon_0}{\ln\epsilon_0}\{\sigma\hat{\tau}\ln\hat{\tau}\}$$

in the outer solution.

(5) To match the term,

$$\tilde{b}_{0,0,0}\hat{\tau}_0\frac{\epsilon_0}{\ln\epsilon_0}\sigma = -\frac{\epsilon_0}{\ln\epsilon_0}\sigma$$

in the inner solution, we may introduce the higher-order outer solution

$$\epsilon_0\bar{\nu}_{*,0}(\epsilon_0)\bar{\Psi}_1(\sigma,\tau) = -\frac{\epsilon_0}{\ln\epsilon_0}\bar{\Psi}_{0,*}(\sigma,\tau),$$

for which, as $\tau\to 0$, we have

$$\epsilon_0\bar{\nu}_{*,0}(\epsilon_0)\bar{\Psi}_1(\sigma,\tau) \sim -\frac{\epsilon_0}{\ln\epsilon_0}\sigma.$$

Up to this point, the remaining unbalanced terms in the inner solution $\hat{\Psi}$ are

$$-2(1-\ln 2)\frac{\epsilon_0}{\ln\epsilon_0}\sigma\hat{\tau}+\cdots \tag{4.175}$$

while the remaining unbalanced terms in the outer solution $\bar{\Psi}$ are

$$-2(\gamma_0-1)\frac{\epsilon_0}{\ln\epsilon_0}\sigma\hat{\tau}+\cdots \tag{4.176}$$

These terms are to be balanced with the higher-order outer solutions:

$$\epsilon_0\bar{\nu}_0(\epsilon_0)\sigma\bar{b}_{0,0,1}\mathcal{F}_0(\tau)L_0^{(1)}(\sigma) = \frac{\epsilon_0}{\ln\epsilon_0}\sigma\Big[2(2-\gamma_0+\ln 2)\Big]E_2(\tau).$$

We finally obtain the outer solution

$$\begin{aligned}\Psi_{\text{outer}}(\sigma,\epsilon_0\tau,\epsilon_0) &= 2\sigma\tau+\frac{2}{\ln\epsilon_0}\sigma\Big[\Psi_{*,0}(\sigma,\tau)-E_2(\tau)\Big]\\ &\quad+\frac{\epsilon_0}{\ln\epsilon_0}\sigma\Big[2(2-\gamma_0+\ln 2)E_2(\tau)-\Psi_{*,0}(\sigma,\tau)\Big]\\ &\quad+O\Big(\frac{\epsilon_0}{\ln\epsilon_0}\Big),\end{aligned} \tag{4.177}$$

$$\zeta_{\text{outer}}(\sigma,\epsilon_0\tau,\epsilon_0) = \frac{2}{\ln\epsilon_0}\,\sigma\sum_{n=0}^{\infty}n!\mathcal{F}_n(\tau)L_n^{(1)}(\sigma)+O\Big(\frac{\epsilon_0}{\ln\epsilon_0}\Big), \tag{4.178}$$

and the inner solution

$$\Psi_{\text{inner}}(\sigma, \hat{\tau}, \epsilon_0) = -2\frac{\epsilon_0}{\ln \epsilon_0} \sigma\Big[\hat{\tau}\ln\hat{\tau} - \hat{\tau}(1+\ln\hat{\tau}_0) + \hat{\tau}_0\Big] + O(\frac{\epsilon_0^2}{\ln \epsilon_0}), \tag{4.179}$$

$$\zeta_{\text{inner}}(\sigma, \hat{\tau}, \epsilon_0) = \frac{2}{\ln \epsilon_0} \sigma \sum_{n=0}^{\infty} \frac{1}{n+1} L_n^{(1)}(\sigma) + O(\frac{\epsilon_0}{\ln \epsilon_0}) = \frac{2}{\ln \epsilon_0} + O(\frac{\epsilon_0}{\ln \epsilon_0}). \tag{4.180}$$

The above procedure may be continued to the higher-order approximations. One of the most important features of the results obtained is that in the leading-order approximation, while the outer solution contains all the Laguerre components, $L_n^{(1)}(\sigma)$, $(n = 0, 1, 2, \ldots)$ the inner solution only contains the Laguerre component $L_0^{(1)}(\sigma)$. Subsequently, in the inner region, the stream function $\Psi_{\text{inner}}(\sigma, \hat{\tau}, \epsilon_0)$ is a linear function of σ, while the vorticity function $\zeta_{\text{inner}}(\sigma, \hat{\tau}, \epsilon_0)$ is a constant.

One can compare the above uniformly valid solution with the Oseen Model solution obtained by Ananth and Gill. As shown in the last subsection, the Oseen Model solution for the stream function obtained by Ananth and Gill can be written in the form

$$\Psi_{AG} = 2\sigma\tau + \sigma\left\{\frac{2}{E_1(\frac{\epsilon_0}{2})}\Big[E_2(\tau) - E_2\Big(\frac{\epsilon_0}{2}\Big)\Big] - \epsilon_0\right\}. \tag{4.181}$$

The qualitative difference between the uniformly valid expansion of the stream function solution (4.177) and the Oseen model solution of the stream function obtained by Ananth and Gill, (4.181) in the outer region, is clear. The former contains all components of Laguerre series and the stream function approaches the given uniform flow in the far field, while the latter is a linear function of the variable σ, and does not approach the given uniform flow in the far field. In the inner field, the comparison between the uniformly valid expansion of the stream function solution (4.179) and the Oseen model solution of Ananth and Gill (4.181), is shown in Fig. 4.10. It it seen that there is a significant discrepancy between the two solutions.

3.6 Skin Friction at the Surface of a Paraboloid

In the previous sections, we derived uniformly valid expansion solution in the leading-order approximation. The procedure can be continued to

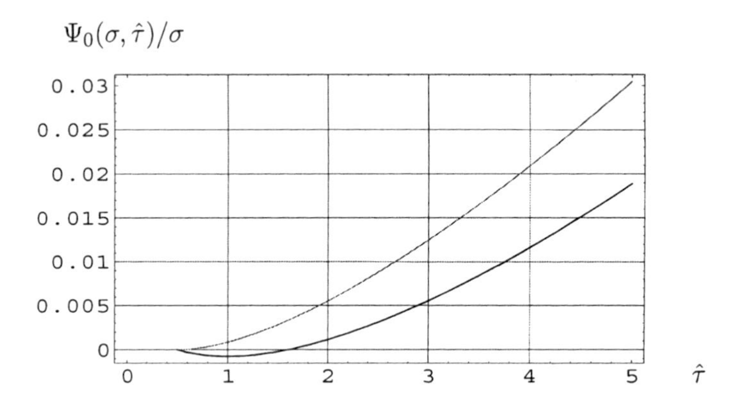

Figure 4.10. The comparison of the uniformly valid solution on top and the Oseen model solution by Ananth and Gill on bottom for the case $\epsilon_0 = 0.01$ in the inner region. The horizontal axis is the inner variable $\hat{\tau} = \frac{\tau}{\epsilon_0}$

systematically derive the higher-order approximate solutions. No principal difficulty seems to occur.

With the results obtained in this section, we are going to derive the skin friction at the surface of a body, which is of interest in many applications.

By using the variables (σ, τ), formula (9.34) can be written in the form

$$\begin{aligned} u &= \frac{1}{2\sqrt{\sigma(\sigma+\tau)}}\frac{\partial\Psi}{\partial\tau}, \\ v &= -\frac{1}{2\sqrt{\tau(\sigma+\tau)}}\frac{\partial\Psi}{\partial\sigma}. \end{aligned} \tag{4.182}$$

Note that along the normal direction of the surface $\eta = 1$, the differential of arc length is

$$\mathrm{d}\ell = \eta_0^2\sqrt{1+\xi^2}\mathrm{d}\eta = \frac{1}{2}\sqrt{\frac{\epsilon_0+2\sigma}{\epsilon_0}}\,\mathrm{d}\tau. \tag{4.183}$$

Let us define the viscous stress coefficient at the surface $\eta = 1$ as

$$C_w = \frac{\partial u}{\partial \ell} = \frac{1}{\sqrt{2}}\sqrt{\frac{\epsilon_0}{\epsilon_0+2\sigma}}\frac{\partial u}{\partial\tau} = \frac{1}{2^{\frac{3}{2}}\epsilon_0^{\frac{3}{2}}(2\sigma+\epsilon_0)\sqrt{\sigma}}\frac{\partial^2\hat{\Psi}}{\partial\hat{\tau}^2}. \tag{4.184}$$

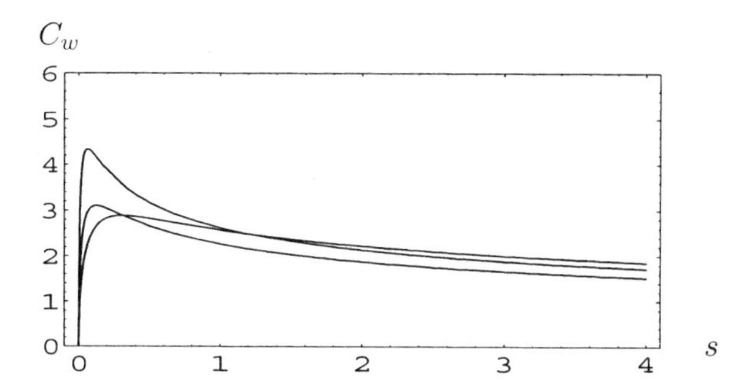

Figure 4.11. The variation stress coefficient C_w with the arc length s for different $\mathrm{Re} = \epsilon_0 = 0.1, 0.2, 0.5$ from top to bottom

From the inner solution obtained in the previous section, we find that at $\hat{\tau} = \frac{1}{2}$,

$$\frac{\partial^2 \hat{\Psi}}{\partial \hat{\tau}^2} = 4 \frac{\epsilon_0}{|\ln \epsilon_0|} \sigma + \cdots, \tag{4.185}$$

so that,

$$C_w = \frac{2^{\frac{3}{2}}}{\epsilon_0^{\frac{1}{2}} |\ln \epsilon_0|} \frac{\sigma^{\frac{1}{2}}}{2\sigma + \epsilon_0} + \cdots, \tag{4.186}$$

Along the tangential direction of the body surface $\eta = 1$, the differential of arc length is

$$\mathrm{d}s = \eta_0^2 \sqrt{1 + \xi^2} \mathrm{d}\xi = \sqrt{\frac{\epsilon_0 + 2\sigma}{\epsilon_0}} \mathrm{d}\sigma. \tag{4.187}$$

The arc length measured along the surface starting from the tip is

$$s = s(\sigma) = \int_0^{\sigma} \sqrt{\frac{\epsilon_0 + 2\sigma'}{\epsilon_0}} \mathrm{d}\sigma'. \tag{4.188}$$

One can derive

$$\sigma = \sigma(s) = \frac{\epsilon_0}{2} \left[\left(\frac{3s}{\epsilon_0} + 1 \right)^{\frac{2}{3}} - 1 \right]. \tag{4.189}$$

We find that the stress coefficient C_w has the maximum

$$C_w = C_{\max} = \frac{1}{\epsilon_0 |\ln \epsilon_0|}, \tag{4.190}$$

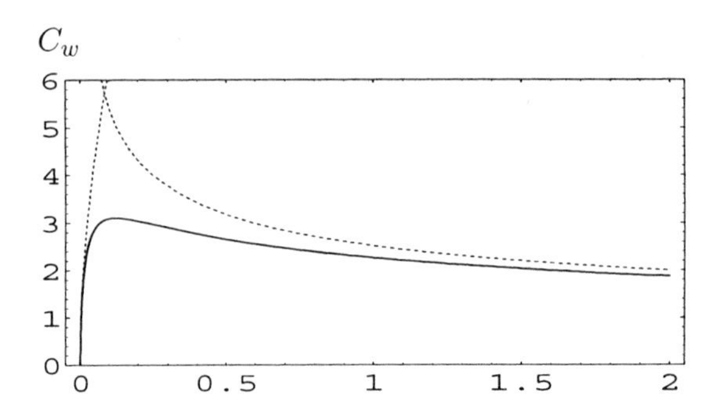

Figure 4.12. The variation of stress coefficient C_w with the arc length s, and its asymptotic behaviors for $\mathrm{Re} = \epsilon_0 = 0.2$

at $\sigma = \sigma_* = \frac{\epsilon_0}{2}$, or $s = s_* = \frac{\sqrt{8}-1}{3}\epsilon_0$ (see Fig. 4.11).

Moreover, near the dendrite's tip,

$$\sigma \sim s, \tag{4.191}$$

so that

$$C_w = \frac{2^{\frac{3}{2}}}{\epsilon_0^{\frac{3}{2}} |\ln \epsilon_0|}\sqrt{s} + \cdots \qquad (s \to 0)\ ; \tag{4.192}$$

whereas away from the tip,

$$\sigma \sim \frac{3^{\frac{2}{3}}\epsilon_0^{\frac{1}{3}}}{2} s^{\frac{2}{3}}, \tag{4.193}$$

so that,

$$C_w = \frac{2}{3^{\frac{1}{3}}} \frac{1}{\epsilon_0^{\frac{2}{3}} |\ln \epsilon_0|} \frac{1}{s^{\frac{1}{3}}} + \cdots \qquad (s \to \infty) \tag{4.194}$$

(see Fig. 4.12).

3.7 Appendix (B)

3.7.1 Asymptotic behavior of the outer solution $\bar{\Psi}_0$ in the limit $\tau \to 0$

With the formulas

$$\begin{aligned} &U\left(n+\tfrac{3}{2},1,x\right) \sim -\tfrac{1}{\Gamma(n+\frac{3}{2})}\ln x, \\ &U\left(n+\tfrac{3}{2},2,x\right) \sim \tfrac{1}{\Gamma\left(n+\frac{3}{2}\right)}x^{-1}, \\ &U\left(n+1,1,x\right) \sim -\tfrac{1}{n!}\ln x, \\ &U\left(n+2,2,x\right) \sim \tfrac{1}{(n+1)!}x^{-1}, \end{aligned} \tag{4.195}$$

as $x \to 0$, it is derived that as $\tau \to 0$,

$$\begin{aligned} &\{\mathcal{A}_{n,k}(\tau), \quad \hat{\mathcal{A}}_{n,k}(\tau)\} \sim O(1), \\ &\{\mathcal{B}_{n,k}(\tau), \quad \hat{\mathcal{B}}_{n,0}(\tau) \sim O(\ln \tau). \end{aligned} \tag{4.196}$$

Hence, by letting $\tau = \epsilon_0 \hat{\tau}$, we derive that as $\epsilon_0 \to 0$, with fixed $\hat{\tau} = O(1)$,

$$\begin{aligned} &\mathcal{A}_{n,k}(\epsilon_0\hat{\tau}) \sim \alpha_{n,k}^{(0)} + \alpha_{n,k}^{(1)}\epsilon_0 \ln \epsilon_0 + \cdots, \\ &\hat{\mathcal{A}}_{n,k}(\epsilon_0\hat{\tau}) \sim \hat{\alpha}_{n,k}^{(0)} + \hat{\alpha}_{n,k}^{(1)}\epsilon_0 \ln \epsilon_0 + \cdots, \\ &\mathcal{B}_{n,k}(\epsilon_0\hat{\tau}) \sim \beta_{n,k}^{(0)} \ln \epsilon_0 + \beta_{n,k}^{(1)} + \cdots, \\ &\hat{\mathcal{B}}_{n,k}(\epsilon_0\hat{\tau}) \sim \hat{\beta}_{n,k}^{(0)} \ln \epsilon_0 + \hat{\beta}_{n,k}^{(1)} + \cdots, \end{aligned} \tag{4.197}$$

where

$$\begin{aligned} &\alpha_{n,0}^{(0)} = \frac{(-1)^n}{4}\frac{\Gamma(n+\frac{1}{2})}{(n+1)!}, \qquad &\hat{\alpha}_{n,0}^{(0)} = \frac{(-1)^n}{4}\frac{1}{n+1}, \\ &\beta_{n,0}^{(0)} = \frac{(-1)^{n+1}}{4}\frac{\left[\Gamma\left(n+\frac{1}{2}\right)\right]^2}{(n+1)!n!}, \qquad &\hat{\beta}_{n,0}^{(0)} = \tfrac{(-1)^{n+1}}{4}\tfrac{n!}{(n+1)\Gamma\left(n+\frac{3}{2}\right)}. \end{aligned} \tag{4.198}$$

The other constants $\left\{\alpha_{n,k}, \hat{\alpha}_{n,k}; \beta_{n,k}, \hat{\beta}_{n,k}\right\}$ $(n = 0, 1, 2, \ldots; k = 1, 2, 3, \ldots)$ can be determined by the recurrence formulas

$$\begin{aligned} &\alpha_{n,1}^{(0)} = 2\alpha_{n,0}^{(0)} - \hat{\alpha}_{n,0}^{(0)}, \\ &\hat{\alpha}_{n,1}^{(0)} = 2\hat{\alpha}_{n,0}^{(0)} - (n+2)\alpha_{n+1,0}^{(0)}, \end{aligned} \tag{4.199}$$

$$\begin{aligned} \alpha_{n,k+1}^{(0)} &= 2\alpha_{n,k}^{(0)} - \alpha_{n,k-1}^{(0)} - \frac{1}{k+1}\hat{\alpha}_{n,k}^{(0)}, \\ \hat{\alpha}_{n,k+1}^{(0)} &= 2\hat{\alpha}_{n,k}^{(0)} - \hat{\alpha}_{n,k-1}^{(0)} - \frac{n+2}{k+1}\alpha_{n+1,k}^{(0)}, \end{aligned} \tag{4.200}$$

$$\begin{aligned} \beta_{n,1}^{(0)} &= 2\beta_{n,0} - \hat{\beta}_{n,0}, \\ \hat{\beta}_{n,1} &= 2\hat{\beta}_{n,0}^{(0)} - (n+2)\beta_{n+1,0}^{(0)}, \end{aligned} \tag{4.201}$$

and

$$\begin{aligned} \beta_{n,k+1}^{(0)} &= 2\beta_{n,k}^{(0)} - \beta_{n,k-1}^{(0)} - \frac{1}{k+1}\hat{\beta}_{n,k}^{(0)} \\ \hat{\beta}_{n,k+1}^{(0)} &= 2\hat{\beta}_{n,k}^{(0)} - \hat{\beta}_{n,k-1}^{(0)} - \frac{n+2}{k+1}\beta_{n+1,k}. \end{aligned} \tag{4.202}$$

On the other hand, we derive that as $\epsilon_0 \to 0$,

$$\begin{aligned} \mathcal{F}_n(\epsilon_0\hat{\tau}) &\sim \kappa_n^{(0)} + \kappa_n^{(1)}\epsilon_0 \ln\epsilon_0 + \cdots, \\ \mathcal{F}_n'(\epsilon_0\hat{\tau}) &\sim \delta_n^{(0)} \ln\epsilon_0 + \delta_n^{(1)} + \delta_n^{(2)}\epsilon_0 \ln\epsilon_0 + \cdots \end{aligned} \tag{4.203}$$

where

$$\kappa_n^{(0)} = \frac{1}{(n+1)!}, \qquad \delta_n^{(0)} = \frac{1}{n!}. \tag{4.204}$$

Therefore, we derive that as $\tau \to 0$, any outer solution

$$\bar{\Psi}_0(\sigma,\tau) = \sigma\mathcal{C}_{0,0}\sum_{n=0}^{N} L_n^{(1)}(\sigma)\left[\sum_{k=0}^{N}\widehat{\mathcal{C}}_{0,k}\mathcal{A}_{n,k} + \widehat{b}_{0,n}\mathcal{F}_n(\tau)\right] \tag{4.205}$$

with

$$\widehat{\mathcal{C}}_{0,0} = 1, \tag{4.206}$$

has the asymptotic form

$$\bar{\Psi}_0(\sigma,\tau) = \bar{\Psi}_0(\sigma,\epsilon_0\hat{\tau}) \sim \sigma\,\mathcal{C}_{0,0}\sum_{n=0}^{\infty}\omega_{0,n}L_n^{(1)}(\sigma) + \cdots, \tag{4.207}$$

and

$$\frac{\partial\bar{\Psi}_0}{\partial\tau}(\sigma,\tau) = \frac{\partial\bar{\Psi}_0}{\partial\tau}(\sigma,\epsilon_0\hat{\tau}) \sim \sigma\,\mathcal{C}_{0,0}\sum_{n=0}^{\infty}\widehat{\omega}_{0,n}L_n^{(1)}(\sigma) + \cdots, \tag{4.208}$$

where we define

$$\omega_{0,n} = \widehat{b}_{0n}\kappa_n^{(0)} + \sum_{k=0}^{\infty} \widehat{\mathcal{C}}_{0,k}\alpha_{n,k}^{(0)}, \tag{4.209}$$

$$\widehat{\omega}_{0,n} = \ln\epsilon_0 \left(\widehat{b}_{0n}\delta_n^{(0)} + \sum_{k=0}^{\infty} \widehat{\mathcal{C}}_{0,k}\widehat{\beta}_{n,k}^{(0)} \right), \tag{4.210}$$

and

$$\widehat{b}_{00} = \frac{\bar{b}_{00}}{\mathcal{C}_{0,0}}. \tag{4.211}$$

3.7.2 Determination of the special outer solution $\bar{\Psi}_{*0}$

The system allows a special outer solution $\Psi_{*,0}(\sigma,\tau)$, such that as $\tau = O(\epsilon_0)$, we have

$$\bar{\Psi}_{*,0}(\sigma,\tau) \sim \sigma L_0^{(1)}(\sigma) + o(1); \quad \frac{\partial \bar{\Psi}_{*,0}}{\partial \tau}(\sigma,\tau) \sim O(1); \tag{4.212}$$

$$\bar{\zeta}_{*,0}(\sigma,\tau) \sim o(1). \tag{4.213}$$

To determine this special solution, one needs to solve the infinite linear system

$$\begin{cases} \mathcal{C}_{*,0}\,\omega_{0,0} = 1, & \widehat{\omega}_{0,0} = 0, \\ \omega_{0,n} = 0, & \widehat{\omega}_{0,0} = 0 \quad (n = 1, 2, \ldots). \end{cases} \tag{4.214}$$

for the coefficients $\{\widehat{\mathcal{C}}_{*,n}; \bar{b}_{*n}\}$ $(n = 0, 1, 2, \ldots)$. In doing so, we follow the same procedure described in the last subsection. We first apply the conditions

$$\omega_{0,n} = 0, \quad \widehat{\omega}_{0,n} = 0 \quad (n = 1, 2, 3, \ldots), \tag{4.215}$$

which may be written in the form

$$\begin{pmatrix} \mathcal{Q}_{1,1} & \cdots & \mathcal{Q}_{1,n} & \cdots \\ & \vdots & & \\ \mathcal{Q}_{n,1} & \cdots & \mathcal{Q}_{1,n} & \cdots \\ & \vdots & & \end{pmatrix} \cdot \begin{pmatrix} \widehat{\mathcal{C}}_{*,1} \\ \vdots \\ \widehat{\mathcal{C}}_{*,n} \\ \vdots \end{pmatrix} = \begin{pmatrix} \mathcal{H}_1 \\ \vdots \\ \mathcal{H}_n \\ \vdots \end{pmatrix}. \tag{4.216}$$

We set

$$\mathcal{Q}_{n,k} = (\ln\epsilon_0)\overline{\mathcal{Q}}_{n,k} \tag{4.217}$$

and

$$\mathcal{H}_n = (\ln \epsilon_0)\overline{\mathcal{H}}_n, \tag{4.218}$$

where $(n, k = 1, 2, 3, \ldots)$. Here, we defined

$$\mathcal{C}_{*,n} = \mathcal{C}_{*,0}\widehat{\mathcal{C}}_{*,n}; \quad \bar{b}_{*n} = \mathcal{C}_{*,0}\widehat{b}_{*n}. \tag{4.219}$$

It is derived that

$$\begin{aligned} \overline{\mathcal{Q}}_{n,k} &= \frac{\alpha_{n,k}^{(0)}}{n!} - \frac{\hat{\beta}_{n,k}^{(0)}}{(n+1)!}, \\ \overline{\mathcal{H}}_n &= -\frac{\alpha_{n,k}^{(0)}}{n!} + \frac{\hat{\beta}_{n,k}^{(0)}}{(n+1)!}. \end{aligned} \tag{4.220}$$

It is seen that as n increases, the elements $\overline{\mathcal{Q}}_{n,k}$ and $\overline{\mathcal{H}}_n$ vanish very rapidly. Thus, we can truncate the sequence $n = 0, 1, \ldots$ at a sufficiently large number $n = N$, and derive that

$$\begin{aligned} \begin{pmatrix} \widehat{\mathcal{C}}_{*,1} \\ \vdots \\ \widehat{\mathcal{C}}_{*,N} \end{pmatrix} &= \begin{pmatrix} \overline{\mathcal{Q}}_{1,1} & \cdots & \overline{\mathcal{Q}}_{1,N} \\ & \vdots & \\ \overline{\mathcal{Q}}_{N,1} & \cdots & \overline{\mathcal{Q}}_{1,N} \end{pmatrix}^{-1} \cdot \begin{pmatrix} \overline{\mathcal{H}}_1 \\ \vdots \\ \overline{\mathcal{H}}_N \end{pmatrix}, \\ \widehat{\mathcal{C}}_{*,n} &= 0 \quad (n > N), \end{aligned} \tag{4.221}$$

and

$$\widehat{b}_{*n} = -\frac{1}{(n+1)!}\left[\alpha_{n,0}^{(0)} + \sum_{k=1}^{N} \widehat{\mathcal{C}}_{*,k}\alpha_{n,k}^{(0)}\right], \quad (n = 1, 2, \ldots, N). \tag{4.222}$$

Up to this point, two constants $\{\mathcal{C}_{*,0}, \bar{b}_{*,0}\}$ remain undetermined. To determine these two constants, we apply the conditions

$$\mathcal{C}_{*,0}\,\omega_{0,0} = 1, \quad \widehat{\omega}_{0,0} = 0. \tag{4.223}$$

It follows that

$$\begin{aligned} \mathcal{C}_{*,0} &\sim \frac{\delta_0^{(0)}}{\sum\limits_{k=0}^{N} \widehat{\mathcal{C}}_{*,k}(\delta_0^{(0)}\alpha_{0,k}^{(0)} - \kappa_0^{(0)}\hat{\beta}_{0,k}^{(0)})}, \\ \widehat{b}_{*,0} &= \frac{\bar{b}_{*,0}}{\mathcal{C}_{*,0}} \sim -\frac{1}{\delta_0^{(0)}}\sum_{k=0}^{N} \widehat{\mathcal{C}}_{*,k}\hat{\beta}_{0,k}. \end{aligned} \tag{4.224}$$

The stream function $\bar{\Psi}_{*,0}$ is finally fully determined. With the relationship

$$L_2[\bar{\Psi}_{*,0}(\sigma,\tau)] = -(\sigma+\tau)\bar{\zeta}_{*,0}(\sigma,\tau), \tag{4.225}$$

it is deduced that for the corresponding vorticity function, we have

$$\bar{\zeta}_{*,0}(\sigma,\epsilon_0\hat{\tau}) \sim 0, \tag{4.226}$$

as $\epsilon_0 \to 0$.

It should be noted that the above results can be extended further to the statement that the system actually allows a special outer solution

$$\bar{\Psi}_{*,m}(\sigma,\tau) = \mathcal{C}^{(m)}_{*,m}\sigma \sum_{n=0}^{N} L^{(1)}_n(\sigma)\left[\sum_{k=0}^{N}\widehat{\mathcal{C}}^{(m)}_{*,k}\mathcal{A}_{n,k} + \widehat{b}^{(m)}_{*,n}\mathcal{F}_n(\tau)\right] \tag{4.227}$$

with any integer $m = 0, 1, 2, \ldots$ and

$$\widehat{\mathcal{C}}^{(m)}_{*,m} = 1, \tag{4.228}$$

such that as $\epsilon_0 \to 0$ the following asymptotic expansions hold:

$$\bar{\Psi}_{*,m}(\sigma,\tau) \sim \sigma L^{(1)}_m(\sigma) + o(1); \quad \frac{\partial\bar{\Psi}_{*,m}}{\partial\tau}(\sigma,\tau) \sim O(1); \tag{4.229}$$

$$\bar{\zeta}_{*,m}(\sigma,\tau) \sim o(1). \tag{4.230}$$

Chapter 5

ASYMPTOTIC SOLUTION OF DENDRITIC GROWTH IN EXTERNAL FLOW (I): THE CASE OF RAPID GROWTH $U \gg U_\infty$

In the following two chapters, we are going to investigate the effect of enforced uniform flow on dendritic growth. During the past several years, this subject has been studied by a number of authors, such as Ananth and Gill; Ben Amar, Bouillou and Pelce; Saville and Beaghton, Xu and Yu, etc., numerically or analytically. The problem, however, was not well resolved. To describe the flow field induced by dendritic growth in the external flow, Ananth and Gill used an Oseen model solution of the uniform flow past a paraboloid (Ananth and Gill, 1989, 1991). Xu made the first attempt to derive an asymptotic expansion solution for the free boundary problem, in terms of the Navier–Stokes model of fluid dynamics for the case of the Prandtl number $\mathrm{Pr} \to \infty$ (Xu, 1994a). Neither Ananth and Gill's solution, nor Xu's solution yields the correct stream function approaching that of the given uniformly flow in the up-stream far-field. Therefore, the problem needs to be reconsidered with the Navier–Stokes model of fluid dynamics and the fully justified mathematical formulation. We shall separately discuss the two limiting cases:

- The case of rapid growth or, weak external flow: $U_\infty/U \ll 1$ with $\mathrm{Pr} = O(1)$;

- The case of large Prandtl number: $\mathrm{Pr} \gg 1$ with $U_\infty/U = O(1)$.

In this chapter, we deal with the first case. The solution for the second case will be presented in the next chapter.

1. Mathematical Formulation of the Problem

Now, the temperature field is no longer treated as uniform. We shall, as usual for dendritic growth, adopt the thermal diffusion length ℓ_{T} as length scale. Furthermore, for the present case, we assume that the system contains no density change and gravity, so that the parameters $\mathrm{Gr} = \alpha = 0$. Moreover, as the surface tension is assumed to be zero, one may set $T_{\mathrm{S}} = 0$.

We now have the following system of governing equations:

1. Kinematic equation:

$$D^2\Psi = -\eta_0^4(\xi^2 + \eta^2)\zeta. \tag{5.1}$$

2. Vorticity equation:

$$\mathrm{Pr}D^2\zeta = \frac{2\zeta}{\eta_0^4\xi^2\eta^2}\frac{\partial(\Psi, \eta_0^2\xi\eta)}{\partial(\xi,\eta)} - \frac{1}{\eta_0^2\xi\eta}\frac{\partial(\Psi,\zeta)}{\partial(\xi,\eta)}. \tag{5.2}$$

3. Heat conduction equation:

$$\nabla^2 T = \frac{1}{\eta_0^2\xi\eta}\left(\frac{\partial\Psi}{\partial\eta}\frac{\partial T}{\partial\xi} - \frac{\partial\Psi}{\partial\xi}\frac{\partial T}{\partial\eta}\right). \tag{5.3}$$

The boundary conditions are:

1. As $\eta \to \infty$, the perturbed flow induced by dendrite vanishes, so that

$$\Psi \sim \frac{1}{2}(1 + \bar{U}_\infty)\eta_0^4\xi^2\eta^2 + o(1); \quad \zeta \to 0, \tag{5.4}$$

where

$$\bar{U}_\infty = \frac{U_\infty}{U} = \epsilon_0 \ll 1, \tag{5.5}$$

and

$$T \to T_\infty. \tag{5.6}$$

2. Axi-symmetrical condition: at the symmetrical axis $\xi = 0$,

$$u = 0, \quad v = O(1), \tag{5.7}$$

or

$$\Psi = \zeta = \frac{\partial\Psi}{\partial\xi} = \frac{\partial\Psi}{\partial\eta} = 0. \tag{5.8}$$

3. The interface condition: at $\eta = \eta_s(\xi)$,

(i) Thermo-dynamical equilibrium condition:

$$T = 0. \tag{5.9}$$

(ii) Enthalpy conservation condition:

$$\frac{\partial T}{\partial \eta} - \eta_s' \frac{\partial T}{\partial \xi} + \eta_0^2 (\xi \eta_s)' = 0. \tag{5.10}$$

(iii) Mass conservation condition:

$$\left(\frac{\partial \Psi}{\partial \xi} + \eta_s' \frac{\partial \Psi}{\partial \eta} \right) = \eta_0^4 (\xi \eta_s)(\xi \eta_s)', \tag{5.11}$$

(iv) Continuity condition of the tangential component of velocity:

$$\left(\frac{\partial \Psi}{\partial \eta} - \eta_s' \frac{\partial \Psi}{\partial \xi} \right) + \eta_0^4 (\xi \eta_s)(\eta_s \eta_s' - \xi) = 0. \tag{5.12}$$

The system involves three independent dimensionless parameters: $\{T_\infty;$ Pr; $\bar{U}_\infty\}$. The parameter η_0^2 is a function of these independent parameters. We further adopt $\epsilon_0 = \bar{U}_\infty$ as the basic small parameter and attempt, using the Laguerre series representation method, to find the regular perturbation expansion solution uniformly valid in the whole physical domain as $\epsilon_0 \to 0$ for the problem.

We denote the uniform flow with unit velocity in the far field as

$$\begin{cases} \bar{\zeta}_* = 0, \\ \bar{\Psi}_* = \dfrac{1}{2} \eta_0^4 \xi^2 \eta^2 \, . \end{cases} \tag{5.13}$$

For convenience, we set

$$\begin{aligned} \Psi(\xi, \eta, \bar{U}_\infty) &= \bar{\Psi}_*(\xi, \eta) + \bar{\Psi}(\xi, \eta, \bar{U}_\infty), \\ \zeta(\xi, \eta, \bar{U}_\infty) &= \frac{1}{\mathrm{Pr}^2} \bar{\zeta}(\xi, \eta, \bar{U}_\infty), \\ T(\xi, \eta, \bar{U}_\infty) &= T_\infty + \mathrm{Pr}\ \bar{T}(\xi, \eta, \bar{U}_\infty). \end{aligned} \tag{5.14}$$

Thus, from system (5.1)–(5.3), one can derive the governing system for the perturbation part, $\{\bar{\Psi}, \bar{\zeta}\}$ as follows:

$$D^2 \bar{\Psi} = -\frac{\eta_0^4}{\mathrm{Pr}^2} (\xi^2 + \eta^2) \bar{\zeta} \, , \tag{5.15}$$

$$\Pr D^2\bar{\zeta} = \eta_0^2\left(\xi\frac{\partial\bar{\zeta}}{\partial\xi} - \eta\frac{\partial\bar{\zeta}}{\partial\eta}\right) + \frac{2\bar{\zeta}}{\eta_0^2\xi^2\eta^2}\left(\eta\frac{\partial\bar{\Psi}}{\partial\eta} - \xi\frac{\partial\bar{\Psi}}{\partial\xi}\right)$$

$$-\frac{1}{\eta_0^2\xi\eta}\frac{\partial(\bar{\Psi},\bar{\zeta})}{\partial(\xi,\eta)}, \tag{5.16}$$

$$\nabla^2\bar{T} = \eta_0^2\left(\xi\frac{\partial\bar{T}}{\partial\xi} - \eta\frac{\partial\bar{T}}{\partial\eta}\right) + \frac{1}{\eta_0^2\xi\eta}\left(\frac{\partial\bar{\Psi}}{\partial\eta}\frac{\partial\bar{T}}{\partial\xi} - \frac{\partial\bar{\Psi}}{\partial\xi}\frac{\partial\bar{T}}{\partial\eta}\right). \tag{5.17}$$

For more convenience, we shall utilize the new variables

$$\sigma = \frac{\eta_0^2}{2\Pr}\xi^2, \qquad \tau = \frac{\eta_0^2}{2\Pr}\eta^2. \tag{5.18}$$

We remark that one should not confuse the new independent variable σ introduced here with the eigenvalue σ introduced for the stability analysis in Chap. 2, as well as later in Chap. 9.

With the variables $(\sigma;\tau)$, the operator D^2 and ∇^2 become

$$D^2 = 2\frac{\eta_0^2}{\Pr}L_2, \qquad \nabla^2 = 2\frac{\eta_0^2}{\Pr}(L_2 + L_1), \tag{5.19}$$

where

$$\begin{cases} L_2 = \left(\sigma\dfrac{\partial^2}{\partial\sigma^2} + \tau\dfrac{\partial^2}{\partial\tau^2}\right), \\ L_1 = \left(\dfrac{\partial}{\partial\sigma} + \dfrac{\partial}{\partial\tau}\right). \end{cases} \tag{5.20}$$

Hence, the system (5.15)– (5.17) is changed to

$$L_2[\bar{\Psi}] = -(\sigma+\tau)\bar{\zeta}, \tag{5.21}$$

$$L_2[\bar{\zeta}] = \left(\sigma\frac{\partial}{\partial\sigma} - \tau\frac{\partial}{\partial\tau}\right)\bar{\zeta} - \frac{\bar{\zeta}}{2\Pr^2\tau\sigma}\left(\tau\frac{\partial\bar{\Psi}}{\partial\tau} - \sigma\frac{\partial\bar{\Psi}}{\partial\sigma}\right)$$

$$-\frac{1}{2\Pr^2}\frac{\partial(\bar{\Psi},\bar{\zeta})}{\partial(\sigma,\tau)}, \tag{5.22}$$

$$(L_2 + L_1)[\bar{T}] = \Pr\left(\sigma\frac{\partial\bar{T}}{\partial\sigma} - \tau\frac{\partial\bar{T}}{\partial\tau}\right) + \frac{1}{2\Pr}\left(\frac{\partial\bar{\Psi}}{\partial\tau}\frac{\partial\bar{T}}{\partial\sigma} - \frac{\partial\bar{\Psi}}{\partial\sigma}\frac{\partial\bar{T}}{\partial\tau}\right). \tag{5.23}$$

With the new variables, the interface can be described as $\tau = \tau_s(\sigma)$. Hence the interface conditions can be written as: at $\tau = \tau_s(\sigma)$,

$$\frac{\partial \bar{\Psi}}{\partial \sigma} + \tau_s' \frac{\partial \bar{\Psi}}{\partial \tau} = 0, \tag{5.24}$$

$$\tau_s \frac{\partial \bar{\Psi}}{\partial \tau} - \sigma \tau_s' \frac{\partial \bar{\Psi}}{\partial \sigma} = 0 \tag{5.25}$$

or

$$\bar{\Psi} = \frac{\partial \bar{\Psi}}{\partial \tau} = 0, \tag{5.26}$$

$$T_\infty + \mathrm{Pr}\, \bar{T} = 0, \tag{5.27}$$

$$\tau_s \frac{\partial \bar{T}}{\partial \tau} - \sigma \tau_s' \frac{\partial \bar{T}}{\partial \sigma} + (\sigma \tau_s' + \tau_s) = 0. \tag{5.28}$$

Note that with the normalization condition (2.2) we have chosen η_0^2, such that the tip of the interface shape locates itself at $\eta = 1$. Thus, let us define

$$\eta_s = 1 + \bar{\eta}_s(\xi, \epsilon_0); \tag{5.29}$$

then we shall have

$$\bar{\eta}_s(0, \epsilon_0) = 0. \tag{5.30}$$

On the other hand, due to the dependence of the normalization parameter η_0^2 on ϵ_0, we can write

$$\eta_0 = \eta_{0,0} + \tilde{\eta}_0(\epsilon_0), \tag{5.31}$$

and assume that as $\epsilon \to 0$,

$$\tilde{\eta}_0 = \epsilon_0 \eta_{0,1} + \cdots. \tag{5.32}$$

Now, with the variables $\{\sigma, \tau\}$, the interface shape can be expressed in the form

$$\tau = \tau_s(\sigma) = \tau_0 + \bar{h}_s(\sigma, \epsilon_0), \tag{5.33}$$

where

$$\tau_0 = \frac{\eta_0^2}{2\mathrm{Pr}}, \quad \bar{h}_s(\sigma, \epsilon_0) = \frac{\eta_0^2}{2\mathrm{Pr}}[2\bar{\eta}_s(\xi, \epsilon_0) + \bar{\eta}_s^2(\xi, \epsilon_0)]. \tag{5.34}$$

Accordingly, we may write

$$\tau_0 = \tau_{0,0} + \tilde{\tau}_0(\epsilon_0) \tag{5.35}$$

and assume

$$\tilde{\tau}_0(\epsilon_0) = \epsilon_0 \tau_{0,1} + \cdots. \tag{5.36}$$

Moreover, due to the normalization condition (2.2), we have

$$\bar{h}_{\rm s}(0, \epsilon_0) = 0. \tag{5.37}$$

One may make a Taylor expansion for the interface conditions around $\tau = \tau_{0,0}$. It is derived that at $\tau = \tau_{0,0}$,

$$\bar{\Psi} + \frac{\partial \bar{\Psi}}{\partial \tau}(\tilde{\tau}_0 + \bar{h}_{\rm s}) + \frac{1}{2!}\frac{\partial^2 \bar{\Psi}}{\partial \tau^2}(\tilde{\tau}_0 + \bar{h}_{\rm s})^2 + \cdots = 0, \tag{5.38}$$

$$\frac{\partial \bar{\Psi}}{\partial \tau} + \frac{\partial^2 \bar{\Psi}}{\partial \tau^2}(\tilde{\tau}_0 + \bar{h}_{\rm s}) + \frac{1}{2!}\frac{\partial^3 \bar{\Psi}}{\partial \tau^3}(\tilde{\tau}_0 + \bar{h}_{\rm s})^2 + \cdots = 0, \tag{5.39}$$

$$T_\infty + \Pr\left[\bar{T} + \frac{\partial \bar{T}}{\partial \tau}(\tilde{\tau}_0 + \bar{h}_{\rm s}) + \frac{1}{2!}\frac{\partial^2 T}{\partial \tau^2}(\tilde{\tau}_0 + \bar{h}_{\rm s})^2 + \cdots\right] = 0, \tag{5.40}$$

$$\begin{aligned}(\tau_0 + \tilde{\tau}_0 + \bar{h}_{\rm s})\Big[\frac{\partial \bar{T}}{\partial \tau} + \frac{\partial^2 \bar{T}}{\partial \tau^2}(\tilde{\tau}_0 + \bar{h}_{\rm s}) + \frac{1}{2!}\frac{\partial^3 \bar{T}}{\partial \tau^3}(\tilde{\tau}_0 + \bar{h}_{\rm s})^2 \\ + \cdots\Big] - \sigma \bar{h}_{\rm s}'\Big[\frac{\partial \bar{T}}{\partial \sigma} + \frac{\partial^2 \bar{T}}{\partial \sigma \partial \tau}(\tilde{\tau}_0 + \bar{h}_{\rm s}) + \cdots\Big] \\ + (\tau_{0,0} + \tilde{\tau}_0 + \bar{h}_{\rm s} + \sigma \bar{h}_{\rm s}') = 0.\end{aligned} \tag{5.41}$$

2. Laguerre Series Representation of Solutions

We follow the approach developed in the last chapter by expanding the solution in the following Laguerre series:

$$\begin{cases} \bar{\zeta}(\sigma, \tau, \epsilon_0) = \sigma \sum_{n=0}^{\infty} A_n(\tau, \epsilon_0) L_n^{(1)}(\sigma), \\ \bar{\Psi}(\sigma, \tau, \epsilon_0) = \sigma \sum_{n=0}^{\infty} B_n(\tau, \epsilon_0) L_n^{(1)}(\sigma), \\ \bar{T}(\sigma, \tau, \epsilon_0) = \sum_{n=0}^{\infty} D_n(\tau, \epsilon_0) L_n^{(1)}(\sigma), \\ \bar{h}_{\rm s}(\sigma, \epsilon_0) = \sum_{n=0}^{\infty} h_n(\epsilon_0) L_n^{(1)}(\sigma). \end{cases} \tag{5.42}$$

By substituting the above expansions into the governing equations, it is derived that

$$\begin{aligned}\frac{\mathrm{d}^2 B_n}{\mathrm{d}\tau^2} &= \frac{n+2}{\tau} B_{n+1} - \left[1 + \frac{2n+2}{\tau}\right] A_n \\ &\quad + \frac{1}{\tau}\Big[(n+2)A_{n+1} + nA_{n-1}\Big],\end{aligned} \tag{5.43}$$

$$\begin{aligned}\frac{\mathrm{d}^2 A_n}{\mathrm{d}\tau^2} + \frac{\mathrm{d}A_n}{\mathrm{d}\tau} - \frac{n+1}{\tau} A_n &= \frac{1}{2\mathrm{Pr}^2}\mathcal{N}\Big\{A_n; B_n\Big\}, \\ &\quad (n = 0, 1, 2, \ldots,)\end{aligned} \tag{5.44}$$

where $\mathcal{N}\Big\{A_n; B_n\Big\}$ is the nonlinear operator defined in the last chapter.

For the temperature field, we have

$$\begin{aligned}\tau\frac{\mathrm{d}^2 D_n}{\mathrm{d}\tau^2} + (1 + \mathrm{Pr}\,\tau)\frac{\mathrm{d}D_n}{\mathrm{d}\tau} - n\mathrm{Pr}D_n &= \Big[(n+1) - (n+2)\mathrm{Pr}\Big] D_{n+1} \\ &\quad - \sum_{m=n+2}^{\infty} D_n + \frac{1}{2\mathrm{Pr}}\mathcal{N}_T\Big\{D_n, B_n\Big\}\end{aligned} \tag{5.45}$$

where

$$\mathcal{N}_T\Big\{D_n, B_n\Big\} = \frac{\partial(T, \Psi)}{\partial(\sigma, \tau)} \tag{5.46}$$

is the nonlinear part of the differential operator of the temperature field, which describes the interaction between flow field and temperature field.

3. Asymptotic Expansion Form of the Solution as $\epsilon_0 \longrightarrow 0$

We further make the following asymptotic expansion in the limit $\epsilon_0 \to 0$:

$$\begin{aligned}A_n(\tau, \epsilon_0) &= \nu_0(\epsilon_0)\Big\{\bar{A}_{n,0}(\tau) + \epsilon_0 \bar{A}_{n,1}(\tau) + \cdots\Big\}, \\ B_n(\tau, \epsilon_0) &= \nu_0(\epsilon_0)\Big\{\bar{B}_{n,0}(\tau) + \epsilon_0 \bar{B}_{n,1}(\tau) + \cdots\Big\}, \\ D_n(\tau, \epsilon_0) &= \gamma_0(\epsilon_0)\Big\{\bar{D}_{n,0}(\tau) + \epsilon_0 \bar{D}_{n,1}(\tau) + \cdots\Big\}, \\ h_n(\epsilon_0) &= \delta_0(\epsilon_0)\Big\{\bar{h}_{n,0} + \epsilon_0 \bar{h}_{n,1} + \cdots\Big\},\end{aligned} \tag{5.47}$$

for $(n = 0, 1, \ldots)$. It is easy to derive from the far-field conditions and interface conditions that the asymptotic factors, $\nu_0(\epsilon_0) = \delta_0(\epsilon_0) = \epsilon_0$ and

$\gamma_0(\epsilon_0) = 1$. Then, with the above asymptotic expansion, the solution $\bar{\Psi}$ will have the following asymptotic structure:

$$\begin{aligned}
\bar{\Psi} &= \epsilon_0\Big(\bar{\Psi}_0 + \epsilon_0\bar{\Psi}_1 + \cdots\Big) \\
&= \epsilon_0\sigma\Big\{\Big[\bar{B}_{0,0} + \epsilon_0\bar{B}_{0,1} + \epsilon_0^2\bar{B}_{0,2} + \cdots\Big]L_0^{(1)}(\sigma) \\
&\quad + \Big[\bar{B}_{1,0} + \epsilon_0\bar{B}_{1,1} + \epsilon_0^2\bar{B}_{1,2} + \cdots\Big]L_1^{(1)}(\sigma) \\
&\quad + \Big[\bar{B}_{2,0} + \epsilon_0\bar{B}_{2,1} + \epsilon_0^2\bar{B}_{2,2} + \cdots\Big]L_2^{(1)}(\sigma) \\
&\quad + \cdots\Big\}.
\end{aligned} \tag{5.48}$$

For the interface function we have the asymptotic expansion

$$\begin{aligned}
\bar{h}_s &= \epsilon_0\Big[\bar{h}_0(\sigma) + \epsilon_0\bar{h}_1(\sigma) + \epsilon_0^2\bar{h}_2(\sigma) + \cdots\Big] \\
&= \epsilon_0\Big\{\Big[\bar{h}_{0,0} + \epsilon_0\bar{h}_{0,1} + \epsilon_0^2\bar{h}_{0,2} + \cdots\Big]L_0^{(1)}(\sigma) \\
&\quad + \Big[\bar{h}_{1,0} + \epsilon_0\bar{h}_{1,1} + \epsilon_0^2\bar{h}_{1,2} + \cdots\Big]L_1^{(1)}(\sigma) \\
&\quad + \cdots\Big\}.
\end{aligned} \tag{5.49}$$

The solutions $\bar{\zeta}$, and $\bar{T}$ have similar asymptotic structure.

3.1 Leading-Order Asymptotic Expansion Solutions of the Flow Field

We seek the flow field solutions first. In the leading-order approximation, the general solution of the flow field is the solution of the Oseen model problem derived in the last chapter. For convenience in further discussion, we write the solution of the stream function in the form

$$\bar{\Psi}_0 = 2\mathrm{Pr}^2\sigma\tau + \bar{\Psi}_0^{(I)} + \bar{\Psi}_0^{(II)} \tag{5.50}$$

where the first term on the right-hand side is from the uniform external flow, while the second and third term represent the perturbed stream function in the laboratory frame due to the growing dendrite. of these two terms,

$$\bar{\Psi}_0^{(I)} = \sigma\left\{\bar{b}_{0,0}\mathcal{F}_0(\tau) + \sum_{k=0}^{\infty} \mathcal{C}_{0,k}\mathcal{A}_{0,k}\right\} \tag{5.51}$$

is the linear part of the perturbed stream function with regard to the variable σ, while

$$\bar{\Psi}_0^{(II)} = \sigma \sum_{n=1}^{N} \left\{ \bar{b}_{n,0}\mathcal{F}_n(\tau) + \sum_{k=0}^{\infty} C_{0,k}\mathcal{A}_{n,k} \right\} L_n^{(1)}(\sigma) \tag{5.52}$$

is the nonlinear part. The corresponding solution of vorticity function can be written as

$$\bar{\zeta}_0 = \bar{\zeta}_0^{(I)} + \bar{\zeta}_0^{(II)} \tag{5.53}$$

where

$$\bar{\zeta}_0^{(I)} = \sigma \bar{a}_{0,0}\mathcal{F}_0(\tau) \tag{5.54}$$

and

$$\bar{\zeta}_0^{(II)} = \sigma \sum_{n=1}^{N} \bar{a}_{n,0}\mathcal{F}_n(\tau) L_n^{(1)}(\sigma). \tag{5.55}$$

The numerical calculations for various cases with different $\tau_0 = 0.01 - 1$ and $\mathrm{Pr} = 0.1 - 10$ have been carried out. To compare the weights of the two parts of stream function: $\bar{\Psi}_0^{(I)}$ and $\bar{\Psi}_0^{(II)}$, we define $\Delta\bar{\Psi}_0 = \bar{\Psi}_0^{(I)} + \bar{\Psi}_0^{(II)}$, and calculate the variations of $\Delta\bar{\Psi}_0/\sigma$ with τ at $\sigma = 0, 3, 6$ for the typical case $\tau_0 = 0.1, \mathrm{Pr} = 5$. The results are shown in Fig. 5.1. It is seen that these three curves are almost identical in the region ($\tau_0 \leq \tau < 3$), or say, the function $\Delta\bar{\Psi}_0/\sigma$ is nearly independent of variable σ. This result implies that the second part of the perturbed stream function $\bar{\Psi}_0^{(II)}(\sigma, \tau)$ and the perturbed vorticity function, $\bar{\zeta}_0^{(II)}(\sigma, \tau)$ are numerically negligible.

3.2 Zeroth-Order Solution of Temperature Field $O(1)$

In the zeroth-order approximation, we have

$$\bar{D}_{n,0} = 0, \tag{5.56}$$

for $n \geq 1$, and

$$\tau \bar{D}_{0,0}'' + (1 + \mathrm{Pr}\,\tau)\,\bar{D}_{0,0}' = 0. \tag{5.57}$$

The above system allows the solutions

$$\bar{D}_{0,0} = I_* + I_0 E_1(\mathrm{Pr}\,\tau), \tag{5.58}$$

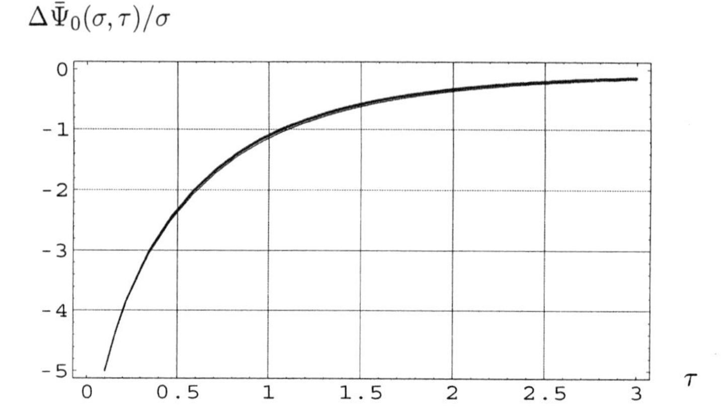

Figure 5.1. The variations of functions $\Delta\bar{\Psi}_0/\sigma$ with τ for the typical case $\tau_{0,0}=0.1$, $\mathrm{Pr}=5$, as $\sigma=0,3,6$

where I_0 and I_* are arbitrary constants. From the far-field condition $\bar{D}_{0,0}\to 0$ as $\tau\to\infty$, we derive

$$I_* = 0. \tag{5.59}$$

The zeroth-order interface conditions contain only the components of $L_0^{(1)}(\sigma)$; they are

$$\begin{aligned} &\bar{\nu}_0(\epsilon_0)\mathrm{Pr}\ \bar{D}_{0,0}(\tau_{0,0}) + T_\infty = 0,\\ &\tau_{0,0}\bar{\nu}_0(\epsilon_0)\bar{D}'_{0,0}(\tau_{0,0}) + \tau_{0,0} = 0. \end{aligned} \tag{5.60}$$

We derive that

$$I_0 = \tau_{0,0}\mathrm{e}^{\mathrm{Pr}\tau_{0,0}}. \tag{5.61}$$

Thus, as we expected, the zeroth-order solution is just the Ivantsov solution:

$$\begin{cases} \bar{D}_{0,0}(\tau) = \tau_{0,0}\mathrm{e}^{\mathrm{Pr}\,\tau_{0,0}}E_1(\mathrm{Pr}\,\tau),\\ T_\infty = -\mathrm{Pr}\ \tau_{0,0}\mathrm{e}^{\mathrm{Pr}\,\tau_{0,0}}E_1(\mathrm{Pr}\,\tau_{0,0}). \end{cases} \tag{5.62}$$

3.3 First Order Asymptotic Expansion Solution of Temperature Field $O(\epsilon_0)$

In the leading-order approximation, the solution $\bar{D}_{n,1}$ may be written in three parts:

$$\bar{D}_{n,1} = \overline{W}_{n,1} + \overline{P}_{n,1} + \overline{Q}_{n,1}. \tag{5.63}$$

The first part of the solution, $\overline{W}_{n,1}$, is the general solution of the associated homogeneous equation,

$$\tau\frac{\mathrm{d}^2\overline{W}_{n,1}}{\mathrm{d}\tau^2}+(1+\mathrm{Pr}\,\tau)\frac{\mathrm{d}\overline{W}_{n,1}}{\mathrm{d}\tau}-n\mathrm{Pr}\overline{W}_{n,1}(\tau)=0; \tag{5.64}$$

the second part of the solution, $\overline{P}_{n,1}$, is the particular solution of the equation with the inhomogeneous term related to the nonlinear part of the stream function solution $\bar{\Psi}_0^{(II)}$; the third part of the solution, $\overline{Q}_{n,1}$, is the particular solution of the equation with the inhomogeneous term related to the total linear part of the stream function solution,

$$\bar{\Psi}_0^{(L)}=2\mathrm{Pr}^2\sigma\tau+\bar{\Psi}_0^{(I)}. \tag{5.65}$$

In terms of the transformation

$$\begin{aligned}&\overline{W}_{n,1}(\tau)=x^{-\frac{1}{2}}\mathrm{e}^{-\frac{x}{2}}W(x),\\&x=\mathrm{Pr}\,\tau,\end{aligned} \tag{5.66}$$

the equation (5.64) can be transformed into the Whittaker equation

$$\frac{\mathrm{d}^2W}{\mathrm{d}x^2}+\left(-\frac{1}{4}+\frac{\kappa}{x}+\frac{\frac{1}{4}-\mu}{x^2}\right)W=0, \tag{5.67}$$

where

$$\kappa=n+\frac{1}{2},\quad \mu=0. \tag{5.68}$$

The fundamental solutions of (5.67) are

$$W(x)=\begin{cases}x^{\frac{1}{2}}\mathrm{e}^{-\frac{x}{2}}M(n+1,1,x),\\x^{\frac{1}{2}}\mathrm{e}^{-\frac{x}{2}}U(n+1,1,x).\end{cases} \tag{5.69}$$

So, the fundamental solutions of (5.64) are

$$\overline{W}_{n,1}(\tau)=\begin{cases}P_n(\tau)=\mathrm{e}^{-\mathrm{Pr}\,\tau}M(n+1,1,\mathrm{Pr}\,\tau),\\\widehat{W}_{n,1}(\tau)=\mathrm{e}^{-\mathrm{Pr}\,\tau}U(n+1,1,\mathrm{Pr}\,\tau).\end{cases} \tag{5.70}$$

Note that the solution

$$P_n(x)=\mathrm{e}^{-x}M(n+1,1,\mathrm{Pr}\,x) \tag{5.71}$$

is a polynomial of degree n.

Taking into account the far-field condition, we have

$$\overline{W}_{n,1} = \bar{d}_{n,1}\widehat{W}_{n,1}(\tau) = \bar{d}_{n,1}e^{-\Pr\tau}U(n+1,1,\Pr\tau). \tag{5.72}$$

Especially, we have

$$\begin{aligned}
&\widehat{W}_{0,1}(\tau) = E_1(\Pr\tau),\\
&\widehat{W}_{1,1}(\tau) = \Big[E_1(\Pr\tau) - E_2(\Pr\tau)\Big],\\
&\vdots\\
&\widehat{W}_{n,1}(\tau) = (-1)^n\sum_{k=0}^{n}\frac{(-1)^k}{(n-k)!k!}E_{k+1}(\Pr\tau).
\end{aligned} \tag{5.73}$$

For the second part of the solution, $\overline{P}_{n,1}$, we need to find the particular solution of the equations

$$\tau\frac{d^2\overline{P}_{n,1}}{d\tau^2} + (1+\Pr\tau)\frac{d\overline{P}_{n,1}}{d\tau} - n\Pr\overline{P}_{n,1}(\tau) = H_n(\tau) \tag{5.74}$$

for $(n = 0, 1, 2, \ldots,)$, where

$$H_n(\tau) = \Big[(n+1) - (n+2)\Pr\Big]\overline{P}_{n+1,1} - \sum_{m=n+2}^{\infty}\overline{P}_{m,1} + \frac{1}{2\Pr}\mathcal{N}_T\Big\{\bar{D}_{0,0}, \bar{\Psi}_0^{(II)}\Big\}. \tag{5.75}$$

We can truncate the system at $n = N$, by approximately setting

$$\overline{P}_{n,1} = 0, \qquad (n \geq N+1). \tag{5.76}$$

By using the method of variation parameter, we derive

$$\overline{P}_{n,1}(\tau) = \overline{\mathcal{L}}_n\Big\{H_n(\tau)\Big\}\widehat{W}_{n,1}(\tau) + \widehat{\mathcal{L}}_n\Big\{H_n(\tau)\Big\}P_n(\tau) \tag{5.77}$$

with the linear operators

$$\begin{aligned}
\overline{\mathcal{L}}_n\Big\{H_n(\tau)\Big\} &= \int_\tau^\infty \frac{P_n(\tilde{\tau})H_n(\tilde{\tau})}{\tilde{\tau}\Delta(\widehat{W}_{n,1}, P_n)}d\tilde{\tau},\\
\widehat{\mathcal{L}}_n\Big\{H_n(\tau)\Big\} &= -\int_\tau^\infty \frac{\widehat{W}_{n,1}(\tilde{\tau})H_n(\tilde{\tau})}{\tilde{\tau}\Delta(\widehat{W}_{n,1}, P_n)}d\tilde{\tau},
\end{aligned} \tag{5.78}$$

where $\Delta(\widehat{W}_{n,1}, P_n)$ is the Wronskian of the solutions $\Big\{\widehat{W}_{n,1}(\tau)$ and $P_n(\tau)\Big\}$. One can find $\overline{P}_{N,1}(\tau)$ first, then consecutively find $\overline{P}_{N-1,1}(\tau)$, $\overline{P}_{N-2,1}(\tau)$, ..., etc. To simplify the formulas to be shown later, one can always set

$$\overline{P}'_{n,1}(\tau_0) = 0, \quad (n = 0, 1, 2, \ldots). \tag{5.79}$$

For the third part of the solution, $\overline{Q}_{n,1}$, we have

$$\overline{Q}_{n,1} = 0 \quad (n = 1, 2, 3, \ldots). \tag{5.80}$$

and

$$\tau \overline{Q}''_{0,1}(\tau) + (1 + \mathrm{Pr}\,\tau)\overline{Q}'_{0,1}(\tau) = H_{0,1}(\tau), \tag{5.81}$$

where

$$\begin{aligned} H_{0,1}(\tau) &= -\frac{1}{2\mathrm{Pr}} \mathcal{N}_T\left\{\bar{D}_{0,0}, \bar{\Psi}_0^{(L)}\right\} \\ &= -\frac{I_0}{2\mathrm{Pr}} \frac{\mathrm{e}^{-\mathrm{Pr}\,\tau}}{\tau} \left\{2\mathrm{Pr}^2\tau + \bar{b}_{0,0}\mathcal{F}_0(\tau) + \sum_{k=0}^{\infty} \mathcal{C}_{0,k}\mathcal{A}_{0,k}(\tau)\right\}. \end{aligned} \tag{5.82}$$

We derive that

$$\overline{Q}'_{0,1}(\tau) = \frac{\mathrm{e}^{-\mathrm{Pr}\,\tau}}{\tau} \int_{\tau_0}^{\tau} H_{0,1}(\tau_1)\mathrm{e}^{-\mathrm{Pr}\,\tau_1} \mathrm{d}\tau_1, \tag{5.83}$$

which satisfies $\overline{Q}'_{0,1}(\tau_0) = 0$, and subsequently,

$$\overline{Q}_{0,1}(\tau) = -\int_{\tau}^{\infty} \frac{\mathrm{e}^{-\mathrm{Pr}\,\tilde{\tau}}}{\tilde{\tau}} \left\{ \int_{\tau_0}^{\tilde{\tau}} H_{0,1}(\tau_1)\mathrm{e}^{-\mathrm{Pr}\,\tau_1}\mathrm{d}\tau_1 \right\} \mathrm{d}\tilde{\tau}. \tag{5.84}$$

In the first-order approximation, the interface conditions for the temperature field can be written in the following forms: at $\tau = \tau_{0,0}$,

$$\begin{aligned} &\bar{T}_1 + \bar{T}'_0(\tau_{0,1} + \bar{h}_0) = 0, \\ &\tau_{0,0}\left[\frac{\partial \bar{T}_1}{\partial \tau} + \bar{T}''_0(\tau_{0,1} + \bar{h}_0)\right] + \bar{T}'_0(\tau_{0,1} + \bar{h}_0) \\ &\qquad + \left[\tau_{0,1} + \bar{h}_0 + \sigma \bar{h}'_0(\sigma)\right] = 0. \end{aligned} \tag{5.85}$$

Noting that

$$\sigma L_n^{(1)'}(\sigma) = n L_n^{(1)}(\sigma) - (n+1)L_{n-1}^{(1)}(\sigma), \tag{5.86}$$

we derive that

$$\begin{aligned} &\bar{D}_{0,1}(\tau_{0,0}) + \bar{D}'_{0,0}(\tau_{0,0})(\tau_{0,1} + \bar{h}_{0,0}) = 0, \\ &\tau_{0,0}\left[\bar{D}'_{0,1}(\tau_{0,0}) + \bar{D}''_{0,0}(\tau_{0,0})(\tau_{0,1} + \bar{h}_{0,0})\right] \\ &\qquad + \left[\bar{D}'_{0,0}(\tau_{0,0}) + 1\right](\tau_{0,1} + \bar{h}_{0,0}) - 2\bar{h}_{1,0} = 0, \end{aligned} \tag{5.87}$$

and

$$\bar{D}_{n,1}(\tau_{0,0}) + \bar{D}'_{0,0}(\tau_{0,0})\bar{h}_{n,0} = 0,$$
$$\tau_{0,0}\Big[\bar{D}'_{n,1}(\tau_{0,0}) + \bar{D}''_{0,0}(\tau_{0,0})\bar{h}_{n,0}\Big] + \Big[\bar{D}'_{0,0}(\tau_{0,0}) + 1\Big]\bar{h}_{n,0} \tag{5.88}$$
$$+nh_{n,0} - (n+2)h_{n+1,0} = 0,$$

for $(n = 1, 2, \cdots, N)$. Moreover, in terms of the formulas: Noting that

$$\bar{D}'_{0,0}(\tau_{0,0}) = -1, \quad \bar{D}''_{0,0}(\tau_{0,0}) = \Big(\mathrm{Pr} + \frac{1}{\tau_{0,0}}\Big), \tag{5.89}$$

one may re-write (5.87)– (5.88) as

$$\begin{aligned} &\bar{d}_{0,1}\widehat{W}_{0,1}(\tau_{0,0}) - (\tau_{0,1} + \bar{h}_{0,0}) = -\Big[\overline{Q}_{0,1}(\tau_{0,0}) + \overline{P}_{0,1}(\tau_{0,0})\Big], \\ &\tau_{0,0}\bar{d}_{0,1}\widehat{W}'_{0,1}(\tau_{0,0}) + (1 + \mathrm{Pr}\tau_{0,0})(\tau_{0,1} + \bar{h}_{0,0}) - 2h_{1,0} = 0, \end{aligned} \tag{5.90}$$

and

$$\bar{d}_{n,1}\widehat{W}_{n,1}(\tau_{0,0}) - \bar{h}_{n,0} = -\overline{P}_{n,1}(\tau_{0,0}),$$
$$\tau_{0,0}\bar{d}_{n,1}\widehat{W}'_{n,1}(\tau_{0,0}) + (n + 1 + \mathrm{Pr}\tau_{0,0})\bar{h}_{n,0} \tag{5.91}$$
$$-(n+2)h_{n+1,0} = 0,$$

for $(n = 1, 2, \ldots, N)$.

Furthermore, from the tip condition, $\bar{h}_s(0, \epsilon_0) = 0$, we have

$$\sum_{n=0}^{n=N} \bar{h}_{n,0} = 0. \tag{5.92}$$

The two sets of interface conditions (5.90)–(5.91) plus the tip condition (5.92) may uniquely determine the $2N + 3$ unknown constants: $\{\bar{d}_{n,1}\}$; $\{\bar{h}_{n,0}\}$ $(n = 0, 1, 2, \ldots, N)$ and $\tau_{0,1}$.

Recall that, as we have shown in the last section, the nonlinear part $\bar{\Psi}_0^{(II)}$ is numerically negligible. Thus, we can consider $\overline{P}_{n,1}(\tau) \approx 0$ for all $n = 0, 1, 2, \ldots$. With these approximations we derive $\bar{h}_{n,0} \approx 0$ for $(n = 0, 1, 2, \ldots)$, and $d_{n,1} \approx 0$ for $(n = 1, 2, \ldots)$. Then the first-order solution then can be written as

$$\bar{T}_1(\tau) \approx \bar{d}_{0,1}\widehat{W}_{0,1}(\tau) + \widehat{Q}_{0,1}(\tau). \tag{5.93}$$

We obtain that

$$\bar{d}_{0,1} \approx \frac{\tau_{0,1} - \overline{Q}_{0,1}(\tau_{0,0})}{\widehat{W}_{0,1}(\tau_{0,0})}, \tag{5.94}$$

$$\tau_{0,1} \approx \tau_{0,0}\mathcal{Z}(\tau_{0,0}, \mathrm{Pr}) \tag{5.95}$$

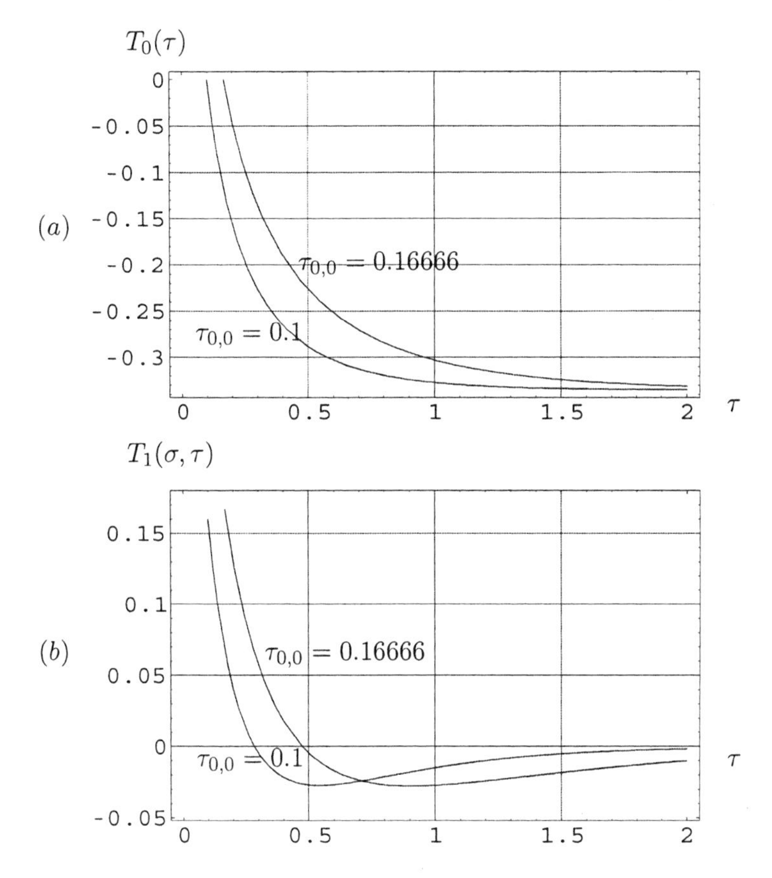

Figure 5.2. The asymptotic solutions of temperature field with $T_\infty = -0.335221$: (a) $T_0(\tau) = T_\infty + \Pr\, D_{0,0}(\tau)$; (b) $T_1(\sigma,\tau) = \Pr\big[\bar{d}_{0,1}\widehat{W}_{0,1}(\Pr\ \tau) + \overline{Q}_{0,1}(\tau)\big]$, for the cases, $\Pr = 1.5$ corresponding to $\tau_{0,0} = 0.16666$, and $\Pr = 2.5$ corresponding to $\tau_{0,0} = 0.1$

where

$$\mathcal{Z}(\tau_{0,0},\Pr) = \frac{\overline{Q}_{0,1}(\tau_{0,0})\widehat{W}'_{0,1}(\tau_{0,0})}{\tau_{0,0}\widehat{W}'_{0,1}(\tau_{0,0}) + (1+\Pr\tau_{0,0})\widehat{W}_{0,1}(\tau_{0,0})}. \tag{5.96}$$

Finally, we can write the interface shape function in the form

$$\begin{aligned}\tau = \tau_{\mathrm{s}}(\sigma) &= (\tau_{0,0} + \epsilon_0\tau_{0,1} + \cdots) + \epsilon_0\sum_{n=0}^{N}\bar{h}_{n,0}L_n^{(1)}(\sigma) + \cdots \\ &= \tau_{0,0}\Big[1 + \epsilon_0\mathcal{Z}(\tau_{0,0},\Pr)\Big] + O(\epsilon_0^2).\end{aligned} \tag{5.97}$$

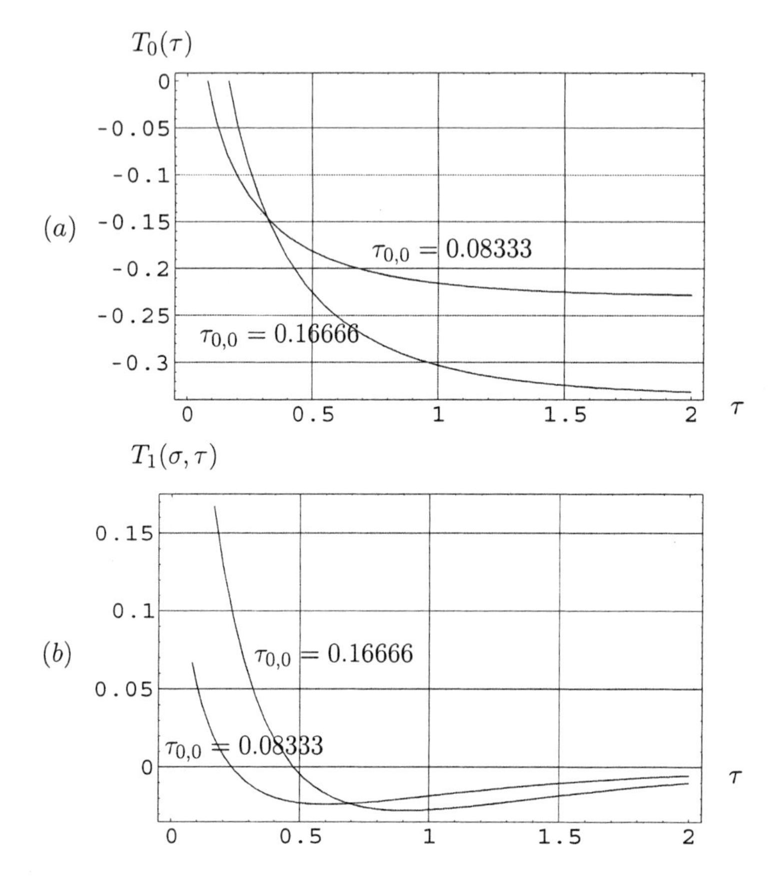

Figure 5.3. The asymptotic solutions of temperature field with $\mathrm{Pr} = 1.5$: (a) $T_0(\tau) = T_\infty + \mathrm{Pr}\ D_{0,0}(\tau)$; (b) $T_1(\sigma, \tau) = \mathrm{Pr}\big[\bar{d}_{0,1}\widehat{W}_{0,1}(\mathrm{Pr}\ \tau) + \overline{Q}_{0,1}(\tau)\big]$, for the cases, $T_\infty = -0.335221$ corresponding to $\tau_{0,0} = 0.16666$, and $T_\infty = -0.229948$ corresponding to $\tau_{0,0} = 0.083333$

In the coordinate system (ξ, η), we have

$$\eta_{\mathrm{s}} = 1 + \epsilon_0 \frac{\mathrm{Pr}}{\eta_0^2} \sum_{n=0}^{N} \bar{h}_{n,0} L_n^{(1)}(\sigma) + O(\epsilon_0^2) = 1 + O(\epsilon_0^2), \tag{5.98}$$

and

$$\begin{aligned} \eta_0^2 &= 2\mathrm{Pr}(\tau_{0,0} + \epsilon_0 \tau_{0,1} + \cdots) \\ &= \eta_{0,0}^2 \Big[1 + \epsilon_0 \mathcal{Z}(\tau_{0,0}, \mathrm{Pr}) + \cdots\Big], \end{aligned} \tag{5.99}$$

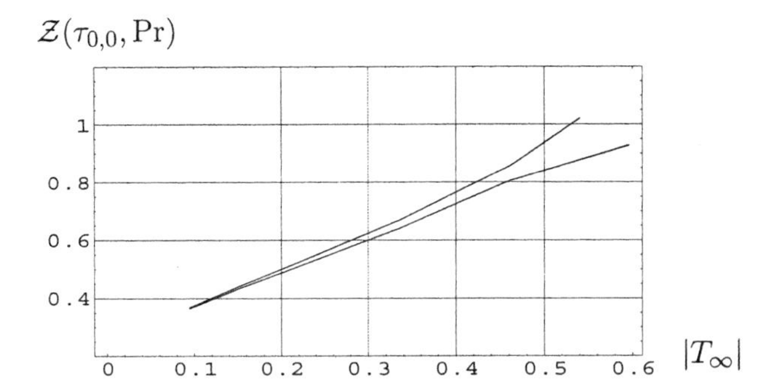

Figure 5.4. The variations of functions $\mathcal{Z}(\tau_{0,0},\mathrm{Pr})$ versus the undercooling $|T_\infty|$ with fixed ϵ_0 for the cases: $\mathrm{Pr} = 1.5, 2.5$, from top to bottom.

where

$$\eta_{0,0}^2 = 2\mathrm{Pr}\tau_{0,0} \tag{5.100}$$

is the Peclet number of dendritic growth without external flow.

It is seen from the above results that in the order of $O(\epsilon_0)$ the external flow does not affect the paraboloidal shape of the dendrite's interface. The external flow only affects the Peclet number of dendritic growth with the factor $1+\epsilon_0\mathcal{Z}(\tau_{0,0},\mathrm{Pr})$, which is related to the Prandtl number Pr, undercooling temperature $|T_\infty|$, as well as the flow parameter ϵ_0.

In Fig. 5.2(a) and (b), we respectively show the temperature profiles of the leading-order approximation, $T_0(\tau) = T_\infty + \mathrm{Pr}\ \bar{D}_{0,0}(\tau)$, and the first-order approximate solution, $T_1(\sigma,\tau) = \mathrm{Pr}[\bar{d}_{0,1}\widehat{W}_{01}(\mathrm{Pr}\ \tau)+\overline{Q}_{0,1}(\tau)]$, for the two cases, which have the same undercooling $T_\infty = -0.335221$, but different Prandtl numbers $\mathrm{Pr} = 1.5, 2.5$. The solution for the case $\mathrm{Pr} = 1.5$ corresponds to $\tau_0 = 0.16666$, while the solution for the case $\mathrm{Pr} = 2.5$ corresponds to $\tau_0 = 0.1$. In Fig. 5.3(a) and (b), we respectively show these temperature profiles for another two cases, which have the same Prandtl numbers $\mathrm{Pr} = 1.5$, but different undercooling temperatures $T_\infty = -0.335221; -0.229948$. The solution for the case $T_\infty = -0.335221$ corresponds to $\tau_0 = 0.16666$, and the case $T_\infty = -0.229948$ corresponds to $\tau_0 = 0.083333$.

In Fig. 5.4, we show the variations of the function $\mathcal{Z}(\tau_{0,0},\mathrm{Pr})$ with the undercooling $|T_\infty|$ for the cases, $\mathrm{Pr} = 1.5, 2.5$. It can be seen that the function $\mathcal{Z}(\tau_{0,0},\mathrm{Pr})$ is always positive, it increases with increasing undercooling $|T_\infty|$ for the given Prandtl number $\mathrm{Pr} = O(1)$, and decreases with increasing Prandtl number Pr for given undercooling $|T_\infty|$.

Chapter 6

ASYMPTOTIC SOLUTION OF DENDRITIC GROWTH IN EXTERNAL FLOW (II): THE CASE OF PR $\to \infty$

For the case of large Prandtl number $\mathrm{Pr} \gg 1$ with $\overline{U}_\infty = \frac{U_\infty}{U} = O(1)$, we adopt

$$\epsilon_2 = \frac{1}{\mathrm{Pr}} \ll 1 \tag{6.1}$$

as the basic small parameter and attempt to find a uniformly valid asymptotic expansion solution for the problem in the limit $\epsilon_2 \to 0$. The governing equations and boundary conditions are the same as given in the previous chapter. The up-stream far-field conditions are re-written in the form: as $\eta \to \infty$,

$$\Psi \sim \frac{1}{2}\lambda_0 \eta_0^4 \xi^2 \eta^2 + o(1), \qquad \zeta \to 0, \tag{6.2}$$

$$T \to T_\infty, \tag{6.3}$$

where

$$\lambda_0 = (1 + \overline{U}_\infty), \qquad \overline{U}_\infty = \frac{U_\infty}{U}. \tag{6.4}$$

We now assume $\{\lambda_0; \overline{U}_\infty\} = O(1)$. To measure the strength of the external flow, instead of the flow parameter $\overline{U}_\infty$, one may use the Reynolds number of flow based on the tip radius of dendrite, ℓ_t, as first defined in Chap. 4,

$$\mathrm{Re} = \frac{U_\infty \ell_t}{\nu}. \tag{6.5}$$

The Reynolds number Re is connected with the flow parameter $\overline{U}_\infty$ by the formula

$$\mathrm{Re} = \overline{U}_\infty \frac{\mathrm{Pe}}{\mathrm{Pr}}, \tag{6.6}$$

where the Peclet number is defined as

$$\mathrm{Pe} = \frac{\ell_{\mathrm{t}}}{\ell_{\mathrm{T}}}. \tag{6.7}$$

In the far field the flow is nearly uniform and approximately described by the functions

$$\begin{cases} \bar{\zeta}_* = 0, \\ \bar{\Psi}_* = \frac{1}{2}\eta_0^4 \lambda_0 \xi^2 \eta^2 \ . \end{cases} \tag{6.8}$$

So, for convenience, we set

$$\begin{aligned} \Psi(\xi,\eta,\lambda_0) &= \bar{\Psi}_*(\xi,\eta) + \bar{\Psi}(\xi,\eta,\lambda_0), \\ \zeta(\xi,\eta,\lambda_0) &= \frac{\lambda_0^2}{\mathrm{Pr}^2}\bar{\zeta}(\xi,\eta,\lambda_0), \\ T(\xi,\eta,\lambda_0) &= T_\infty + \mathrm{Pr}\ \bar{T}(\xi,\eta,\lambda_0). \end{aligned} \tag{6.9}$$

On the other hand, we define the new variables

$$\begin{aligned} \bar{\sigma} &= \frac{\eta_0^2 \lambda_0}{2\mathrm{Pr}} \xi^2, \\ \bar{\tau} &= \frac{\eta_0^2 \lambda_0}{2\mathrm{Pr}} \eta^2 \ . \end{aligned} \tag{6.10}$$

With the variables $(\bar{\sigma};\bar{\tau})$, the operator D^2 and ∇^2 become

$$D^2 = \frac{2\eta_0^2\lambda_0}{\mathrm{Pr}}\bar{L}_2, \qquad \nabla^2 = \frac{2\eta_0^2\lambda_0}{\mathrm{Pr}}(\bar{L}_2 + \bar{L}_1), \tag{6.11}$$

where

$$\begin{cases} \bar{L}_2 = \left(\bar{\sigma}\dfrac{\partial^2}{\partial\bar{\sigma}^2} + \bar{\tau}\dfrac{\partial^2}{\partial\bar{\tau}^2}\right), \\ \bar{L}_1 = \left(\dfrac{\partial}{\partial\bar{\sigma}} + \dfrac{\partial}{\partial\bar{\tau}}\right), \end{cases} \tag{6.12}$$

the stream function at the far field becomes

$$\bar{\Psi}_* = \frac{2\mathrm{Pr}^2}{\lambda_0}\bar{\sigma}\bar{\tau} = \frac{2}{\epsilon_2^2\lambda_0}\bar{\sigma}\bar{\tau}, \tag{6.13}$$

while the interface shape function $\eta = \eta_s(\xi, \epsilon_2)$ is transformed into $\bar{\tau} = \bar{\tau}_s(\bar{\sigma}, \epsilon_2)$. We then have the following governing equations:

$$\bar{L}_2[\bar{\Psi}] = -(\bar{\sigma} + \bar{\tau})\bar{\zeta}, \tag{6.14}$$

$$\bar{L}_2[\bar{\zeta}] = \left(\bar{\sigma}\frac{\partial}{\partial\bar{\sigma}} - \bar{\tau}\frac{\partial}{\partial\bar{\tau}}\right)\bar{\zeta} - \frac{\lambda_0\bar{\zeta}}{2\mathrm{Pr}^2\bar{\tau}\bar{\sigma}}\left(\bar{\tau}\frac{\partial\bar{\Psi}}{\partial\bar{\tau}} - \bar{\sigma}\frac{\partial\bar{\Psi}}{\partial\bar{\sigma}}\right)$$
$$-\frac{\lambda_0}{2\mathrm{Pr}^2}\frac{\partial(\bar{\Psi},\bar{\zeta})}{\partial(\bar{\sigma},\bar{\tau})}, \tag{6.15}$$

$$(\bar{L}_2 + \bar{L}_1)[\bar{T}] = \mathrm{Pr}\left(\bar{\sigma}\frac{\partial\bar{T}}{\partial\bar{\sigma}} - \bar{\tau}\frac{\partial\bar{T}}{\partial\bar{\tau}}\right)$$
$$+\frac{1}{2\mathrm{Pr}\lambda_0}\left(\frac{\partial\bar{\Psi}}{\partial\bar{\tau}}\frac{\partial\bar{T}}{\partial\bar{\sigma}} - \frac{\partial\bar{\Psi}}{\partial\bar{\sigma}}\frac{\partial\bar{T}}{\partial\bar{\tau}}\right). \tag{6.16}$$

The interface conditions now have the following forms: at $\bar{\tau} = \bar{\tau}_s(\bar{\sigma})$,

$$T_\infty + \mathrm{Pr}\,\bar{T} = 0, \tag{6.17}$$

$$\bar{\tau}_s\frac{\partial\bar{T}}{\partial\bar{\tau}} - \bar{\sigma}\bar{\tau}_s'\frac{\partial\bar{T}}{\partial\bar{\sigma}} + \frac{1}{\lambda_0}(\bar{\sigma}\bar{\tau}_s' + \bar{\tau}_s) = 0, \tag{6.18}$$

$$\frac{\partial\bar{\Psi}}{\partial\bar{\sigma}} + \bar{\tau}_s'\frac{\partial\bar{\Psi}}{\partial\bar{\tau}} - 2\mathrm{Pr}^2\frac{\lambda_0 - 1}{\lambda_0^2}(\bar{\tau}_s + \bar{\sigma}\bar{\tau}_s') = 0, \tag{6.19}$$

$$\bar{\tau}_s\frac{\partial\bar{\Psi}}{\partial\bar{\tau}} - \bar{\sigma}\bar{\tau}_s'\frac{\partial\bar{\Psi}}{\partial\bar{\sigma}} + 2\mathrm{Pr}^2\frac{\lambda_0 - 1}{\lambda_0^2}\bar{\sigma}\bar{\tau}_s(1 - \bar{\tau}_s') = 0. \tag{6.20}$$

Note that in the coordinate system (ξ, η), the interface shape function $\eta_s = O(1)$; however, in the new coordinate system $(\bar{\sigma}, \bar{\tau})$, the interface shape function now becomes

$$\bar{\tau}_s(\bar{\sigma}, \epsilon_2) = \frac{\eta_0^2\lambda_0}{2\mathrm{Pr}}\eta_s^2 = \epsilon_2\frac{\eta_0^2\lambda_0}{2}\eta_s^2 = O(\epsilon_2). \tag{6.21}$$

At the dendrite's tip, $\xi = \bar{\sigma} = 0$, from the normalization condition (2.2) we derive

$$\bar{\tau}_s(0) = \bar{\tau}_0 = \epsilon_2\frac{\eta_0^2\lambda_0}{2}. \tag{6.22}$$

It is, therefore, seen that the problem under study is the so-called *singular boundary problem.* To solve this problem, by following the approach developed in Chap. 4, we divide the whole physical region into two sub-regions:

- the outer region away from the interface, $\bar{\tau} = O(1)$,
- the inner region near the interface, $\bar{\tau} \ll 1$,

and derive the outer and inner expansion solutions in each sub-region, then match them in the intermediate region.

In what follows, we shall derive the outer expansion solution first.

1. Laguerre Series Representation of Solutions

To proceed, as in the last chapter, we expand the solution in the Laguerre series

$$\begin{cases} \bar{\zeta}(\bar{\sigma},\bar{\tau}) = \bar{\sigma} \sum\limits_{n=0}^{\infty} \bar{A}_n(\bar{\tau},\epsilon_2) L_n^{(1)}(\bar{\sigma}), \\ \bar{\Psi}(\bar{\sigma},\bar{\tau}) = \bar{\sigma} \sum\limits_{n=0}^{\infty} \bar{B}_n(\bar{\tau},\epsilon_2) L_n^{(1)}(\bar{\sigma}), \\ \bar{T}(\bar{\sigma},\bar{\tau}) = \sum\limits_{n=0}^{\infty} \bar{D}_n(\bar{\tau},\epsilon_2) L_n^{(1)}(\bar{\sigma}). \end{cases} \tag{6.23}$$

Furthermore, we derive that

$$\begin{aligned} \frac{\mathrm{d}^2 \bar{B}_n}{\mathrm{d}\bar{\tau}^2} &= \frac{n+2}{\bar{\tau}} \bar{B}_{n+1} - \left[1 + \frac{2(n+1)}{\bar{\tau}}\right] \bar{A}_n \\ &\quad + \frac{1}{\bar{\tau}} \Big[(n+2)\bar{A}_{n+1} + n\bar{A}_{n-1}\Big], \\ \frac{\mathrm{d}^2 \bar{A}_n}{\mathrm{d}\bar{\tau}^2} &+ \frac{\mathrm{d}\bar{A}_n}{\mathrm{d}\bar{\tau}} - \frac{n+1}{\bar{\tau}} \bar{A}_n = \frac{\epsilon_2^2 \lambda_0}{2} \mathcal{N}\Big\{\bar{A}_n; \bar{B}_n\Big\} \\ &(n = 0, 1, 2, \ldots,) \end{aligned} \tag{6.24}$$

where $\mathcal{N}\Big\{\bar{A}_n; \bar{B}_n\Big\} = \dfrac{\partial(\bar{\zeta}, \bar{\Psi})}{\partial(\bar{\sigma}, \bar{\tau})}$ is the nonlinear part of the differential operator for the flow field.

For the temperature field, we have

$$\begin{aligned} \epsilon_2 \bar{\tau} \frac{\mathrm{d}^2 \bar{D}_n}{\mathrm{d}\bar{\tau}^2} + (\epsilon_2 + \bar{\tau}) \frac{\mathrm{d}\bar{D}_n}{\mathrm{d}\bar{\tau}} - n\bar{D}_n &= \Big[\epsilon_2(n+1) - (n+2)\Big] \bar{D}_{n+1} \\ &\quad - \epsilon_2 \sum_{m=n+2}^{\infty} \bar{D}_n + \frac{\epsilon_2^2 \lambda_0}{2} \mathcal{N}_T\Big\{\bar{D}_n, \bar{B}_n\Big\} \end{aligned} \tag{6.25}$$

where $\mathcal{N}_T\Big\{\bar{D}_n, \bar{B}_n\Big\} = \dfrac{\partial(\bar{T}, \bar{\Psi})}{\partial(\bar{\sigma}, \bar{\tau})}$ is the nonlinear part of the differential operator for the temperature field, which describes the interaction between flow field and temperature field. The equation of the temperature field itself is a singular perturbation equation, as the small parameter ϵ_2 appears in the front of $\dfrac{d^2\bar{D}_n}{d\bar{\tau}^2}$.

In what follows, we shall find the solution for the flow field first, then find the solution for the temperature field.

2. Asymptotic Expansion Forms of the Solution for the Flow Field

2.1 Outer Expansion Form of the Solution

In the outer region $\tau = O(1)$, we can make the following outer asymptotic expansion in the limit $\epsilon_2 \to 0$ for the solution of the flow field:

$$\begin{aligned} \bar{A}_n(\bar{\tau}, \epsilon_2) &= \bar{\nu}(\epsilon_2)\Big\{\bar{A}_{n,0}(\bar{\tau}) + \epsilon_2 \bar{A}_{n,1}(\bar{\tau}) + \cdots\Big\}, \\ \bar{B}_n(\bar{\tau}, \epsilon_2) &= \bar{\nu}(\epsilon_2)\Big\{\bar{B}_{n,0}(\bar{\tau}) + \epsilon_2 \bar{B}_{n,1}(\bar{\tau}) + \cdots\Big\}, \end{aligned} \tag{6.26}$$

for $(n = 0, 1, \ldots)$.

The leading asymptotic factor, $\bar{\nu}(\epsilon_2)$ is to be determined later. Note that due to the presence of the nonlinear terms in the vorticity equation, the leading asymptotic factor $\bar{\nu}(\epsilon_2)$ may have different orders of magnitude, such as

$$\bar{\nu}(\epsilon_2) = \bar{\nu}_0(\epsilon_2) \gg \bar{\nu}_1(\epsilon_2) \gg \bar{\nu}_2(\epsilon_2) \gg \cdots .$$

Thus, the outer solution will have the following general form of asymptotic expansion:

$$\begin{aligned} \zeta &= \nu_0(\epsilon_2)\zeta_0 + \bar{\nu}_1(\epsilon_2)\bar{\zeta}_1 + \bar{\nu}_2(\epsilon_2)\bar{\zeta}_2 + \cdots, \\ \bar{\Psi} &= \bar{\nu}_0(\epsilon_2)\bar{\Psi}_0 + \bar{\nu}_1(\epsilon_2)\bar{\Psi}_1 + \bar{\nu}_2(\epsilon_2)\bar{\Psi}_2 + \cdots . \end{aligned} \tag{6.27}$$

With the factor $\bar{\nu}_m(\epsilon_2)$, the solution $\bar{\Psi}_m$ will have the asymptotic structure

$$\begin{aligned} \bar{\Psi}_m = & \Big[\bar{B}_{m,0,0} + \epsilon_2 \bar{B}_{m,0,1} + \epsilon_2^2 \bar{B}_{m,0,2} + \cdots\Big] L_0^{(1)}(\bar{\sigma}) \\ & + \Big[\bar{B}_{m,1,0} + \epsilon_2 \bar{B}_{m,1,1} + \epsilon_2^2 \bar{B}_{m,1,2} + \cdots\Big] L_1^{(1)}(\bar{\sigma}) \\ & \vdots \end{aligned}$$

The solution of vorticity function has similar asymptotic structure.

2.2 Inner Expansion Form of the Solution

Noting that the parameter $\bar{\tau}_0$ can be expressed as

$$\bar{\tau}_0 = \epsilon_2 \hat{\tau}_0, \quad \hat{\tau}_0 = \frac{\lambda_0 \eta_0^2}{2} = O(1), \tag{6.28}$$

in the inner region near the interface, $\bar{\tau} = O(\epsilon_2)$, we introduce the inner variables $(\bar{\sigma}, \hat{\tau})$, where

$$\hat{\tau} = \frac{\bar{\tau}}{\epsilon_2} \, . \tag{6.29}$$

With the inner variables, we have

$$\begin{aligned} \Psi &= \bar{\Psi}_*(\bar{\sigma}, \hat{\tau}) + \hat{\Psi}(\bar{\sigma}, \hat{\tau}, \epsilon_2), \\ \zeta &= \frac{\lambda_0^2}{\mathrm{Pr}^2} \hat{\zeta}(\bar{\sigma}, \hat{\tau}, \epsilon_2), \end{aligned} \tag{6.30}$$

where

$$\bar{\Psi}_* = \frac{2}{\epsilon_2 \lambda_0} \bar{\sigma} \hat{\tau} \tag{6.31}$$

and

$$\begin{cases} \hat{\Psi}(\bar{\sigma}, \hat{\tau}, \epsilon_2) = \bar{\Psi}(\bar{\sigma}, \epsilon_2 \hat{\tau}, \epsilon_2), \\ \hat{\zeta}(\bar{\sigma}, \hat{\tau}, \epsilon_2) = \bar{\zeta}(\bar{\sigma}, \epsilon_2 \hat{\tau}, \epsilon_2). \end{cases} \tag{6.32}$$

We expand the inner solution in the Laguerre series

$$\begin{cases} \hat{\zeta}(\bar{\sigma}, \hat{\tau}) = \bar{\sigma} \sum\limits_{n=0}^{\infty} \hat{A}_n(\hat{\tau}, \epsilon_2) L_n^{(1)}(\bar{\sigma}), \\ \hat{\Psi}(\bar{\sigma}, \hat{\tau}) = \bar{\sigma} \sum\limits_{n=0}^{\infty} \hat{B}_n(\hat{\tau}, \epsilon_2) L_n^{(1)}(\bar{\sigma}). \end{cases} \tag{6.33}$$

Then, the inner system can be obtained from (6.24) as follows:

$$\begin{aligned} \frac{\mathrm{d}^2 \hat{B}_n}{\mathrm{d}\hat{\tau}^2} &= \epsilon_2 \frac{n+2}{\hat{\tau}} \hat{B}_{n+1} - \left[\epsilon_2^2 + \epsilon_2 \frac{2(n+1)}{\hat{\tau}} \right] \hat{A}_n \\ &\quad + \frac{\epsilon_2}{\hat{\tau}} \Big[(n+2) \hat{A}_{n+1} + n \hat{A}_{n-1} \Big], \\ \frac{\mathrm{d}^2 \hat{A}_n}{\mathrm{d}\hat{\tau}^2} &+ \epsilon_2 \frac{\mathrm{d}\hat{A}_n}{\mathrm{d}\hat{\tau}} - \epsilon_2 \frac{n+1}{\hat{\tau}} \hat{A}_n = \frac{\epsilon_2^3 \lambda_0}{2} \hat{\mathcal{N}} \Big\{ \hat{A}_n ; \hat{B}_n \Big\}, \\ &(n = 0, 1, 2, \ldots) \end{aligned} \tag{6.34}$$

where $\mathcal{N}\left\{\hat{A}_n; \hat{B}_n\right\} = \dfrac{\partial(\hat{\zeta}, \hat{\Psi})}{\partial(\bar{\sigma}, \hat{\tau})}$ is the nonlinear differential operator.

We make the following inner asymptotic expansion in the limit $\epsilon_2 \to 0$:

$$\begin{aligned}\hat{A}_n(\hat{\tau}, \epsilon_2) &= \hat{\nu}(\epsilon_2)\left\{\hat{A}_{n,0}(\hat{\tau}) + \epsilon_2 \hat{A}_{n,1}(\hat{\tau}) + \cdots\right\}, \\ \hat{B}_n(\hat{\tau}, \epsilon_2) &= \epsilon_2 \hat{\nu}(\epsilon_2)\left\{\hat{B}_{n,0}(\hat{\tau}) + \epsilon_2 \hat{B}_{n,1}(\hat{\tau}) + \cdots\right\}\end{aligned} \tag{6.35}$$

for $(n = 0, 1, \ldots)$.

The leading asymptotic factors $\hat{\nu}(\epsilon_2)$ are to be determined later, and they may have different orders of magnitude:

$$\hat{\nu}(\epsilon_2) = \hat{\nu}_0(\epsilon_2) \gg \hat{\nu}_1(\epsilon_2) \gg \hat{\nu}_2(\epsilon_2) \gg \cdots .$$

So, the inner solution have the following general form of expansion:

$$\begin{aligned}\hat{\zeta} &= \hat{\nu}_0(\epsilon_2)\hat{\zeta}_0 + \hat{\nu}_1(\epsilon_2)\hat{\zeta}_1 + \hat{\nu}_2(\epsilon_2)\hat{\zeta}_2 + \cdots \\ \hat{\Psi} &= \epsilon_2\left[\hat{\nu}_0(\epsilon_2)\hat{\Psi}_0 + \hat{\nu}_1(\epsilon_2)\hat{\Psi}_1 + \hat{\nu}_2(\epsilon_2)\hat{\Psi}_2 + \cdots\right] .\end{aligned} \tag{6.36}$$

With the primary factor $\hat{\nu}_m(\epsilon_2)$, the solution $\hat{\Psi}_m$ will have the asymptotic structure

$$\begin{aligned}\hat{\Psi}_m &= \left[\hat{B}_{m,0,0} + \epsilon_2 \hat{B}_{m,0,1} + \epsilon_2^2 \hat{B}_{m,0,2} + \cdots\right] L_0^{(1)}(\hat{\sigma}) \\ &+ \left[\hat{B}_{m,1,0} + \epsilon_2 \hat{B}_{m,1,1} + \epsilon_2^2 \hat{B}_{m,1,2} + \cdots\right] L_1^{(1)}(\hat{\sigma}) \\ &+ \cdots .\end{aligned} \tag{6.37}$$

The solution of the vorticity function has similar asymptotic structure.

With the inner variable, the interface shape $\eta = \eta_s(\xi)$ is changed to the form $\hat{\tau} = \hat{\tau}_s(\bar{\sigma}) = \hat{\tau}_0 + \hat{h}_s(\bar{\sigma})$, where $|\hat{h}_s| \ll \hat{\tau}_0$. From the normalization condition (6.22), it follows that

$$\hat{h}_s(0) = 0. \tag{6.38}$$

Similarly, we make the Laguerre expansion for the function $\hat{h}_s(\bar{\sigma})$:

$$\hat{h}_s(\bar{\sigma}) = \sum_{n=0}^{\infty} h_n(\epsilon_2) L_n^{(1)}(\bar{\sigma}) \tag{6.39}$$

and assume that, as $\epsilon_2 \to 0$,

$$h_n(\epsilon_2) = \hat{\delta}(\epsilon_2)\left\{\hat{h}_{n,0} + \epsilon_2 \hat{h}_{n,1} + \cdots\right\}. \tag{6.40}$$

The leading asymptotic factor $\hat{\delta}(\epsilon_2)$ is to be determined later, and it may have different orders of magnitude:

$$\hat{\delta}(\epsilon_2) = \hat{\delta}_0(\epsilon_2) \gg \hat{\delta}_1(\epsilon_2) \gg \hat{\delta}_2(\epsilon_2) \gg \cdots .$$

So, the inner solution $\hat{h}_s$ has the general form of expansion

$$\hat{h}_s = \hat{\delta}_0(\epsilon_2)\hat{h}_0 + \hat{\delta}_1(\epsilon_2)\hat{h}_1 + \hat{\delta}_2(\epsilon_2)\hat{h}_2 + \cdots . \tag{6.41}$$

The normalization condition (6.22) then leads to the conditions

$$\hat{h}_n(0) = 0, \qquad (n = 0, 1, 2, \cdots). \tag{6.42}$$

With the primary factor $\hat{\delta}_m(\epsilon_2)$, the solution $\hat{h}_m$ will have the asymptotic structure

$$\begin{aligned} \hat{h}_m &= \Big[\hat{h}_{m,0,0} + \epsilon_2 \hat{h}_{m,0,1} + \epsilon_2^2 \hat{h}_{m,02} + \cdots\Big] L_0^{(1)}(\hat{\sigma}) \\ &+ \Big[\hat{h}_{m,1,0} + \epsilon_2 \hat{h}_{m,11} + \epsilon_2^2 \hat{h}_{m,1,2} + \cdots\Big] L_1^{(1)}(\hat{\sigma}) \\ &+ \cdots . \end{aligned} \tag{6.43}$$

Moreover, noting that the parameter $\hat{\tau}_0$, as well as η_0^2, depend on ϵ_2 and other physical parameters, we may write

$$\begin{aligned} \eta_0(\epsilon_2) &= \eta_{0,0} + \tilde{\eta}_0(\epsilon_2), \\ \tau_0(\epsilon_2) &= \tau_{0,0} + \tilde{\tau}_0(\epsilon_2), \end{aligned} \tag{6.44}$$

and make the asymptotic expansion

$$\begin{aligned} \tilde{\eta}_0(\epsilon_2) &= \hat{\delta}(\epsilon_2)\Big\{\eta_{0,1} + \epsilon_2 \eta_{0,2} + \cdots\Big\}, \\ \tilde{\tau}_0(\epsilon_2) &= \hat{\delta}(\epsilon_2)\Big\{\hat{\tau}_{0,1} + \epsilon_2 \hat{\tau}_{0,2} + \cdots\Big\}. \end{aligned} \tag{6.45}$$

Since $(\tilde{\tau}_0 + \hat{h}_s) \ll 1$, we can make a Taylor expansion for the interface conditions around $\hat{\tau} = \hat{\tau}_{0,0}$ as follows: at $\hat{\tau} = \hat{\tau}_{0,0}$,

$$\begin{aligned} &\frac{\partial \hat{\Psi}}{\partial \bar{\sigma}} + \frac{\partial^2 \hat{\Psi}}{\partial \bar{\sigma} \partial \hat{\tau}}(\tilde{\tau}_0 + \hat{h}_s) + \frac{1}{2!}\frac{\partial^3 \hat{\Psi}}{\partial \bar{\sigma} \partial \hat{\tau}^2}(\tilde{\tau}_0 + \hat{h}_s)^2 + \cdots \\ &+ \hat{h}_s' \Big(\frac{\partial \hat{\Psi}}{\partial \hat{\tau}} + \frac{\partial^2 \hat{\Psi}}{\partial^2 \hat{\tau}}(\tilde{\tau}_0 + \hat{h}_s) + \frac{1}{2!}\frac{\partial^3 \hat{\Psi}}{\partial \hat{\tau}^3}(\tilde{\tau}_0 + \hat{h}_s)^2 + \cdots \Big) \\ &\qquad - \frac{2(\lambda_0 - 1)}{\epsilon_2 \lambda_0^2}(\hat{\tau}_{0,0} + \tilde{\tau}_0 + \hat{h}_s + \bar{\sigma}\hat{h}_s') = 0, \end{aligned} \tag{6.46}$$

$$(\hat{\tau}_{0,0}+\tilde{\tau}_0+\hat{h}_s)\left(\frac{\partial\hat{\Psi}}{\partial\hat{\tau}}+\frac{\partial^2\hat{\Psi}}{\partial\hat{\tau}^2}(\tilde{\tau}_0+\hat{h}_s)+\frac{1}{2!}\frac{\partial^3\hat{\Psi}}{\partial\hat{\tau}^3}(\tilde{\tau}_0+\hat{h}_s)^2+\cdots\right)$$

$$-\epsilon_2\bar{\sigma}\hat{h}_s'\left(\frac{\partial\hat{\Psi}}{\partial\bar{\sigma}}+\frac{\partial^2\hat{\Psi}}{\partial\bar{\sigma}\partial\hat{\tau}}(\tilde{\tau}_0+\hat{h}_s)+\frac{1}{2!}\frac{\partial^3\hat{\Psi}}{\partial\bar{\sigma}\partial\hat{\tau}^2}(\tilde{\tau}_0+\hat{h}_s)^2+\cdots\right)$$

$$+\frac{2(\lambda_0-1)}{\epsilon_2\lambda_0^2}\bar{\sigma}(\hat{\tau}_{0,0}+\tilde{\tau}_0+\hat{h}_s)(1-\epsilon_2\hat{h}_s')=0. \quad (6.47)$$

3. Leading-Order Asymptotic Expansion Solutions of the Flow Field

To obtain a uniformly valid expansion solution for the flow field, one needs to derive the outer expansion solution and the inner expansion solution, then match these solutions. In what follows, we shall not give the detailed derivations, but only the results.

3.1 Zeroth-Order Outer Solution of the Velocity Field

The leading-order approximation of the flow field is $O(\bar{\nu}_0(\epsilon_2))$, which is just the Oseen model solution and subject to the following system:

$$\begin{aligned}\frac{\mathrm{d}^2\bar{B}_{0,n,0}}{\mathrm{d}\bar{\tau}^2}&=\frac{n+2}{\bar{\tau}}\bar{B}_{0,n+1,0}-\left[1+\frac{2(n+1)}{\bar{\tau}}\right]\bar{A}_{0,n,0}\\&\quad+\frac{1}{\bar{\tau}}\Big[(n+2)\bar{A}_{0,n+1,0}+n\bar{A}_{0,n-1,0}\Big],\\\frac{\mathrm{d}^2\bar{A}_{0,n,0}}{\mathrm{d}\bar{\tau}^2}&+\frac{\mathrm{d}\bar{A}_{0,n,0}}{\mathrm{d}\bar{\tau}}-\frac{n+1}{\bar{\tau}}\bar{A}_{0,n,0}=0\\&(n=0,1,2,\ldots.)\end{aligned} \quad (6.48)$$

This system is entirely the same as we studied in the last chapter. Therefore, we can write the general solution of the stream function as

$$\bar{\nu}_0(\epsilon_2)\bar{\Psi}_0=\bar{\nu}_{0,0}(\epsilon_2)\psi_0(\bar{\sigma},\bar{\tau})+\bar{\nu}_0(\epsilon_2)\bar{\sigma}\sum_{n=0}^{\infty}\bar{b}_{0,n,0}\mathcal{F}_n(\bar{\tau})L_n^{(1)}(\bar{\sigma}), \quad (6.49)$$

while we write the general solution of the vorticity function as

$$\bar{\nu}_0(\epsilon_2)\bar{\zeta}_0=\bar{\nu}_0(\epsilon_2)\bar{\sigma}\left\{\sum_{n=0}^{\infty}\bar{a}_{0,n,0}\mathcal{F}_n(\bar{\tau})L_n^{(1)}(\bar{\sigma})\right\}, \quad (6.50)$$

where

$$\psi_0 = \bar{\sigma} \sum_{n=0}^{\infty} L_n^{(1)}(\bar{\sigma}) \sum_{k=0}^{\infty} \mathcal{C}_{0,k} \mathcal{A}_{n,k}(\bar{\tau}), \tag{6.51}$$

$\mathcal{C}_{0,k}$ and $\bar{a}_{0,n,0}$ are arbitrary constants and

$$\bar{b}_{0,n,0} = -\bar{a}_{0,n,0} + n\bar{a}_{0,n-1,0}. \tag{6.52}$$

The outer solution then has the form

$$\begin{aligned} \bar{\nu}_0(\epsilon_2)\bar{\Psi}_0 &= \bar{\nu}_{0,0}(\epsilon_2)\bar{\sigma} \sum_{n=0}^{N} L_n^{(1)}(\bar{\sigma}) \sum_{k=0}^{N} \mathcal{C}_{0,k} \mathcal{A}_{n,k} \\ &+ \bar{\nu}_0(\epsilon_2)\bar{\sigma} \sum_{n=0}^{\infty} \bar{b}_{0,n,0} \mathcal{F}_n(\bar{\tau}) L_n^{(1)}(\bar{\sigma}). \end{aligned} \tag{6.53}$$

This solution involves two sequences of unknowns, $\{\bar{b}_{0,n,0}\ \mathcal{C}_{0,k}(n, k = 0, 1, 2, \ldots)\}$ to be determined by matching conditions with the inner solution. On the other hand, we recall that, as indicated in Appendix (B) of Chap. 4, the system also allows the following special solutions for stream function:

$$\Psi_{*m}(\bar{\sigma}, \bar{\tau}) = \bar{\sigma} \sum_{n=0}^{\infty} \omega_{*,n}^{(m)}(\bar{\tau}) L_n^{(1)}(\bar{\sigma}), \quad (m = 0, 1, 2, \ldots), \tag{6.54}$$

where $\omega_{*,n}^{(m)}(\bar{\tau})$ are some determined functions, such that as $\bar{\tau} \to \infty$, $\psi_{*m}(\bar{\sigma}, \bar{\tau}) \to 0$, whereas as $\bar{\tau} \to 0$,

$$\bar{\Psi}_{*,m}(\bar{\sigma}, \bar{\tau}) \sim \sigma L_m^{(1)}(\bar{\sigma}) + o(1); \quad \bar{\zeta}_{*,m}(\bar{\sigma}, \bar{\tau}) \sim o(1). \tag{6.55}$$

These solutions play important roles in the matching procedure.

3.2 First Sequence of Inner Solutions of the Velocity Field

From the inhomogeneous boundary conditions (6.46) and (6.47), we can derive the first asymptotic factor,

$$\hat{\nu}_0(\epsilon_2) = \frac{1}{\epsilon_2^2}. \tag{6.56}$$

From the system of inner equations, one can obtain the solutions $\{\hat{A}_{0,n,0}, \hat{B}_{0,n,0}\}$, hence derive the following first sequence of inner solutions:

$$\epsilon_2 \hat{\nu}_0(\epsilon_2)\hat{\Psi}_0(\bar{\sigma}, \hat{\tau}) = -\frac{1}{\epsilon_2} \frac{2(\lambda_0 - 1)}{\lambda_0^2} \bar{\sigma}\hat{\tau} - \bar{\sigma}\Big\{2\hat{b}_{0,0,0}[\hat{\tau}(\ln \hat{\tau} - \ln \hat{\tau}_{0,0})$$

$$-(\hat{\tau}-\hat{\tau}_{0,0})]+\hat{a}_{0,0,0}(\hat{\tau}-\hat{\tau}_{0,0})^2\Big\}+\cdots \tag{6.57}$$

$$\hat{\nu}_0(\epsilon_2)\hat{\zeta}_0(\bar{\sigma},\hat{\tau}) = \frac{1}{\epsilon_2^2}\Big[\hat{a}_{0,0,0}\hat{\tau}+\hat{b}_{0,0,0}\Big]+\cdots, \tag{6.58}$$

in which $(\hat{a}_{0,0,0}, \hat{b}_{0,0,0})$ are arbitrary constants to be determined.

3.3 Second Sequence of Inner Solutions of the Velocity Field

With the factor $\hat{\nu}_1(\epsilon_2) \neq O(\epsilon_2^k)(k=1,2,\ldots)$, one can also from the inner system solve $\{\hat{A}_{1,n,0}, \hat{B}_{1,n,0}\}$, hence obtain the following second sequence of inner solutions of the velocity field:

$$\begin{aligned}\epsilon_2\hat{\nu}_1(\epsilon_2)\hat{\Psi}_1(\bar{\sigma},\hat{\tau},\epsilon_2) = {}& \epsilon_2\hat{\nu}_1(\epsilon_2)\,\bar{\sigma}\sum_{n=0}^{\infty}\Big[\tilde{b}_{1,n,0}(\hat{\tau}\ln\hat{\tau}-\hat{\tau}(1+\ln\hat{\tau}_{0,0})\\ &+\hat{\tau}_{0,0})+\tilde{a}_{1,n,0}(\hat{\tau}-\hat{\tau}_{0,0})^2\Big]L_n^{(1)}(\bar{\sigma})\\ &+O(\epsilon_2\hat{\nu}_1(\epsilon_2)),\end{aligned} \tag{6.59}$$

$$\begin{aligned}\hat{\nu}_1(\epsilon_2)\ \hat{\zeta}_1(\bar{\sigma},\hat{\tau},\epsilon_2) &= \hat{\nu}_1(\epsilon_2)\ \bar{\sigma}\sum_{n=0}^{\infty}[\hat{a}_{1,n,0}\hat{\tau}+\hat{b}_{1,n,0}]L_n^{(1)}(\bar{\sigma})\\ &\quad+O(\epsilon_2\hat{\nu}_1(\epsilon_2)),\end{aligned}$$

where

$$\begin{cases}\tilde{b}_{1,n,0} = -2(n+1)\hat{b}_{1,n,0}+(n+2)\hat{b}_{1,n+1,0}+n\hat{b}_{1,n-1,0},\\ \tilde{a}_{1,n,0} = -2(n+1)\hat{a}_{1,n,0}+(n+1)\,\hat{a}_{1,n+1,0}+n\hat{a}_{1,n-1,0},\end{cases} \tag{6.60}$$

and $(\hat{a}_{1,n,0}, \hat{b}_{1,n,0})$ are arbitrary constants. The inner solutions obtained satisfy all the interface conditions, but do not satisfy the far-field conditions.

3.4 Matching Conditions for Leading-Order Solutions of the Flow Field

We now turn to matching the outer solutions

$$\begin{aligned}\bar{\nu}_0(\epsilon_2)\bar{\Psi}_0 &= \bar{\nu}_{0,0}(\epsilon_2)\gamma_{0,0}\Psi_{*0}(\bar{\sigma},\bar{\tau})\\ &\quad+\bar{\nu}_0(\epsilon_2)\bar{\sigma}\sum_{n=0}^{\infty}\bar{b}_{0,n,0}\mathcal{F}_n(\bar{\tau})L_n^{(1)}(\bar{\sigma}),\end{aligned} \tag{6.61}$$

$$\bar{\nu}_0(\epsilon_2)\bar{\zeta}_0 = \bar{\nu}_{0,0}(\epsilon_2)\gamma_{0,0}\zeta_{*0}(\bar{\sigma},\bar{\tau})$$

$$+\bar{\nu}_0(\epsilon_2)\bar{\sigma}\left\{\sum_{n=0}^{\infty}\bar{a}_{0,n,0}\mathcal{F}_n(\bar{\tau})L_n^{(1)}(\bar{\sigma})\right\}, \tag{6.62}$$

with the inner solutions

$$\begin{aligned}\hat{\Psi}(\bar{\sigma},\hat{\tau},\epsilon_2) &= -\frac{1}{\epsilon_2}\frac{2(\lambda_0-1)}{\lambda_0^2}\bar{\sigma}\hat{\tau}+\epsilon_2\hat{\nu}_1(\epsilon_2)\bar{\sigma}\\ &\times\sum_{n=0}^{\infty}\left\{\tilde{b}_{1,n,0}\left[\hat{\tau}\ln\hat{\tau}-\hat{\tau}(1+\ln\hat{\tau}_{0,0})+\hat{\tau}_{0,0}\right]\right.\\ &\left.+\tilde{a}_{1,n,0}(\hat{\tau}-\hat{\tau}_{0,0})^2\right\}L_n^{(1)}(\bar{\sigma})\\ &-\bar{\sigma}\left\{2\hat{b}_{0,0,0}[\hat{\tau}(\ln\hat{\tau}-\ln\hat{\tau}_{0,0})-(\hat{\tau}-\hat{\tau}_{0,0})]\right.\\ &\left.+\hat{a}_{0,0,0}(\hat{\tau}-\hat{\tau}_{0,0})^2\right\}+\cdots,\end{aligned} \tag{6.63}$$

$$\begin{aligned}\hat{\zeta}(\bar{\sigma},\hat{\tau},\epsilon_2) &= \frac{\bar{\sigma}}{\epsilon_2^2}(\hat{a}_{0,0,0}\hat{\tau}+\hat{b}_{0,0,0})\\ &+\hat{\nu}_1(\epsilon_2)\;\bar{\sigma}\sum_{n=0}^{\infty}(\hat{a}_{1,n,0}\hat{\tau}+\hat{b}_{1,n,0})L_n^{(1)}(\bar{\sigma})+\cdots.\end{aligned} \tag{6.64}$$

For this purpose, one may first rewrite the outer solution (6.61–6.62) with inner variables $\{\bar{\sigma},\bar{\tau}\}=\{\bar{\sigma},\ \epsilon_2\hat{\tau}\}$. Namely,

$$\begin{aligned}\bar{\Psi}(\bar{\sigma},\epsilon_2\hat{\tau},\epsilon_2) &= \bar{\nu}_{0,0}(\epsilon_2)\sigma\gamma_{0,0}+\bar{\nu}_0(\epsilon_2)\bar{\sigma}\Big]\\ &\times\left\{\bar{b}_{0,0,0}\Big[1+\epsilon_2\ln\epsilon_2\hat{\tau}+\epsilon_2(\gamma_0-1)\hat{\tau}+\epsilon_2\hat{\tau}\ln\hat{\tau}\Big]\right.\\ &\left.\sum_{n=1}^{\infty}\frac{\bar{b}_{0,n,0}}{(n+1)!}L_n^{(1)}(\bar{\sigma})\right\}+O(\epsilon_2\nu_0(\epsilon_2))\cdots,\end{aligned} \tag{6.65}$$

$$\begin{aligned}\bar{\zeta}(\bar{\sigma},\epsilon_2\hat{\tau},\epsilon_2) &= \bar{\nu}_0(\epsilon_2)\;\sigma\sum_{n=0}^{\infty}\bar{a}_{0,n,0}\mathcal{F}_n(\epsilon_2\hat{\tau})L_n^{(1)}(\sigma)+\cdots\\ &= \bar{\nu}_0(\epsilon_2)\;\bar{\sigma}\sum_{n=0}^{\infty}\frac{\bar{a}_{0,n,0}}{(n+1)!}L_n^{(1)}(\bar{\sigma})+O(\epsilon_2\nu_0(\epsilon_2)).\end{aligned} \tag{6.66}$$

We now compare these terms in the inner and outer solutions that have the same orders of magnitudes. Thus, one can determine the arbitrary constants involved in the inner and outer solution. We first match the solutions of stream functions (6.65) and (6.63).

(1) By matching the terms of $\{\bar{\sigma}\hat{\tau}\}$, we obtain

$$\begin{aligned} \bar{\nu}_0(\epsilon_2) &= \frac{1}{\epsilon_2^2 \ln \epsilon_2}, \\ \bar{b}_{0,0,0} &= -\frac{2(\lambda_0 - 1)}{\lambda_0^2}. \end{aligned} \tag{6.67}$$

(2) By matching the terms of $\{\bar{\sigma}\}$, we obtain

$$\begin{aligned} &\bar{\nu}_{0,0}(\epsilon_2) = \bar{\nu}_0(\epsilon_2) = \frac{1}{\epsilon_2^2 \ln \epsilon_2}, \\ &\gamma_{0,0} + \bar{b}_{0,0,0} = 0, \\ &\bar{b}_{0,n,0} = 0 \qquad (n = 1, 2, \ldots). \end{aligned} \tag{6.68}$$

From the above, we derive

$$\gamma_{0,0} = \frac{2(\lambda_0 - 1)}{\lambda_0^2}. \tag{6.69}$$

Furthermore, from (6.52), we derive that

$$\begin{aligned} \bar{a}_{0,0,0} &= -\bar{b}_{0,0,0} = \tfrac{2(\lambda_0 - 1)}{\lambda_0^2}, \\ \bar{a}_{0,n,0} &= n! \bar{a}_{0,0,0} = \tfrac{2(\lambda_0 - 1)}{\lambda_0^2} n!\,. \end{aligned} \tag{6.70}$$

(3). By matching the solutions of vorticity functions (6.66) and (6.64), it is found that

$$\hat{a}_{0,0,0} = \hat{b}_{0,0,0,} = 0, \tag{6.71}$$

and

$$\begin{aligned} &\hat{\nu}_1(\epsilon_2) = \bar{\nu}_0(\epsilon_2) = \tfrac{1}{\epsilon_2^2 \ln \epsilon_2}, \\ &\hat{a}_{1,n,0} = 0, \\ &\hat{b}_{1,n,0} = \tfrac{\bar{a}_{0,n,0}}{(n+1)!} = \tfrac{2(\lambda_0 - 1)}{\lambda_0^2} \tfrac{1}{n+1}. \end{aligned} \tag{6.72}$$

From (6.60), we derive

$$\begin{aligned} &\tilde{b}_{1,0,0} = -\tfrac{2(\lambda_0 - 1)}{\lambda_0^2}, \\ &\tilde{b}_{1,n,0} = 0, \qquad (n = 1, 2, \ldots), \\ &\tilde{a}_{1,n,0} = 0, \qquad (n = 0, 1, 2, \ldots). \end{aligned} \tag{6.73}$$

Now, we match the higher-order solutions of stream functions.

(4). Note that the term of the inner solution

$$\epsilon_2 \hat{\nu}_1(\epsilon_2)\tilde{b}_{0,0,0}\{\bar{\sigma}\hat{\tau}\ln\hat{\tau}\} = \frac{1}{\epsilon_2 \ln \epsilon_2}\tilde{b}_{0,0,0}\{\bar{\sigma}\hat{\tau}\ln\hat{\tau}\}$$

automatically matches with the term of the outer solution

$$\frac{1}{\epsilon_2 \ln \epsilon_2}\bar{b}_{0,0,0}\{\bar{\sigma}\hat{\tau}\ln\hat{\tau}\}.$$

(5) To match the term of the inner solution

$$\epsilon_2 \hat{\nu}_1(\epsilon_2)\tilde{b}_{0,0,0}\hat{\tau}_{0,0}\bar{\sigma} = -\frac{1}{\epsilon_2 \ln \epsilon_2}\frac{2(\lambda_0 - 1)}{\lambda_0^2}\hat{\tau}_{0,0}\bar{\sigma},$$

we may introduce the higher-order outer solution

$$\bar{\nu}_{1,0}(\epsilon_2)\gamma_{1,0}\Psi_{*0}(\bar{\sigma},\bar{\tau}) = -\frac{1}{\epsilon_2 \ln \epsilon_2}\frac{2(\lambda_0 - 1)}{\lambda_0^2}\hat{\tau}_{0,0}\Psi_{*0}(\bar{\sigma},\bar{\tau}),$$

for which, as $\bar{\tau} \to 0$, we have

$$\bar{\nu}_{1,0}(\epsilon_2)\Psi_{*0}(\bar{\sigma},\bar{\tau}) \sim -\frac{1}{\epsilon_2 \ln \epsilon_2}\frac{2(\lambda_0 - 1)}{\lambda_0^2}\hat{\tau}_{0,0}\bar{\sigma}.$$

Up to this point, the remaining unbalanced terms in the inner solution $\hat{\Psi}$ are

$$-2(1 + \ln\hat{\tau}_{0,0})\frac{1}{\epsilon_2 \ln \epsilon_2}\frac{\lambda_0 - 1}{\lambda_0^2}\bar{\sigma}\hat{\tau} + \cdots \tag{6.74}$$

while the remaining unbalanced terms in outer solution $\bar{\Psi}$ are

$$-2(\gamma_0 - 1)\frac{1}{\epsilon_2 \ln \epsilon_2}\frac{\lambda_0 - 1}{\lambda_0^2}\bar{\sigma}\hat{\tau} + \cdots. \tag{6.75}$$

These terms are to be balanced with the higher-order outer solutions

$$\epsilon_2 \bar{\nu}_0(\epsilon_2)\bar{\Psi}_{0,1}(\bar{\sigma},\bar{\tau}) = \epsilon_2 \bar{\nu}_0(\epsilon_2)\bar{\sigma}\sum_{n=0}^{\infty}\bar{B}_{0,n,1}(\bar{\sigma},\bar{\tau})L_n^{(1)}(\bar{\sigma}), \tag{6.76}$$

where

$$\begin{cases} \epsilon_2 \bar{\nu}_0(\epsilon_2) & = \dfrac{1}{\epsilon_2 \ln \epsilon_2}, \\ \bar{B}_{0,0,1}(\bar{\sigma},\bar{\tau}) & = \bar{b}_{0,0,1}\mathcal{F}_0(\bar{\tau}) \\ & = \left[2(2 - \gamma_0 + \ln\hat{\tau}_{0,0})\right]\left(\frac{\lambda_0 - 1}{\lambda_0^2}\right)\mathcal{F}_0(\bar{\tau}), \\ \bar{B}_{0,n,1}(\bar{\sigma},\bar{\tau}) & = \bar{b}_{0,n,1}\mathcal{F}_n(\bar{\tau}) = 0\,. \end{cases} \tag{6.77}$$

Therefore, the asymptotic factors are found as

$$\bar{\nu}_{0,0}(\epsilon_2) = \bar{\nu}_0(\epsilon_2) = \hat{\nu}_1(\epsilon_2) = \frac{1}{\epsilon_2^2 \ln \epsilon_2}. \tag{6.78}$$

The outer solutions of the total stream function and the vorticity function are found to be

$$\begin{aligned}\Psi_{\text{outer}}(\bar{\sigma}, \bar{\tau}, \epsilon_2) = {} & \frac{2}{\epsilon_2^2 \lambda_0} \bar{\sigma}\bar{\tau} + \frac{2}{\epsilon_2^2 \ln \epsilon_2} \left(\frac{\lambda_0 - 1}{\lambda_0^2} \right) \bar{\sigma} \Big[\Psi_{*0}(\bar{\sigma}, \bar{\tau}) - \mathcal{F}_0(\bar{\tau}) \Big] \\ & + \frac{1}{\epsilon_2 \ln \epsilon_2} \left(\frac{\lambda_0 - 1}{\lambda_0^2} \right) \bar{\sigma} \Big[2(2 - \gamma_0 + \ln \hat{\tau}_{0,0}) \mathcal{F}_0(\bar{\tau}) - \Psi_{*0}(\bar{\sigma}, \bar{\tau}) \Big] \\ & + \cdots, \end{aligned} \tag{6.79}$$

$$\begin{aligned}\zeta_{\text{outer}}(\bar{\sigma}, \bar{\tau}, \epsilon_2) & = \frac{2}{\epsilon_2^2 \ln \epsilon_2} \left(\frac{\lambda_0 - 1}{\lambda_0^2} \right) \bar{\sigma} \sum_{n=0}^{\infty} n! \mathcal{F}_n(\bar{\tau}) L_n^{(1)}(\bar{\sigma}) + \cdots \\ & = \frac{2}{\epsilon_2^2 \ln \epsilon_2} \left(\frac{\lambda_0 - 1}{\lambda_0^2} \right) \frac{\sigma}{\sigma + \tau} \mathrm{e}^{-\tau} + \cdots . \end{aligned} \tag{6.80}$$

Accordingly, the inner solutions are found to be

$$\begin{aligned}\Psi_{\text{inner}}(\bar{\sigma}, \hat{\tau}, \epsilon_2) & = \frac{2}{\epsilon_2 \lambda_0^2} \bar{\sigma}\hat{\tau} - \frac{2}{\epsilon_2 \ln \epsilon_2} \left(\frac{\lambda_0 - 1}{\lambda_0^2} \right) \bar{\sigma} \\ & \quad \times \Big[\hat{\tau} \ln \hat{\tau} - \hat{\tau}(1 + \ln \hat{\tau}_{0,0}) + \hat{\tau}_{0,0} \Big] + \cdots , \end{aligned} \tag{6.81}$$

$$\begin{aligned}\zeta_{\text{inner}}(\bar{\sigma}, \hat{\tau}, \epsilon_2) & = \frac{2}{\epsilon_2^2 \ln \epsilon_2} \left(\frac{\lambda_0 - 1}{\lambda_0^2} \right) \bar{\sigma} \sum_{n=0}^{\infty} \frac{1}{n+1} L_n^{(1)}(\bar{\sigma}) + \cdots \\ & = \frac{2}{\epsilon_2^2 \ln \epsilon_2} \left(\frac{\lambda_0 - 1}{\lambda_0^2} \right) + \cdots . \end{aligned} \tag{6.82}$$

The inner solution of the perturbed stream function is

$$\begin{aligned}\hat{\Psi}(\bar{\sigma}, \hat{\tau}, \epsilon_2) & = -\frac{2}{\epsilon_2} \frac{\lambda_0 - 1}{\lambda_0^2} \bar{\sigma}\hat{\tau} - \frac{2}{\epsilon_2 \ln \epsilon_2} \left(\frac{\lambda_0 - 1}{\lambda_0^2} \right) \bar{\sigma} \\ & \quad \times \Big[\hat{\tau} \ln \hat{\tau} - \hat{\tau}(1 + \ln \hat{\tau}_{0,0}) + \hat{\tau}_{0,0} \Big] + \cdots . \end{aligned} \tag{6.83}$$

4. Asymptotic Expansion Solution of the Temperature Field

We now turn to finding the solutions for the temperature field. As we have pointed out before, this is a singular perturbation problem. Hence, we need to use the inner variables: $(\bar{\sigma}, \hat{\tau})$, and write the solution in the form:

$$T = T_\infty + \Pr \hat{T}(\bar{\sigma}, \hat{\tau}, \epsilon_2), \tag{6.84}$$

where

$$\hat{T}(\bar{\sigma}, \hat{\tau}, \epsilon_2) = \bar{T}(\bar{\sigma}, \epsilon_2 \hat{\tau}, \epsilon_2). \tag{6.85}$$

On the other hand, with the inner variables, as the interface shape has the form: $\hat{\tau} = \hat{\tau}_s(\bar{\sigma}) = \hat{\tau}_{0,0} + \tilde{\tau}_0 + \hat{h}_s(\bar{\sigma})$, the interface conditions for the temperature field can be written as follows: at $\hat{\tau} = \hat{\tau}_{0,0}$,

$$\epsilon_2 T_\infty + \left(\hat{T} + \frac{\partial \hat{T}}{\partial \hat{\tau}} (\tilde{\tau}_0 + \hat{h}_s) + \frac{1}{2!} \frac{\partial^2 \hat{T}}{\partial \hat{\tau}^2} (\tilde{\tau}_0 + \hat{h}_s)^2 + \cdots \right) = 0, \tag{6.86}$$

$$\begin{aligned} &(\hat{\tau}_{0,0} + \tilde{\tau}_0 + \hat{h}_s) \left(\frac{\partial \hat{T}}{\partial \hat{\tau}} + \frac{\partial^2 \hat{T}}{\partial \hat{\tau}^2} (\tilde{\tau}_0 + \hat{h}_s) + \frac{1}{2!} \frac{\partial^3 \hat{T}}{\partial \hat{\tau}^3} (\tilde{\tau}_0 + \hat{h}_s)^2 + \cdots \right) \\ &- \epsilon_2 \bar{\sigma} \hat{h}_s' \left(\frac{\partial \hat{T}}{\partial \bar{\sigma}} + \frac{\partial^2 \hat{T}}{\partial \bar{\sigma} \partial \hat{\tau}} (\tilde{\tau}_0 + \hat{h}_s) + \frac{1}{2!} \frac{\partial^3 \hat{T}}{\partial \bar{\sigma} \partial \hat{\tau}^2} (\tilde{\tau}_0 + \hat{h}_s)^2 + \cdots \right) \\ &+ \frac{\epsilon_2}{\lambda_0} (\hat{\tau}_{0,0} + \tilde{\tau}_0 + \hat{h}_s + \bar{\sigma} \hat{h}_s') = 0. \end{aligned} \tag{6.87}$$

We now make the Laguerre series expansion

$$\hat{T}(\bar{\sigma}, \hat{\tau}) = \sum_{n=0}^{\infty} \hat{D}_n(\hat{\tau}, \epsilon_2) L_n^{(1)}(\bar{\sigma}) . \tag{6.88}$$

Then, from (5.45), we derive

$$\begin{aligned} \hat{\tau} \frac{\mathrm{d}^2 \hat{D}_n}{\mathrm{d}\hat{\tau}^2} + (1 + \hat{\tau}) \frac{\mathrm{d}\hat{D}_n}{\mathrm{d}\hat{\tau}} - n\hat{D}_n &= \Big[\epsilon_2(n+1) - (n+2) \Big] \hat{D}_{n+1} \\ &- \epsilon_2 \sum_{m=n+2}^{\infty} \hat{D}_n + \frac{\epsilon_2 \lambda_0}{2} \hat{\mathcal{N}}_T \Big\{ \hat{D}_n, \hat{B}_n \Big\}, \end{aligned} \tag{6.89}$$

where $\hat{\mathcal{N}}_T \Big\{ \hat{D}_n, \hat{B}_n \Big\} = \dfrac{\partial(\hat{T}, \hat{\Psi})}{\partial(\bar{\sigma}, \hat{\tau})}$.

It is important to notice that the variation of the temperature is restricted in the boundary layer temperature. The boundary layer of temperature and the boundary layer of flow have the same thickness of $O(\epsilon_2)$. Thus, to determine the temperature field, one only needs to use the inner solution of the flow field.

We further make the following asymptotic expansion for the solution of the temperature field:

$$\hat{T} = \hat{\gamma}_0(\epsilon_2)\hat{T}_0 + \hat{\gamma}_1(\epsilon_2)\hat{T}_1 + \hat{\gamma}_1(\epsilon_2)\hat{T}_2 + \cdots, \tag{6.90}$$

in the limit $\epsilon_2 \to 0$, where $\hat{\gamma}_0 \gg \hat{\gamma}_1 \gg \hat{\gamma}_2 \gg \cdots$. With the factor $\hat{\gamma}_m(\epsilon_2)$, the m-th sequence of solutions $\hat{T}_m$ has the asymptotic structure

$$\begin{aligned}\hat{T}_m &= \Big[\hat{D}_{m,0,0} + \epsilon_2\hat{D}_{m,0,1} + \epsilon_2^2\hat{D}_{m,0,2} + \cdots\Big]L_0^{(1)}(\hat{\sigma}) \\ &+ \Big[\hat{D}_{m,1,0} + \epsilon_2\hat{D}_{m,1,1} + \epsilon_2^2\hat{D}_{m,1,2} + \cdots\Big]L_1^{(1)}(\hat{\sigma}) \\ &+ \cdots .\end{aligned} \tag{6.91}$$

In the last section, we derived the first sequence of inner solutions of the stream function

$$\epsilon_2\hat{\nu}_0(\epsilon_2)\hat{\Psi}_0(\bar{\sigma},\hat{\tau},\epsilon_2) = -\frac{2}{\epsilon_2}\frac{\lambda_0 - 1}{\lambda_0^2}\bar{\sigma}\hat{\tau} + \cdots.$$

So, it follows that the equation of the $(m+1)$-th sequence of solutions, $\{\hat{\gamma}_m(\epsilon_2)\hat{T}_m\}$ will always involve the interaction term

$$\frac{\epsilon_2\lambda_0}{2}\hat{\mathcal{N}}_T\Big\{\gamma_m\hat{T}_m, \epsilon_2\hat{\nu}_0\hat{\Psi}_0\Big\} = O\Big(\hat{\gamma}_m(\epsilon_2)\hat{T}_m\Big).$$

4.1 First Sequence of Solutions of the Temperature Field

For the first sequence of solutions $\hat{\gamma}_0(\epsilon_2)\hat{T}_0$, the leading factor $\hat{\gamma}_0(\epsilon_2)$ is only related to the first sequence of inner solutions of the flow field $\epsilon_2\hat{\nu}_0(\epsilon_2)\hat{\Psi}_0$, which only involves the component $L_0^{(1)}(\bar{\sigma})$. Thus, it can be assumed that $\hat{\gamma}_0(\epsilon_2)\hat{T}_0$ is a similar solution, only depending on the variable $\hat{\tau}$. Namely, we have

$$\hat{D}_{0,n,0} = 0 \quad \text{for} \quad n = 1, 2, \ldots, \tag{6.92}$$

and

$$\hat{T}_0 = \Big[\hat{D}_{0,0,0}(\bar{\tau}) + \epsilon_2\hat{D}_{0,0,1}(\bar{\tau}) + \cdots\Big]L_0^{(1)}(\bar{\sigma}). \tag{6.93}$$

The governing equation of $\hat{D}_{0,0,0}(\bar{\tau})$ is:

$$\frac{\mathrm{d}^2\hat{D}_{0,0,0}}{\mathrm{d}\hat{\tau}^2} + \left(1+\frac{1}{\hat{\tau}}\right)\frac{\mathrm{d}\hat{D}_{0,0,0}}{\mathrm{d}\hat{\tau}} = -\left(\frac{\lambda_0 - 1}{\lambda_0}\right)\frac{\mathrm{d}\hat{D}_{0,0,0}}{\mathrm{d}\hat{\tau}} \tag{6.94}$$

or

$$\frac{\mathrm{d}^2\hat{D}_{0,0,0}}{\mathrm{d}\hat{\tau}^2} + \left(\hat{\lambda}_0+\frac{1}{\hat{\tau}}\right)\frac{\mathrm{d}\hat{D}_{0,0,0}}{\mathrm{d}\hat{\tau}} = 0, \tag{6.95}$$

where we have defined

$$\hat{\lambda}_0 = \frac{2\lambda_0 - 1}{\lambda_0} > 0. \tag{6.96}$$

From the leading-order approximation of the interface condition (6.86), we derive

$$\hat{\gamma}_0(\epsilon_2) = \epsilon_2, \tag{6.97}$$

and at $\hat{\tau} = \hat{\tau}_{0,0}$,

$$\hat{D}_{0,0,0}(\hat{\tau}_{0,0}) + T_\infty = 0, \tag{6.98}$$

$$\hat{D}'_{0,0,0}(\hat{\tau}_{0,0}) + \frac{1}{\lambda_0} = 0\ . \tag{6.99}$$

The solution is obtained as

$$\hat{D}_{0,0,0} = I_* + I_{0,0}E_1(\hat{\lambda}_0\hat{\tau}). \tag{6.100}$$

From the far-field condition,

$$\hat{D}_{0,0,0} \to 0, \qquad \text{as} \quad \hat{\tau} \to \infty, \tag{6.101}$$

we derive

$$I_* = 0. \tag{6.102}$$

Furthermore, from interface condition (6.99), we derive

$$I_{0,0} = \frac{\hat{\tau}_{0,0}}{\lambda_0}\mathrm{e}^{\hat{\lambda}_0\hat{\tau}_{0,0}}, \tag{6.103}$$

so that we have an Ivantsov-like solution as

$$\hat{D}_{0,0,0} = \frac{\hat{\tau}_{0,0}}{\lambda_0}\mathrm{e}^{\hat{\lambda}_0\hat{\tau}_{0,0}}E_1(\hat{\lambda}_0\hat{\tau}). \tag{6.104}$$

Finally, from interface condition (6.98), we derive

$$\begin{aligned} T_\infty &= -\frac{\hat{\tau}_{0,0}}{\lambda_0} e^{\hat{\lambda}_0 \hat{\tau}_{0,0}} E_1(\hat{\lambda}_0 \hat{\tau}_{0,0}) \\ &= -\frac{\eta_{0,0}^2}{2} e^{\frac{(2\lambda_0 - 1)\eta_{0,0}^2}{2}} E_1\Big(\frac{2\lambda_0 - 1}{2}\eta_{0,0}^2\Big). \end{aligned} \tag{6.105}$$

This formula is the generalization of the Ivantsov solution to the case of dendritic growth in external flow with Pr $= \infty$. For the case with no external flow, $\lambda_0 = 1$, we regain the Ivantsov solution:

$$T_\infty = -\frac{\eta_{0,0}^2}{2} e^{\frac{\eta_{0,0}^2}{2}} E_1\Big(\frac{\eta_{0,0}^2}{2}\Big). \tag{6.106}$$

In Fig 6.1, we show the solution $\hat{T}_0(\hat{\tau}) = T_\infty + \hat{D}_{0,0,0}(\tau)$ for the cases $\hat{\tau}_{0,0} = 0.2$ and $\lambda_0 = 1.0, 1.5, 3.0$. In Fig 6.2, we show the variations of undercooling temperature T_∞ with the parameter $\hat{\tau}_{0,0} = 0.2$ for different values of flow parameter, $\lambda_0 = 1.0, 1.25, 1.5, 3.0, 6.0$, while in Fig 6.3, we show the 3D graphics of T_∞ as a function of the variables, τ_0 and λ_0. It is seen that, like a steady dendritic growth without external flow, which can be described the Ivantsov similar solution, the dendritic growth in uniform flow with Pr $= \infty$ can be also described by a similar solution $\hat{T}_0(\hat{\tau})$. The interface of dendrite is also paraboloidal, given by $\eta = 1$. The Peclet number Pe $= \eta_0^2 = \frac{2}{\lambda_0}\hat{\tau}_{0,0}$, however, is now dependent of T_∞ as well as flow parameter λ_0. It is noted that for Ivantsov's solution of dendritic growth without external flow ($\lambda_0 = 1$), the maximum of undercooling temperature must be $|T_\infty| = 1$. Now, with the inclusion of external flow we find that the maximum of undercooling temperature must be $|T_\infty| < 1$.

4.2 Second Sequence of Solutions of the Temperature Field

The second sequence of solutions $\hat{\gamma}_1(\epsilon_2)\hat{T}_1$ is related to the first and second sequence of inner solutions of flow field $\epsilon_2\hat{\nu}_0(\epsilon_2)\hat{\Psi}_0$ and $\epsilon_2\hat{\nu}_1(\epsilon_2)\hat{\Psi}_1$. Since both $\hat{\Psi}_0$ and $\hat{\Psi}_1$ only involve the component $L_0^{(1)}(\hat{\sigma})$, it is implied that $\hat{T}_1$ is also a similar solutions, namely,

$$\hat{D}_{1,n,0} = \hat{h}_{1,n,0} = 0, \qquad \text{for} \quad n = 1, 2, \cdots . \tag{6.107}$$

Hence, we have

$$\hat{T}_1 = \Big[\hat{D}_{1,0,0}(\hat{\tau}) + \epsilon_2 \hat{D}_{1,0,1}(\hat{\tau}) + \cdots\Big] L_0^{(1)}(\bar{\sigma}). \tag{6.108}$$

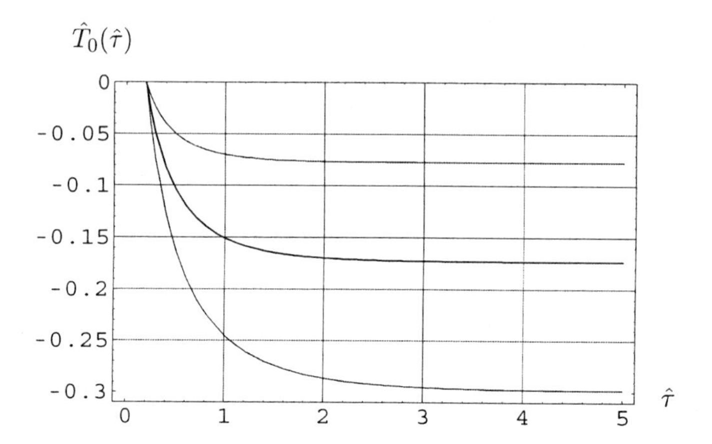

Figure 6.1. The solutions of temperature field for the case: $\hat{\tau}_{0,0} = 0.2$, $\lambda_0 = 1.0$, 1.5, 3.0 from bottom to top.

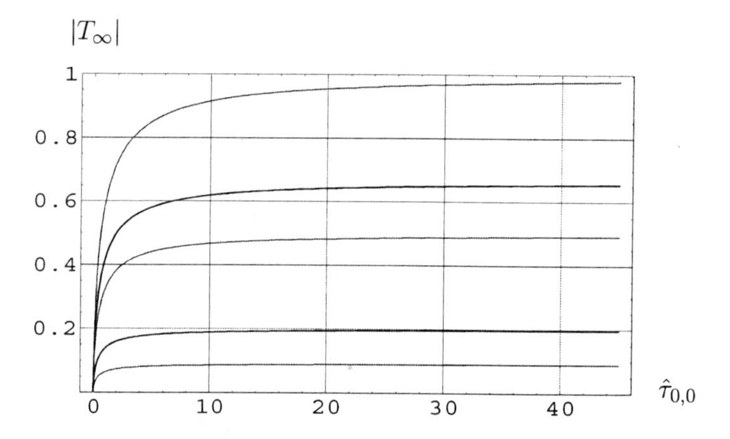

Figure 6.2. The variations of undercooling temperature T_∞ with variable $\hat{\tau}_{0,0}$ for different values of $\lambda_0 = 1.0, 1.25, 1.5, 3.0, 6.0$ from top to bottom.

Furthermore, due to

$$\epsilon_2 \hat{\nu}_1(\epsilon_2) \hat{\Psi}_1 = O\Big(\frac{1}{\epsilon_2 \ln \epsilon_2}\Big),$$

the interaction term has the order of magnitude

$$\frac{\epsilon_2 \lambda_0}{2} \hat{\mathcal{N}}_T \Big\{ \hat{T}_0, \hat{\Psi}_1 \Big\} = O(\epsilon_2^2 \hat{\nu}_1(\epsilon_2)) = O\Big(\frac{\epsilon_2}{\ln \epsilon_2}\Big),$$

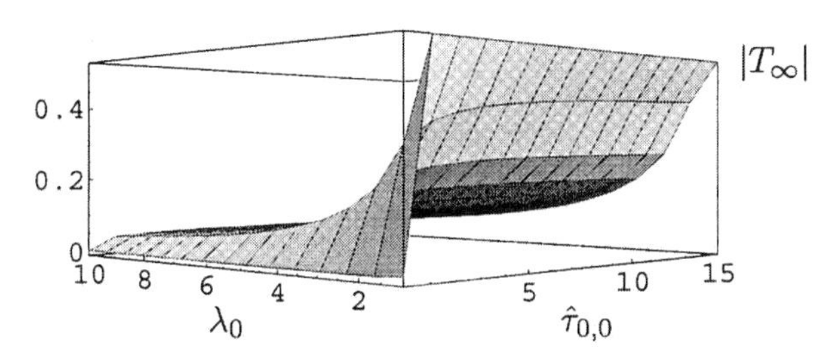

Figure 6.3. The graphics of undercooling temperature T_∞ as a function of $\hat{\tau}_{0,0}$ and λ_0

and we derive that

$$\hat{\gamma}_1(\epsilon_2) = \frac{\epsilon_2}{\ln \epsilon_2}. \tag{6.109}$$

The governing equation for $\hat{D}_{1,0,0}(\bar{\tau})$ is

$$\frac{\mathrm{d}^2\hat{\mathrm{D}}_{1,0,0}}{\mathrm{d}\hat{\tau}^2} + \left(\hat{\lambda}_0 + \frac{1}{\hat{\tau}}\right)\frac{\mathrm{d}\hat{\mathrm{D}}_{1,0,0}}{\mathrm{d}\hat{\tau}} = \hat{H}_{1,0,0}, \tag{6.110}$$

where

$$\hat{H}_{1,0,0} = \left(\frac{\lambda_0 - 1}{\lambda_0}\right)\left[\ln\hat{\tau} - (1 + \ln\hat{\tau}_{0,0}) + \frac{\hat{\tau}_{0,0}}{\hat{\tau}}\right]\hat{D}'_{0,0,0}$$

$$= -\hat{\tau}_{0,0}\left(\frac{\lambda_0 - 1}{\lambda_0^2}\right)\frac{\mathrm{e}^{-\hat{\lambda}_0(\hat{\tau}-\hat{\tau}_{0,0})}}{\hat{\tau}}\left[\ln\hat{\tau} - (1 + \ln\hat{\tau}_{0,0}) + \frac{\hat{\tau}_{0,0}}{\hat{\tau}}\right]. \tag{6.111}$$

The general solution can be expressed in the form

$$\hat{D}_{1,0,0}(\hat{\tau}) = I_{1,0}E_1(\hat{\lambda}_0\hat{\tau}) + \hat{\tau}_{0,0}\widehat{R}_{1,0,0}(\hat{\tau}); \tag{6.112}$$

then we derive that

$$\begin{aligned}\widehat{R}'_{1,0,0}(\hat{\tau}) &= -\left(\frac{\lambda_0 - 1}{\lambda_0^2}\right)\frac{\mathrm{e}^{-\hat{\lambda}_0(\hat{\tau}-\hat{\tau}_{0,0})}}{\hat{\tau}}\int_{\tau_{0,0}}^{\hat{\tau}}\left[\ln\hat{\tau}' - (1 + \ln\hat{\tau}_{0,0}) + \frac{\hat{\tau}_{0,0}}{\hat{\tau}'}\right]\mathrm{d}\hat{\tau}' \\ &= -\left(\frac{\lambda_0 - 1}{\lambda_0^2}\right)\mathrm{e}^{-\hat{\lambda}_0(\hat{\tau}-\hat{\tau}_{0,0})}\left[\left(1 + \frac{\hat{\tau}_{0,0}}{\hat{\tau}}\right)\ln\hat{\tau}\right. \\ &\quad \left. + (2\hat{\tau}_{0,0} - \hat{\tau}_{0,0}\ln\hat{\tau}_{0,0})\frac{1}{\hat{\tau}} - (2 + \ln\hat{\tau}_{0,0})\right],\end{aligned} \tag{6.113}$$

with $\widehat{R}'_{1,0,0}(\hat{\tau}_{0,0}) = 0$, and

$$\widehat{R}_{1,0,0}(\hat{\tau}) = \left(\frac{\lambda_0 - 1}{\lambda_0^2}\right) e^{\hat{\lambda}_0 \hat{\tau}_{0,0}} \left\{ \hat{\tau}_{0,0} E_1^{(2)}(\hat{\lambda}_0 \hat{\tau}) + \frac{e^{-\hat{\lambda}_0 \hat{\tau}}}{\hat{\lambda}_0} (\ln \hat{\tau} - \ln \hat{\tau}_{0,0} - 2) \right.$$
$$\left. + \left[\frac{1}{\hat{\lambda}_0} + \hat{\tau}_{0,0}(\ln \hat{\tau} - \ln \hat{\tau}_{0,0}) + 2\hat{\tau}_{0,0}\right] E_1(\hat{\lambda}_0 \hat{\tau}) \right\}, \tag{6.114}$$

where we defined

$$E_1^{(2)}(x) = \int_x^\infty \frac{E_1(t)}{t} \mathrm{d}t. \tag{6.115}$$

Furthermore, from the interface condition (6.86), we derive

$$\hat{\delta}_0(\epsilon_2) = \hat{\gamma}_1(\epsilon_2)/\hat{\gamma}_0(\epsilon_2) = \frac{1}{\ln \epsilon_2}. \tag{6.116}$$

The interface conditions turn out to be: at $\hat{\tau} = \hat{\tau}_{0,0}$,

$$\begin{aligned} &\hat{D}_{1,0,0}(\hat{\tau}_{0,0}) + \hat{D}'_{0,0,0}\ (\hat{\tau}_{1,0} + \hat{h}_{0,0,0}) = 0, \\ &\hat{\tau}_{0,0}\left[\hat{D}'_{1,0,0}(\hat{\tau}_{0,0}) + \hat{D}''_{0,0,0}\ (\hat{\tau}_{1,0} + \hat{h}_{0,0,0})\right] \\ &\quad + \left(D'_{0,0,0,} + \frac{1}{\lambda_0}\right)\ (\hat{\tau}_{1,0} + \hat{h}_{0,0,0}) = 0. \end{aligned} \tag{6.117}$$

Noting that

$$\hat{D}'_{0,0,0} = -\frac{1}{\lambda_0}, \qquad \hat{D}''_{0,0,0} = \frac{1}{\lambda_0}\left(\hat{\lambda}_0 + \frac{1}{\hat{\tau}_{0,0}}\right), \tag{6.118}$$

one can re-write (6.117) in the form:

$$\begin{aligned} &\hat{D}_{1,0,0}(\hat{\tau}_{0,0}) - \frac{1}{\lambda_0}\ (\hat{\tau}_{1,0} + \hat{h}_{0,0,0}) = 0 \\ &\hat{\tau}_{0,0}\hat{D}'_{1,0,0}(\hat{\tau}_{0,0}) + \frac{1}{\lambda_0}\left(1 + \hat{\lambda}_0\hat{\tau}_{0,0}\right)(\hat{\tau}_{1,0} + \hat{h}_{0,0,0}) = 0. \end{aligned} \tag{6.119}$$

From the above, we derive that $\hat{h}_{0,0,0}$ must be a constant. Moreover, from the normalization condition (6.22), we further derive

$$\hat{h}_{0,0,0} = 0. \tag{6.120}$$

Thus, the two interface conditions (6.119) can be used to uniquely determine two constants, $\{I_{1,0};\ \hat{\tau}_{1,0}\}$. The results are as follows:

$$\begin{aligned} &\frac{\hat{\tau}_{1,0}}{\hat{\tau}_{0,0}} = \mathcal{Z}_e(\hat{\tau}_{0,0}, \lambda_0) = \lambda_0\left[E_1(\hat{\lambda}_0\hat{\tau}_0)\tilde{I}_0 + \widehat{R}_{1,0,0}\right], \\ &I_{1,0} = \hat{\tau}_{0,0}\tilde{I}_0, \end{aligned} \tag{6.121}$$

where we have defined

$$\tilde{I}_0 = \frac{(1+\hat{\lambda}_0\hat{\tau}_{0,0})\widehat{R}_{1,0,0}(\hat{\tau}_{0,0})}{e^{-\hat{\lambda}_0\hat{\tau}_{0,0}} - (1+\hat{\lambda}_0\hat{\tau}_{0,0})E_1(\hat{\lambda}_0\hat{\tau}_{0,0})}. \tag{6.122}$$

We finally obtain the solution of the interface shape function:

$$\hat{\tau} = \hat{\tau}_s = \hat{\tau}_{0,0} + \frac{1}{\ln\epsilon_2}\hat{\tau}_{1,0} + O\,(\text{h.r.t}) \tag{6.123}$$

and

$$\eta_s = \sqrt{\frac{\hat{\tau}_s}{\hat{\tau}_0}} = 1 + O\,(\text{h.r.t})\ . \tag{6.124}$$

The normalization parameter is obtained as

$$\begin{aligned}\eta_0^2 &= \eta_{0,0}^2 - \frac{2}{\lambda_0\ln\text{Pr}}\hat{\tau}_{1,0} + O\,(\text{h.r.t}) \\ &= \eta_{0,0}^2\left[1 - \frac{1}{\ln\text{Pr}}\mathcal{Z}_e(\hat{\tau}_{0,0},\lambda_0) + O\,(\text{h.r.t})\right].\end{aligned} \tag{6.125}$$

Thus, the Peclet number of growth is calculated as

$$\text{Pe} = \frac{\ell_t}{\ell_T} = \eta_{0,0}^2\left[1 - \frac{1}{\ln\text{Pr}}\mathcal{Z}_e(\hat{\tau}_{0,0},\lambda_0) + O\,(\text{h.r.t})\right], \tag{6.126}$$

where the parameter $\eta_{0,0}$ is determined by the undercooling parameter T_∞ and flow parameter λ_0 through (6.105).

In Fig. 6.4, we show the solutions $\hat{T}_1(\hat{\tau}) = \left[I_{10}E_1(\hat{\lambda}_0\hat{\tau}) + \hat{\tau}_{0,0}\widehat{R}_{1,0,0}(\hat{\tau})\right]$ for the cases of $\hat{\tau}_{0,0} = 0.2$ and $\lambda_0 = 1.0, 1.05, 1.2, 1.5, 3.0$.

In Fig. 6.5, we plot $\mathcal{Z}_e$ as a function of $\hat{\tau}_{0,0}$ for fixed $\lambda_0 = 1.0, 2.5, 4.0, 5.5$. It is seen that $\mathcal{Z}_e$ is always negative as $\lambda_0 > 1$; with a fixed λ_0, its absolute value decreases with increasing the parameter $\hat{\tau}_{0,0}$.

5. A Brief Summary

So far, we have studied the steady dendritic growth in external flow with the Laguerre series representation and find uniformly valid asymptotic expansion of the solution in the entire flow field for the case of $\overline{U}_\infty \ll 1; \text{Pr} = O(1)$, as well as the case of $\text{Pr} \gg 1, \overline{U}_\infty = O(1)$. The asymptotic analysis presented here provides an approach to systematically finding each order approximation of the solution for the problem. The solution of the stream function generally contains all the components of the Laguerre functions, $L_n^{(1)}(\bar{\sigma})$ $(n = 1, 2, \ldots)$. However, from the results obtained, we can draw the following conclusions:

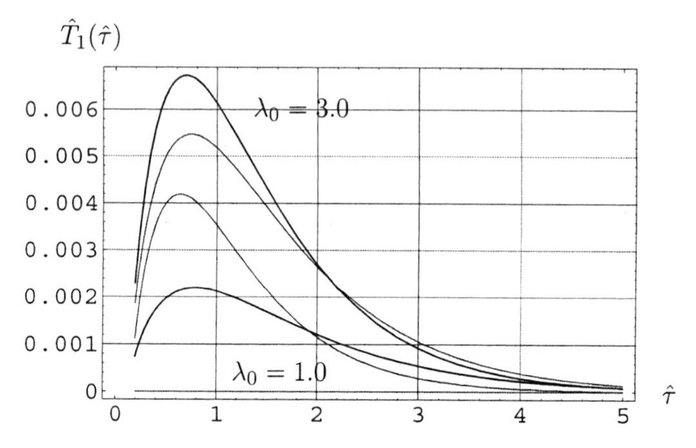

Figure 6.4. The solutions of $T_1(\hat{\tau}) = \left[I_{10}E_1(\hat{\lambda}_0\hat{\tau}) + \hat{\tau}_{0,0}\widehat{R}_{1,0,0}(\hat{\tau})\right]$, for the cases: $\hat{\tau}_{0,0} = 0.2$, $\lambda_0 = 1.0, 1.05, 1.2, 1.5, 3.0$

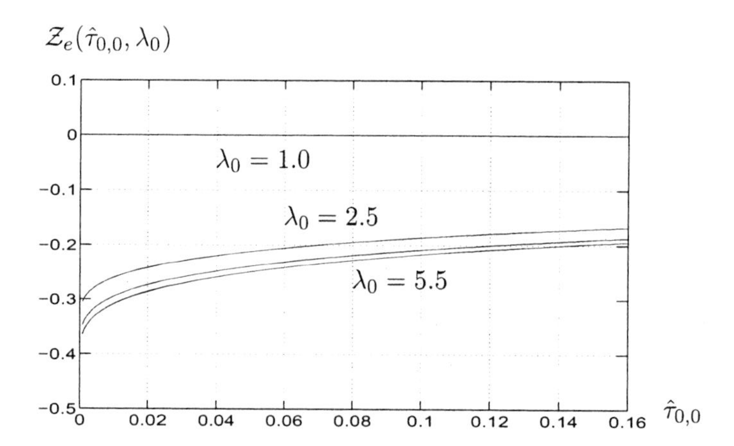

Figure 6.5. The variations of functions $\mathcal{Z}_e(\hat{\tau}_{0,0}, \lambda_0)$ versus $\hat{\tau}_{0,0}$ with $\lambda_0 = 1.0, 2.5, 4.0, 5.5$ from top to bottom

1. For the case of $\overline{U}_\infty \ll 1; \mathrm{Pr} = O(1)$ only the component of $L_0^{(1)}(\bar{\sigma})$ is important, all other components are numerically very small. Therefore, the solution of the stream function is approximately proportional to the variable σ in the whole flow field. In accordance with the flow field, the solution of the temperature field is nearly similar, while the interface shape is nearly paraboloid.

2. For the case of $\mathrm{Pr} \gg 1, \lambda_0 = 1 + \overline{U}_\infty = O(1)$, in the inner region, the solution of the stream function is proportional to the variable $\bar{\sigma}$, as it only contains the basic component of the Laguerre function, $L_0^{(1)}(\bar{\sigma})$. Furthermore, there is a temperature boundary layer near the interface, whose thickness has the same order of magnitude as the inner region of the flow field. The solution of the temperature field, therefore, is only affected by the inner solution of the flow field. In general, due to the presence of the external flow, the temperature field with finite Pr is described by a 'nearly' similar solution and the interface shape of dendrite is 'nearly' paraboloid, described by the shape function $\eta_s \approx 1$. The Peclet number Pe increases from $\eta_{0,0}^2$ with a factor $\left(1 + \frac{1}{\ln \mathrm{Pr}} \mathcal{Z}_e\right)$, as a function of undercooling temperature T_∞, Prandtl number Pr, and flow parameter λ_0.

3. One of the most important results obtained is that for the system with $\mathrm{Pr} = \infty$, dendritic growth in a uniform flow will have a paraboloidal interface $\eta = 1$, and its temperature field can be described by an exact similar solution, like the Ivantsov solution for the case without external flow. The flow parameter λ_0 affects the Peclet number of growth via affecting the normalization parameter $\eta_0^2 = \eta_{0,0}^2(\lambda_0, T_\infty)$ (see 6.105).

Chapter 7

STEADY DENDRITIC GROWTH WITH NATURAL CONVECTION (I): THE CASE OF PR $= O(1)$ AND $\overline{\mathrm{G}} \ll 1$

In the following two chapters, we shall study the cases that the convection in the melt is induced by the buoyancy effect. The effect of buoyancy-driven convection on dendritic growth has been an important topic in micro-gravity research. The first important experimental work on this case was performed by Huang and Glicksman in 1981. They found that the buoyancy-induced convection has a profound effect on dendritic growth, especially in the small undercooling regime (see Fig. 1.3). Huang and Glicksman's observation was later confirmed by Lee and Gill (1991). The later development of the experimental investigations on this subject was reviewed by Lee et al. (1996). The analytical work on this problem was first carried out by Canright and Davis in 1991 with perturbation method. Canright et al. obtained a local solution valid near the interface for the case of small Grashof number.

Similar to the system with external flow studied in the previous chapters, for the present system, one can also discuss two different limit cases:

- $\mathrm{Pr} = O(1), \overline{G} \ll 1$,
- $\mathrm{Pr} \gg 1, \overline{G} = O(1)$,

where we have defined the modified Grashof number

$$\overline{G} = \frac{2\mathrm{Pr}^3\mathrm{Gr}}{|T_\infty|}. \tag{7.1}$$

In this chapter we are going to discuss the first case, while in the next chapter we study the second case.

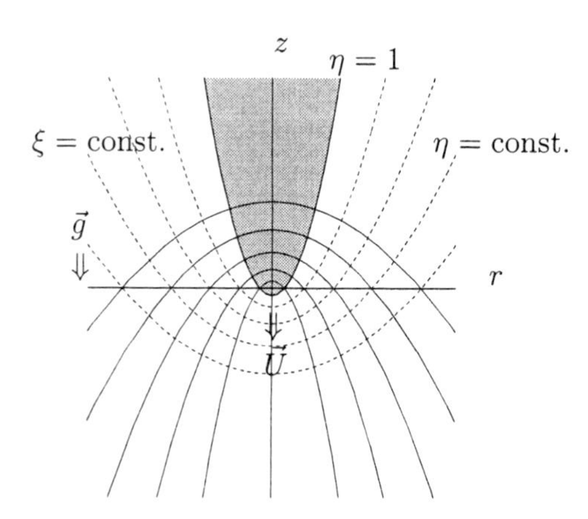

Figure 7.1. A sketch of dendritic growth from a melt with buoyancy effect

1. Mathematical Formulation of The Problem

Assuming $U_\infty = \alpha = 0$, let us consider a single dendrite, with a constant velocity U, steadily growing into an undercooled pure melt with temperature $(T_\infty)_{\rm D}$, in the negative z-axis direction as shown in Fig. 7.1. The gravity is along the growth direction, and surface tension at the interface is zero. Hence, the dendrite is axi-symmetrical and isothermal. The non-dimensional governing equations for this case are as follows:

1 Kinematic equation:

$$D^2\Psi = -\eta_0^4(\xi^2 + \eta^2)\zeta. \tag{7.2}$$

2 Vorticity equation:

$$\begin{aligned}\mathrm{Pr}D^2\zeta = {} & \frac{2\zeta}{\eta_0^4\xi^2\eta^2}\frac{\partial(\Psi, \eta_0^2\xi\eta)}{\partial(\xi,\eta)} - \frac{1}{\eta_0^2\xi\eta}\frac{\partial(\Psi,\zeta)}{\partial(\xi,\eta)} \\ & + \frac{\eta_0^4\mathrm{Gr}}{\mathrm{Pr}|T_\infty|}\xi\eta\left(\eta\frac{\partial T}{\partial \xi} + \xi\frac{\partial T}{\partial \eta}\right).\end{aligned} \tag{7.3}$$

3 Heat conduction equation in liquid phase:

$$\nabla^2 T = \frac{1}{\eta_0^2\xi\eta}\left(\frac{\partial\Psi}{\partial\eta}\frac{\partial T}{\partial\xi} - \frac{\partial\Psi}{\partial\xi}\frac{\partial T}{\partial\eta}\right). \tag{7.4}$$

Here, the differentiation operators ∇^2 and D^2 are defined as

$$\nabla^2 = \left\{ \frac{\partial^2}{\partial \xi^2} + \frac{\partial^2}{\partial \eta^2} + \frac{1}{\xi}\frac{\partial}{\partial \xi} + \frac{1}{\eta}\frac{\partial}{\partial \eta} \right\}, \tag{7.5}$$

$$D^2 = \left\{ \frac{\partial}{\partial \xi^2} + \frac{\partial^2}{\partial \eta^2} - \frac{1}{\xi}\frac{\partial}{\partial \xi} - \frac{1}{\eta}\frac{\partial}{\partial \eta} \right\}. \tag{7.6}$$

The boundary conditions are:

1 . The up-stream far-field conditions: as $\eta \to \infty$,

$$\Psi \sim \frac{1}{2}\eta_0^4 \xi^2 \eta^2 + o(1), \qquad \zeta \to 0; \tag{7.7}$$

$$T \to T_\infty. \tag{7.8}$$

2 . Axi-symmetrical condition: at the symmetrical axis $\xi = 0, \eta > 1$,

$$\Psi = \zeta = \frac{\partial \Psi}{\partial \xi} = \frac{\partial \zeta}{\partial \xi} = \frac{\partial \Psi}{\partial \eta} = \frac{\partial \zeta}{\partial \eta} = \frac{\partial T}{\partial \eta} = 0. \tag{7.9}$$

3 . The interface condition: at $\eta = \eta_s(\xi)$,

(i) Thermo-dynamical equilibrium condition:

$$T = 0, \tag{7.10}$$

(ii) Enthalpy conservation condition:

$$\left(\frac{\partial T}{\partial \eta} - \eta_s' \frac{\partial T}{\partial \xi} \right) + \eta_0^2 \left(\xi \eta_s \right)' = 0, \tag{7.11}$$

(iii) Mass conservation condition:

$$\left(\frac{\partial \Psi}{\partial \xi} + \eta_s' \frac{\partial \Psi}{\partial \eta} \right) = \eta_0^4 (\xi \eta_s)(\xi \eta_s)', \tag{7.12}$$

(iv) Continuity condition of the tangential component of velocity:

$$\left(\frac{\partial \Psi}{\partial \eta} - \eta_s' \frac{\partial \Psi}{\partial \xi} \right) + \eta_0^4 (\xi \eta_s)(\eta_s \eta_s' - \xi) = 0. \tag{7.13}$$

System (7.2)–(7.13) involves four independent dimensionless parameters: $\{T_\infty,\ \mathrm{Pr},\ \mathrm{Gr}\}$. The parameter η_0^2 in the paraboloid coordinate system (2.1) is dependent of these parameters.

In the far field, the velocity field for this case is approximately described by the functions

$$\begin{cases} \bar{\zeta}_* = 0, \\ \bar{\Psi}_* = \frac{1}{2}\eta_0^4\xi^2\eta^2 . \end{cases} \tag{7.14}$$

For convenience, we set

$$\begin{aligned} \Psi(\xi,\eta) &= \bar{\Psi}_*(\xi,\eta) + \bar{\Psi}(\xi,\eta), \\ \zeta(\xi,\eta) &= \frac{1}{\mathrm{Pr}^2}\bar{\zeta}(\xi,\eta), \\ T(\xi,\eta) &= T_\infty + \mathrm{Pr}\,\bar{T}(\xi,\eta). \end{aligned} \tag{7.15}$$

The perturbations of the flow field and temperature field are now subject to the following system:

$$D^2\bar{\Psi} = -\frac{\eta_0^4}{\mathrm{Pr}^2}(\xi^2+\eta^2)\bar{\zeta}\ , \tag{7.16}$$

$$\begin{aligned} \mathrm{Pr}D^2\bar{\zeta} = \eta_0^2\left(\xi\frac{\partial\bar{\zeta}}{\partial\xi} - \eta\frac{\partial\bar{\zeta}}{\partial\eta}\right) + \frac{2\bar{\zeta}}{\eta_0^2\xi^2\eta^2}\left(\eta\frac{\partial\bar{\Psi}}{\partial\eta} - \xi\frac{\partial\bar{\Psi}}{\partial\xi}\right) - \frac{1}{\eta_0^2\xi\eta}\frac{\partial(\bar{\Psi},\bar{\zeta})}{\partial(\xi,\eta)} \\ + \frac{\eta_0^4\mathrm{Pr}^2\mathrm{Gr}}{|T_\infty|}\xi\eta\left(\eta\frac{\partial\bar{T}}{\partial\xi} + \xi\frac{\partial\bar{T}}{\partial\eta}\right)\ , \end{aligned} \tag{7.17}$$

$$\nabla^2\bar{T} = \eta_0^2\left(\xi\frac{\partial\bar{T}}{\partial\xi} - \eta\frac{\partial\bar{T}}{\partial\eta}\right) + \frac{1}{\eta_0^2\xi\eta}\left(\frac{\partial\bar{\Psi}}{\partial\eta}\frac{\partial\bar{T}}{\partial\xi} - \frac{\partial\bar{\Psi}}{\partial\xi}\frac{\partial\bar{T}}{\partial\eta}\right) . \tag{7.18}$$

We utilize the new variables

$$\begin{aligned} \bar{\sigma} &= \frac{\eta_0^2}{2\mathrm{Pr}}\xi^2, \\ \bar{\tau} &= \frac{\eta_0^2}{2\mathrm{Pr}}\eta^2 . \end{aligned} \tag{7.19}$$

With the variables $(\bar{\sigma};\bar{\tau})$, the stream function at the far field, becomes

$$\bar{\Psi}_* = 2\mathrm{Pr}^2\bar{\sigma}\bar{\tau} = \frac{2}{\epsilon_2^2}\bar{\sigma}\bar{\tau}, \tag{7.20}$$

while the differential operator D^2 and ∇^2 become

$$D^2 = 2\frac{\eta_0^2}{\mathrm{Pr}}\bar{L}_2, \qquad \nabla^2 = 2\frac{\eta_0^2}{\mathrm{Pr}}(\bar{L}_2 + \bar{L}_1), \tag{7.21}$$

where

$$\begin{cases} \bar{L}_2 = \left(\bar{\sigma}\dfrac{\partial^2}{\partial\bar{\sigma}^2} + \bar{\tau}\dfrac{\partial^2}{\partial\bar{\tau}^2}\right), \\ \bar{L}_1 = \left(\dfrac{\partial}{\partial\bar{\sigma}} + \dfrac{\partial}{\partial\bar{\tau}}\right). \end{cases} \tag{7.22}$$

Hence, the system (7.16)– (7.18) is changed to

$$\bar{L}_2[\bar{\Psi}] = -(\bar{\sigma} + \bar{\tau})\bar{\zeta}, \tag{7.23}$$

$$\begin{aligned} \bar{L}_2[\bar{\zeta}] = {} & \left(\bar{\sigma}\frac{\partial}{\partial\bar{\sigma}} - \bar{\tau}\frac{\partial}{\partial\bar{\tau}}\right)\bar{\zeta} - \frac{\bar{\zeta}}{2\mathrm{Pr}^2\bar{\tau}\bar{\sigma}}\left(\bar{\tau}\frac{\partial\bar{\Psi}}{\partial\bar{\tau}} - \bar{\sigma}\frac{\partial\bar{\Psi}}{\partial\bar{\sigma}}\right) \\ & - \frac{1}{2\mathrm{Pr}^2}\frac{\partial(\bar{\Psi},\bar{\zeta})}{\partial(\bar{\sigma},\bar{\tau})} + \overline{\mathrm{G}}\bar{\sigma}\bar{\tau}\left(\frac{\partial\bar{T}}{\partial\bar{\tau}} + \frac{\partial\bar{T}}{\partial\bar{\sigma}}\right), \end{aligned} \tag{7.24}$$

$$(\bar{L}_2 + \bar{L}_1)[\bar{T}] = \mathrm{Pr}\left(\bar{\sigma}\frac{\partial\bar{T}}{\partial\bar{\sigma}} - \bar{\tau}\frac{\partial\bar{T}}{\partial\bar{\tau}}\right) + \frac{1}{2\mathrm{Pr}}\left(\frac{\partial\bar{\Psi}}{\partial\bar{\tau}}\frac{\partial\bar{T}}{\partial\bar{\sigma}} - \frac{\partial\bar{\Psi}}{\partial\bar{\sigma}}\frac{\partial\bar{T}}{\partial\bar{\tau}}\right) \tag{7.25}$$

where we have used the modified Grashof number $\overline{\mathrm{G}}$, instead of Gr.

Furthermore, in terms of the variables $\{\bar{\sigma}, \bar{\tau}\}$, the interface shape function $\eta = \eta_s(\xi)$ is transformed into the form $\bar{\tau} = \bar{\tau}_s(\bar{\sigma})$. Then the interface conditions can be written as: at $\bar{\tau} = \bar{\tau}_s(\bar{\sigma})$,

$$\mathrm{Pr}\,\bar{T} + T_\infty = 0, \tag{7.26}$$

$$\bar{\tau}_s\frac{\partial\bar{T}}{\partial\bar{\tau}} - \bar{\sigma}\bar{\tau}_s'\frac{\partial\bar{T}}{\partial\bar{\sigma}} + (\bar{\sigma}\bar{\tau}_s' + \bar{\tau}_s) = 0, \tag{7.27}$$

$$\frac{\partial\bar{\Psi}}{\partial\bar{\sigma}} + \bar{\tau}_s'\frac{\partial\bar{\Psi}}{\partial\bar{\tau}} = 0, \tag{7.28}$$

$$\bar{\tau}_s\frac{\partial\bar{\Psi}}{\partial\bar{\tau}} - \bar{\sigma}\bar{\tau}_s'\frac{\partial\bar{\Psi}}{\partial\bar{\sigma}} = 0. \tag{7.29}$$

2. Laguerre Series Representation of Solutions

As in Chap. 5, we expand the solution in the Laguerre series

$$\begin{cases} \bar{\zeta} = \bar{\sigma} \sum_{n=0}^{\infty} A_n(\bar{\tau}, \overline{\mathrm{G}}) L_n^{(1)}(\bar{\sigma}), \\ \bar{\Psi} = \bar{\sigma} \sum_{n=0}^{\infty} B_n(\bar{\tau}, \overline{\mathrm{G}}) L_n^{(1)}(\bar{\sigma}), \\ \bar{T} = \sum_{n=0}^{\infty} D_n(\bar{\tau}, \overline{\mathrm{G}}) L_n^{(1)}(\bar{\sigma}). \end{cases} \tag{7.30}$$

We then derive

$$\begin{aligned} &\frac{\mathrm{d}^2 B_n}{\mathrm{d}\bar{\tau}^2} + A_n = \frac{n+2}{\bar{\tau}} B_{n+1} - \frac{1}{\bar{\tau}}\Big[2(n+1)A_n - (n+2)A_{n+1} - nA_{n-1}\Big], \\ &\frac{\mathrm{d}^2 A_n}{\mathrm{d}\bar{\tau}^2} + \frac{\mathrm{d}A_n}{\mathrm{d}\bar{\tau}} - \frac{n+1}{\bar{\tau}} A_n = \epsilon_0\Big(\frac{\mathrm{d}D_n}{\mathrm{d}\bar{\tau}} - \sum_{m=n+1}^{\infty} D_m\Big) \\ &\qquad - \frac{1}{2\mathrm{Pr}^2}\mathcal{N}\Big\{A_n; B_n\Big\}, \qquad (n = 0, 1, 2, \ldots,) \end{aligned} \tag{7.31}$$

where $\mathcal{N}\Big\{A_n; B_n\Big\}$ is the nonlinear differential operator. For the temperature field, we have

$$\begin{aligned} \bar{\tau}\frac{\mathrm{d}^2 D_n}{\mathrm{d}\bar{\tau}^2} + (1 + \mathrm{Pr}\,\bar{\tau})\frac{\mathrm{d}D_n}{\mathrm{d}\bar{\tau}} - n\mathrm{Pr}D_n(\bar{\tau}) &= \Big[(n+1) - (n+2)\mathrm{Pr}\Big]D_{n+1} \\ &\quad - \sum_{m=n+2}^{\infty} D_m + \frac{1}{2\mathrm{Pr}}\mathcal{N}_T\Big\{D_n, B_n\Big\} \end{aligned} \tag{7.32}$$

where $\mathcal{N}_T\Big\{D_n, B_n\Big\}$ is the nonlinear operator, which describes the interaction between flow field and temperature field.

3. Asymptotic Expansion Solution with Small Buoyancy Effect

In this chapter, we discuss the case of $\mathrm{Pr} = O(1)$, but $\overline{\mathrm{G}} \ll 1$. Thus, we may set $\overline{\mathrm{G}} = \epsilon_0$ as the basic small parameter, and look for the asymptotic expansion solution in the limit $\epsilon_0 \to 0$:

$$\begin{aligned} A_n(\bar{\tau}, \epsilon_0) &= \epsilon_0\Big\{\bar{A}_{n,0}(\bar{\tau}) + \epsilon_0 \bar{A}_{n,1}(\bar{\tau}) + \cdots\Big\}, \\ B_n(\bar{\tau}, \epsilon_0) &= \epsilon_0\Big\{\bar{B}_{n,0}(\bar{\tau}) + \epsilon_0 \bar{B}_{n,1}(\bar{\tau}) + \cdots\Big\}, \\ D_n(\bar{\tau}, \epsilon_0) &= \bar{D}_{n,0}(\bar{\tau}) + \epsilon_0 \bar{D}_{n,1}(\bar{\tau}) + \cdots \end{aligned} \tag{7.33}$$

for $(n = 0, 1, \dots,)$. Here, we have set the leading-order asymptotic factors for D_n as $O(1)$. This can be verified by the later derivations. With the primary factor ϵ_0, the solution $\bar{\Psi}$ will have the asymptotic structure:

$$\begin{aligned}\bar{\Psi} = \epsilon_0\sigma\Big\{&\Big[\bar{B}_{0,0} + \epsilon_0\bar{B}_{0,1} + \epsilon_0^2\bar{B}_{0,2} + \cdots\Big]L_0^{(1)}(\bar{\sigma})\\ &+\Big[\bar{B}_{1,0} + \epsilon_0\bar{B}_{1,1} + \epsilon_0^2\bar{B}_{1,2} + \cdots\Big]L_1^{(1)}(\bar{\sigma})\\ &+\Big[\bar{B}_{2,0} + \epsilon_0\bar{B}_{2,1} + \epsilon_0^2\bar{B}_{2,2} + \cdots\Big]L_2^{(1)}(\bar{\sigma})\\ &+\Big[\bar{B}_{3,0} + \epsilon_0\bar{B}_{3,1} + \epsilon_0^2\bar{B}_{3,2} + \cdots\Big]L_3^{(1)}(\bar{\sigma})\\ &+\cdots\Big\}.\end{aligned} \tag{7.34}$$

The solutions $\bar{\zeta}$, and $\bar{T}$ have similar asymptotic structure.

With the new variables $\bar{\sigma}, \bar{\tau}$, the interface shape can be written in the form $\bar{\tau} = \bar{\tau}_s(\bar{\sigma}) = \bar{\tau}_0 + \bar{h}_s(\bar{\sigma})$, where $|\bar{h}_s| \ll \bar{\tau}_0$, with

$$\bar{\tau}_0 = \frac{\eta_0^2}{2\mathrm{Pr}}. \tag{7.35}$$

From the normalization condition (2.2), it follows that

$$\bar{h}_s(0) = 0. \tag{7.36}$$

Similar to the Chap. 5, we make the regular asymptotic expansion, as well as the Laguerre expansion for the interface function $\bar{h}_s(\bar{\sigma})$:

$$\begin{aligned}\bar{h}_s(\bar{\sigma}) &= \epsilon_0\Big[\bar{h}_0(\sigma) + \epsilon_0\bar{h}_1(\sigma) + \epsilon_0^2\bar{h}_2(\sigma) + \cdots\Big]\\ &= \epsilon_0\Big\{\Big[\bar{h}_{0,0} + \epsilon_0\bar{h}_{0,1} + \epsilon_0^2\bar{h}_{0,2} + \cdots\Big]L_0^{(1)}(\sigma)\\ &\quad+\Big[\bar{h}_{1,0} + \epsilon_0\bar{h}_{1,1} + \epsilon_0^2\bar{h}_{1,2} + \cdots\Big]L_1^{(1)}(\sigma)\\ &\quad+\cdots\Big\}.\end{aligned} \tag{7.37}$$

Hence, we may write that

$$\bar{h}_s(\bar{\sigma}) = \sum_{n=0}^{\infty} h_n(\epsilon_0)L_n^{(1)}(\bar{\sigma}), \tag{7.38}$$

and as $\epsilon_0 \to 0$, one has

$$h_n(\epsilon_0) = \epsilon_0\Big[\bar{h}_{n,0} + \epsilon_0\bar{h}_{n,1} + \epsilon_0^2\bar{h}_{n,2} + \cdots\Big]. \tag{7.39}$$

Moreover, note that $\bar{\tau}_0$, as well as the parameter η_0^2, depends on ϵ_0. We write that

$$\begin{aligned}\eta_0^2(\epsilon_0) &= \eta_{0,0}^2 + \tilde{\eta}_0(\epsilon_0),\\ \bar{\tau}_0(\epsilon_0) &= \bar{\tau}_{0,0} + \tilde{\tau}_0(\epsilon_0),\end{aligned} \tag{7.40}$$

and assume as $\epsilon_0 \to 0$,

$$\begin{aligned}\tilde{\eta}_0(\epsilon_0) &= \epsilon_0(\eta_{0,1} + \epsilon_0\eta_{0,2} + \cdots),\\ \tilde{\tau}_0(\epsilon) &= \epsilon_0(\bar{\tau}_{0,1} + \epsilon_0\bar{\tau}_{0,2} + \cdots).\end{aligned} \tag{7.41}$$

Since $(\tilde{\tau}_0 + \bar{h}_s) \ll 1$, we can make a Taylor expansion for the interface conditions (7.26)–(7.29) around $\bar{\tau} = \bar{\tau}_{0,0}$ as follows: at $\bar{\tau} = \bar{\tau}_{0,0}$,

$$\bar{\Psi} + \frac{\partial\bar{\Psi}}{\partial\bar{\tau}}(\tilde{\tau}_0 + \bar{h}_s) + \frac{1}{2!}\frac{\partial^2\bar{\Psi}}{\partial\bar{\tau}^2}(\tilde{\tau}_0 + \bar{h}_s)^2 + \cdots = 0, \tag{7.42}$$

$$\frac{\partial\bar{\Psi}}{\partial\bar{\tau}} + \frac{\partial^2\bar{\Psi}}{\partial\bar{\tau}^2}(\tilde{\tau}_0 + \bar{h}_s) + \frac{1}{2!}\frac{\partial^3\bar{\Psi}}{\partial\bar{\tau}^3}(\tilde{\tau}_0 + \bar{h}_s)^2 + \cdots = 0, \tag{7.43}$$

$$T_\infty + \mathrm{Pr}\Big[\bar{T} + \frac{\partial\bar{T}}{\partial\bar{\tau}}(\tilde{\tau}_0 + \bar{h}_s) + \frac{1}{2!}\frac{\partial^2 T}{\partial\bar{\tau}^2}(\tilde{\tau}_0 + \bar{h}_s)^2 + \cdots\Big] = 0, \tag{7.44}$$

$$\begin{aligned}&(\bar{\tau}_{0,0} + \tilde{\tau}_0 + \bar{h}_s))\left[\frac{\partial\bar{T}}{\partial\bar{\tau}} + \frac{\partial^2\bar{T}}{\partial\bar{\tau}^2}(\tilde{\tau}_0 + \bar{h}_s) + \frac{1}{2!}\frac{\partial^3\bar{T}}{\partial\bar{\tau}^3}(\tilde{\tau}_0 + \bar{h}_s)^2 + \cdots\right]\\ &-\bar{\sigma}\bar{h}_s'\left[\frac{\partial\bar{T}}{\partial\bar{\sigma}} + \frac{\partial^2\bar{T}}{\partial\bar{\sigma}\partial\bar{\tau}}(\tilde{\tau}_0 + \bar{h}_s) + \cdots\right] + (\bar{\tau}_{0,0} + \tilde{\tau}_0 + \bar{h}_s + \bar{\sigma}\bar{h}_s') = 0.\end{aligned} \tag{7.45}$$

The driving force for the convection is the inhomogeneity of the temperature field. Hence, in order to find the solution for the velocity field, we must solve the temperature field first.

3.1 Zeroth-Order Solution of the Temperature Field $O(1)$

In the zeroth-order approximation, we have

$$\bar{D}_{n,0} = 0 \tag{7.46}$$

for $n \geq 1$, and

$$\bar{\tau}\bar{D}_{0,0}'' + (1 + \Pr\bar{\tau})\,\bar{D}_{0,0}' = 0. \tag{7.47}$$

The above system allows the solutions

$$\bar{D}_{0,0} = I_* + I_0 E_1(\Pr\bar{\tau}), \tag{7.48}$$

where I_0 and I_* are arbitrary constants. The zeroth-order interface conditions only contain the component of $L_0^{(1)}(\bar{\sigma})$, which are

$$\begin{aligned} &\Pr\bar{D}_{0,0}(\bar{\tau}_{0,0}) + T_\infty = 0,\\ &\bar{\tau}_{0,0}\bar{D}_{0,0}'(\bar{\tau}_{0,0}) + \bar{\tau}_{0,0} = 0. \end{aligned} \tag{7.49}$$

One derives

$$I_0 = \bar{\tau}_{0,0}\mathrm{e}^{\Pr\,\bar{\tau}_{0,0}}. \tag{7.50}$$

Furthermore, from the far-field condition, one derives

$$I_* = \frac{T_\infty}{\Pr}. \tag{7.51}$$

Thus, as we expected, the zeroth-order solution is just the Ivantsov solution

$$\begin{cases} \bar{D}_{0,0}(\bar{\tau}) = \frac{T_\infty}{\Pr} + \bar{\tau}_{0,0}\mathrm{e}^{\Pr\bar{\tau}_{0,0}}E_1(\Pr\bar{\tau}),\\ T_\infty = -\Pr\bar{\tau}_{0,0}\mathrm{e}^{\Pr\bar{\tau}_{0,0}}E_1(\Pr\bar{\tau}_{0,0}). \end{cases} \tag{7.52}$$

In Fig 7.2, we show the distributions of the temperature $\hat{T}_0(\bar{\tau})$ along the $\bar{\tau}$-axis for the cases, $\Pr = 2.5$ and $\bar{\tau}_{0,0} = 0.5, 1.0, 1.5$, while in Fig 7.3, we show the variation of undercooling temperature T_∞ with the parameter $\bar{\tau}_{0,0}$, for the cases for $\Pr = 0.5, 1.5, 5.0$.

3.2 Zeroth-Order Solution of the Velocity Field $O(\epsilon_0)$

The zeroth-order approximation of the flow field induced by the buoyancy effect is of $O(\epsilon_0)$. We have the system

$$\begin{aligned} \frac{\mathrm{d}^2\bar{B}_{n,0}}{\mathrm{d}\bar{\tau}^2} = \frac{n+2}{\bar{\tau}}\bar{B}_{n+1,0} - \left[1 + \frac{2(n+1)}{\bar{\tau}}\right]\bar{A}_{n,0}\\ +\frac{1}{\bar{\tau}}\Big[(n+2)\bar{A}_{n+1,0} + n\bar{A}_{n-1,0}\Big], \end{aligned} \tag{7.53}$$

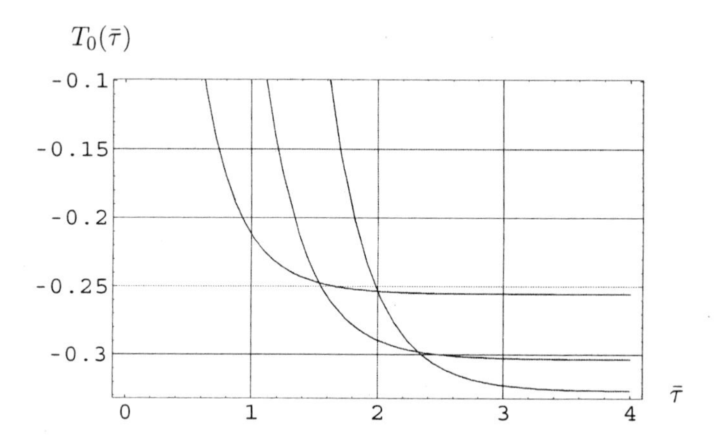

Figure 7.2. The graphs of the solution $T_0(\bar{\tau})$ versus $\bar{\tau}$ for Pr $= 2.5$ and $\bar{\tau}_{0,0} = 0.5, 1.0, 1.5$ from top to bottom.

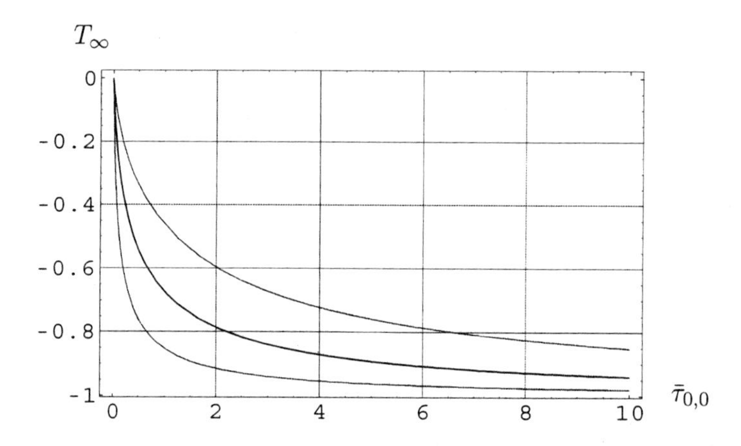

Figure 7.3. The graphics of T_∞ versus $\bar{\tau}_{0,0}$, for Pr $= 0.5, 1.5, 5.0$ from top to bottom.

$$\frac{\mathrm{d}^2 \bar{A}_{n,0}}{\mathrm{d}\bar{\tau}^2} + \frac{\mathrm{d}\bar{A}_{n,0}}{\mathrm{d}\bar{\tau}} - \frac{n+1}{\bar{\tau}} \bar{A}_{n,0} = 0 \qquad (n = 1, 2, \ldots.), \tag{7.54}$$

and

$$\frac{\mathrm{d}^2 \bar{B}_{0,0}}{\mathrm{d}\bar{\tau}^2} = \frac{2}{\bar{\tau}} \bar{B}_{1,0} - \left(1 + \frac{2}{\bar{\tau}}\right) \bar{A}_{0,0} + \frac{2}{\bar{\tau}} \bar{A}_{1,0}, \tag{7.55}$$

$$\frac{\mathrm{d}^2 \bar{A}_{0,0}}{\mathrm{d}\bar{\tau}^2} + \frac{\mathrm{d}\bar{A}_{0,0}}{\mathrm{d}\bar{\tau}} - \frac{1}{\bar{\tau}} \bar{A}_{0,0} = D'_{0,0} = -\frac{\bar{\tau}_{0,0}}{\bar{\tau}} \mathrm{e}^{-\mathrm{Pr}(\bar{\tau} - \bar{\tau}_{0,0})}. \tag{7.56}$$

The fundamental solutions of the homogeneous equation of (7.54) are

$$\left\{\bar{\tau}e^{-\bar{\tau}}M(n+2,2,\bar{\tau});\quad \bar{\tau}e^{-\bar{\tau}}U(n+2,2,\bar{\tau})\right\}. \tag{7.57}$$

Note that the solution

$$V_n(\bar{\tau}) = \bar{\tau}e^{-\bar{\tau}}M(n+2,2,\bar{\tau}) \tag{7.58}$$

is a polynomial of degree n, subject to the recurrence formula

$$\begin{aligned} &V_0(x) = \bar{\tau}, \\ &V_{n+1}(\bar{\tau}) = \frac{n+1}{n+2}V_n(\bar{\tau}) + \frac{\bar{\tau}}{n+2}\Big[V_n(\bar{\tau}) + V_n'(\bar{\tau})\Big]. \end{aligned} \tag{7.59}$$

Consequently, we have

$$\begin{cases} V_1(\bar{\tau}) = \bar{\tau} + \frac{1}{2}\bar{\tau}^2, \\ V_2(x) = \frac{4}{3}\bar{\tau} + \bar{\tau}^2 + \frac{1}{6}\bar{\tau}^3, \\ \vdots . \end{cases} \tag{7.60}$$

The particular solutions (7.56) can be obtained with the method of variation of parameter, since the two fundamental solutions of the associated homogeneous equation of (7.56) is known as

$$\left\{\bar{\tau};\ E_2(\bar{\tau})\right\}. \tag{7.61}$$

We thus obtain the particular solutions

$$\bar{A}_{0,0} = I_0\widehat{A}_0, \tag{7.62}$$

where

$$\widehat{A}_0 = u_1(\bar{\tau})\bar{\tau} + u_2(\bar{\tau})E_2(\bar{\tau}), \tag{7.63}$$

and

$$\begin{aligned} u_1(\bar{\tau}) &= -\int_{\bar{\tau}}^{\infty} \frac{E_2(t)e^{-\mathrm{Pr}t}}{t[E_2(t) - tE_1(t)]}dt, \\ u_2(\bar{\tau}) &= \int_{\bar{\tau}}^{\infty} \frac{e^{-\mathrm{Pr}t}}{E_2(t) - tE_1(t)}dt. \end{aligned} \tag{7.64}$$

Taking into account the far-field condition at $\bar{\tau} = \infty$, we write the general solution of (7.56) and (7.54) as

$$\bar{A}_{0,0}(\bar{\tau}) = \bar{a}_{0,0}\mathcal{F}_0(\bar{\tau}) + I_0\widehat{A}_0 = \bar{a}_{0,0}E_2(\bar{\tau}) + I_0\widehat{A}_0(\bar{\tau}), \tag{7.65}$$

and

$$\bar{A}_{n,0}(\bar{\tau}) = \bar{a}_{n,0}\mathcal{F}_n(\bar{\tau}), \quad (n = 1, 2, \ldots) \tag{7.66}$$

where $\bar{a}_{n,0}$, $(n = 0, 1, \ldots)$ are arbitrary constants.

Furthermore, the particular solution of (7.53) with the inhomogeneous terms produced by $\bar{A}_{n,0}(\bar{\tau}) = \bar{a}_{n,0}\mathcal{F}_n(\bar{\tau})$ can be found in the form

$$\bar{B}_{n,0}(\bar{\tau}) = \bar{b}_{n,0}\mathcal{F}_n(\bar{\tau}), \qquad (n = 1, 2, \ldots), \tag{7.67}$$

where one can derive

$$\bar{b}_{n,0} = -\bar{a}_{n,0} + n\bar{a}_{n-1,0}. \tag{7.68}$$

The particular solutions of (7.53), with the inhomogeneous terms produced by $\bar{A}_{0,0}(\bar{\tau}) = I_0\widehat{A}_0$, are written in the form

$$\bar{B}_{n,0}(\bar{\tau}) = I_0\widehat{B}_n(\bar{\tau}), \quad (n = 0, 1), \tag{7.69}$$

which are subject to the equations

$$\frac{\mathrm{d}^2\widehat{B}_1}{\mathrm{d}\bar{\tau}^2} = \frac{1}{\bar{\tau}}\widehat{A}_0 \tag{7.70}$$

and

$$\frac{\mathrm{d}^2\widehat{B}_0}{\mathrm{d}\bar{\tau}^2} = \frac{2}{\bar{\tau}}\widehat{B}_1 - \left(1 + \frac{2}{\bar{\tau}}\right)\widehat{A}_0, \tag{7.71}$$

respectively. It is found that

$$\begin{aligned} \widehat{B}_1(\bar{\tau}) &= \int_{\bar{\tau}}^{\infty} (t - \bar{\tau})\frac{\widehat{A}_0(t)}{t}\mathrm{d}t, \\ \widehat{B}_0(\bar{\tau}) &= 2\int_{\bar{\tau}} (t - \bar{\tau})\frac{\widehat{B}_1(t)}{t}\mathrm{d}t - \int_{\bar{\tau}}^{\infty} (t - \bar{\tau})\left(1 + \frac{2}{t}\right)\widehat{A}_0(t)\mathrm{d}t. \end{aligned} \tag{7.72}$$

In order to find the general solution of the stream function, one needs to include the general solution for the associated homogeneous equation of stream function, $\psi_0(\bar{\sigma}, \bar{\tau})$. The general solution $\psi_0(\bar{\sigma}, \bar{\tau})$ has been derived in the Appendix. We quote the results as follows:

$$\psi_0 = \bar{\sigma}\sum_{n=0}^{\infty} L_n^{(1)}(\bar{\sigma})\sum_{k=0}^{\infty} C_{0,k}\mathcal{A}_{n,k}(\bar{\tau}), \tag{7.73}$$

and

$$\frac{\partial\psi_0}{\partial\bar{\tau}} = \bar{\sigma}\sum_{n=0}^{\infty} L_n^{(1)}(\bar{\sigma})\sum_{k=0}^{\infty} C_{0,k}\hat{\mathcal{B}}_{n,k}(\bar{\tau}), \tag{7.74}$$

where $\mathcal{C}_{0,k}$ $(k = 0, 1, 2, \ldots)$ are arbitrary constants. With these results, we write the general solution of the stream function as

$$\bar{\Psi}_0 = \psi_0(\bar{\sigma}, \bar{\tau}) + \bar{\sigma} \sum_{n=0}^{\infty} \left[\bar{b}_{n,0}\mathcal{F}_n(\bar{\tau}) + I_0 \widehat{B}_n(\bar{\tau})\right] L_n^{(1)}(\bar{\sigma}), \tag{7.75}$$

where

$$\widehat{B}_n(\bar{\tau}) = 0 \qquad (n \geq 2). \tag{7.76}$$

The first part of this result was missed in the solution of Canright and Davis (1991). The general solution for the vorticity function is

$$\epsilon_0 \bar{\zeta}_0 = \epsilon_0 \bar{\sigma} \left\{ \sum_{n=0}^{\infty} \left[\bar{a}_{n,0}\mathcal{F}_n(\bar{\tau}) + I_0 \widehat{A}_n\right] L_n^{(1)}(\bar{\sigma}) \right\}, \tag{7.77}$$

where

$$\widehat{A}_n(\bar{\tau}) = 0 \qquad (n \geq 1). \tag{7.78}$$

The interface conditions in the zeroth-order approximation are: at $\bar{\tau} = \bar{\tau}_{0,0}$,

1 .

$$\bar{\Psi}_0 = 0, \tag{7.79}$$

2 .

$$\frac{\partial \bar{\Psi}_0}{\partial \bar{\tau}} = 0. \tag{7.80}$$

These interface conditions fully determine the two sequences of unknowns, $\{\bar{b}_{n,0}\ \mathcal{C}_{0,k}(n, k = 0, 1, 2, \ldots)\}$.

To carry out numerical calculations, one can truncate the series at $n = N$ and $k = N$. Thus, from the interface conditions (7.79)–(7.80), one derives that, for $n = 0, 1$,

$$\begin{aligned} \sum_{k=0}^{N} \mathcal{C}_{0,k}\mathcal{A}_{0,k}(\bar{\tau}_{0,0}) &= -I_0\widehat{B}_0(\bar{\tau}_{0,0}) - \bar{b}_{0,0}\mathcal{F}_0(\bar{\tau}_{0,0}), \\ \sum_{k=0}^{N} \mathcal{C}_{0,k}\hat{\mathcal{B}}_{0,k}(\bar{\tau}_{0,0}) &= -I_0\widehat{B}_0(\bar{\tau}_{0,0}) - \bar{b}_{0,0}\mathcal{F}_0'(\bar{\tau}_{0,0}), \end{aligned} \tag{7.81}$$

and

$$\begin{aligned} \sum_{k=0}^{N} \mathcal{C}_{0,k}\mathcal{A}_{1,k}(\bar{\tau}_{0,0}) &= -I_0\widehat{B}_1(\bar{\tau}_{0,0}) - \bar{b}_{1,0}\mathcal{F}_1(\bar{\tau}_{0,0}), \\ \sum_{k=0}^{N} \mathcal{C}_{0,k}\hat{\mathcal{B}}_{1,k}(\bar{\tau}_{0,0}) &= -I_0\widehat{B}_1(\bar{\tau}_{0,0}) - \bar{b}_{1,0}\mathcal{F}_1'(\bar{\tau}_{0,0}), \end{aligned} \tag{7.82}$$

On the other hand, for $(n = 2, 3, \ldots, N)$, we have

$$\begin{aligned}
&\sum_{k=2}^{N} \mathcal{C}_{0,k}\mathcal{A}_{n,k}(\bar{\tau}_{0,0}) + \bar{b}_{n,0}\mathcal{F}_n(\bar{\tau}_{0,0}) = -\mathcal{C}_{0,0}\mathcal{A}_{n,0}(\bar{\tau}_{0,0}) - \mathcal{C}_{0,1}\mathcal{A}_{n,1}(\bar{\tau}_{0,0}), \\
&\sum_{k=2}^{N} \mathcal{C}_{0,k}\hat{\mathcal{B}}_{n,k}(\bar{\tau}_{0,0}) + \bar{b}_{n,0}\mathcal{F}'_n(\bar{\tau}_{0,0}) = -\mathcal{C}_{0,0}\hat{\mathcal{B}}_{n,0}(\bar{\tau}_{0,0}) - \mathcal{C}_{0,1}\hat{\mathcal{B}}_{n,1}(\bar{\tau}_{0,0}).
\end{aligned} \tag{7.83}$$

The system (7.83) determines the constants $\mathcal{C}_{0,k}, \bar{b}_{n,0}$ $(n, k = 2, 3, \ldots, N)$ as the functions of the constants $\mathcal{C}_{0,0}$ and $\mathcal{C}_{0,1}$:

$$\begin{pmatrix} \mathcal{C}_{0,2} \\ \mathcal{C}_{0,3} \\ \vdots \\ \mathcal{C}_{0,N} \end{pmatrix} = \left(\mathcal{Q}_{n,k}\right)^{-1} \begin{pmatrix} \mathcal{H}_2 \\ \mathcal{H}_3 \\ \vdots \\ \mathcal{H}_N \end{pmatrix} \tag{7.84}$$

and

$$\bar{b}_{n,0} = -\frac{1}{\mathcal{F}_n(\bar{\tau}_{0,0})}\left[\mathcal{C}_{0,0}\mathcal{A}_{n,0}(\bar{\tau}_{0,0}) + \mathcal{C}_{0,1}\mathcal{A}_{n,1}(\bar{\tau}_{0,0}) + \sum_{k=2}^{N}\mathcal{C}_{0,k}\mathcal{A}_{n,k}(\bar{\tau}_{0,0})\right], \tag{7.85}$$

where the matrix $\left(\mathcal{Q}_{n,k}\right)$ is defined as

$$\mathcal{Q}_{n,k} = \mathcal{F}'_n(\bar{\tau}_{0,0})\mathcal{A}_{n,k}(\bar{\tau}_{0,0}) - \mathcal{F}_n(\bar{\tau}_{0,0})\hat{\mathcal{B}}_{n,k}(\bar{\tau}_{0,0}) \tag{7.86}$$

and

$$\begin{aligned}
\mathcal{H}_n = {} & \mathcal{C}_{0,0}\Big[\mathcal{F}_n(\bar{\tau}_{0,0})\hat{\mathcal{B}}_{n,0}(\bar{\tau}_{0,0}) - \mathcal{F}'_n(\bar{\tau}_{0,0})\mathcal{A}_{n,0}(\bar{\tau}_{0,0})\Big] \\
& + \mathcal{C}_{0,1}\Big[\mathcal{F}_n(\bar{\tau}_{0,0})\hat{\mathcal{B}}_{n,1}(\bar{\tau}_{0,0}) - \mathcal{F}'_n(\bar{\tau}_{0,0})\mathcal{A}_{n,1}(\bar{\tau}_{0,0})\Big].
\end{aligned} \tag{7.87}$$

To proceed, let us denote

$$\mathcal{H}_n = \mathcal{C}_{0,0}\widehat{\mathcal{H}}_n^{(0)} + \mathcal{C}_{0,1}\widehat{\mathcal{H}}_n^{(1)}, \tag{7.88}$$

where

$$\begin{aligned}
\widehat{\mathcal{H}}_n^{(0)} &= \Big[\mathcal{F}_n(\bar{\tau}_{0,0})\hat{\mathcal{B}}_{n,0}(\bar{\tau}_{0,0}) - \mathcal{F}'_n(\bar{\tau}_{0,0})\mathcal{A}_{n,0}(\bar{\tau}_{0,0})\Big], \\
\widehat{\mathcal{H}}_n^{(1)} &= \Big[\mathcal{F}_n(\bar{\tau}_{0,0})\hat{\mathcal{B}}_{n,1}(\bar{\tau}_{0,0}) - \mathcal{F}'_n(\bar{\tau}_{0,0})\mathcal{A}_{n,1}(\bar{\tau}_{0,0})\Big].
\end{aligned}$$

Thus, we can express the solution $\mathcal{C}_{0,k}$, $(k = 2, 3, 4, \ldots)$ as follows:

$$\mathcal{C}_{0,k} = \mathcal{C}_{0,0}\widehat{\mathcal{C}}_{0,k}^{(0)} + \mathcal{C}_{0,1}\widehat{\mathcal{C}}_{0,k}^{(1)}, \tag{7.89}$$

where

$$\begin{pmatrix} \widehat{\mathcal{C}}_{0,2}^{(i)} \\ \widehat{\mathcal{C}}_{0,3}^{(i)} \\ \vdots \\ \widehat{\mathcal{C}}_{0,N}^{(i)} \end{pmatrix} = \left(\mathcal{Q}_{n,k}\right)^{-1} \begin{pmatrix} \widehat{\mathcal{H}}_{2}^{(i)} \\ \widehat{\mathcal{H}}_{3}^{(i)} \\ \vdots \\ \widehat{\mathcal{H}}_{N}^{(i)} \end{pmatrix}, \qquad (i = 0, 1). \tag{7.90}$$

are fully determined by the parameter $\bar{\tau}_{0,0}$. Now, by substituting $\mathcal{C}_{0,k}$, $(k = 2, 3, 4, \ldots)$ into (7.81)–(7.82), one can uniquely determine the four unknowns: $\{\mathcal{C}_{0,0}, \mathcal{C}_{0,1}, \bar{b}_{0,0}, \bar{b}_{1,0}\}$. The zeroth-order solution of the flow field is then obtained.

For convenience in further discussion, we write the solution of the stream function in the form

$$\bar{\Psi}_0 = \bar{\Psi}_0^{(I)} + \bar{\Psi}_0^{(II)} \tag{7.91}$$

where

$$\begin{aligned} \bar{\Psi}_0^{(I)} &= \bar{\sigma} \left\{ \bar{b}_{0,0} E_2(\bar{\tau}) + I_0 \widehat{B}_0(\bar{\tau}) + \sum_{k=0}^{N} \mathcal{C}_{0,k} \mathcal{A}_{n,k}(\bar{\tau}) \right\}, \\ \bar{\Psi}_0^{(II)} &= \bar{\sigma} \sum_{n=1}^{N} \left\{ \bar{b}_{n,0} \mathcal{F}_n(\bar{\tau}) + I_0 \widehat{B}_n(\bar{\tau}) + \sum_{k=0}^{N} \mathcal{C}_{0,k} \mathcal{A}_{n,k}(\bar{\tau}) \right\} L_n^{(1)}(\bar{\sigma}). \end{aligned} \tag{7.92}$$

It is different from the case of uniform external flow studied in Chap. 5, that for present case, the second part of solution $\bar{\Psi}_0^{(II)}$ numerically may be not small.

3.3 First-Order Solution of the Temperature Field $O(\epsilon_0)$

The first-order approximation is of $O(\epsilon_0)$. The solution $\bar{D}_{n,1}$ may be written in three parts:

$$\bar{D}_{n,1} = \overline{W}_{n,1} + \overline{P}_{n,1} + \overline{Q}_{n,1}. \tag{7.93}$$

The first part of the solution, $\overline{W}_{n,1}$, is the general solution of the associated homogeneous equation,

$$\bar{\tau} \frac{\mathrm{d}^2 \overline{W}_{n,1}}{\mathrm{d}\bar{\tau}^2} + (1 + \mathrm{Pr}\, \bar{\tau}) \frac{\mathrm{d}\overline{W}_{n,1}}{\mathrm{d}\bar{\tau}} - n\mathrm{Pr}\overline{W}_{n,1}(\bar{\tau}) = 0; \tag{7.94}$$

the second part of the solution, $\overline{P}_{n,1}$, is the particular solution of the equation with the inhomogeneous term related to the solution $\bar{\Psi}_0^{(II)}$; the third part of the solution, $\overline{Q}_{n,1}$, is the particular solution of the equation with the inhomogeneous term related to the solution $\bar{\Psi}_0^{(I)}$.

In terms of the transformation

$$\begin{aligned} \overline{W}_{n,1}(\bar{\tau}) &= x^{-\frac{1}{2}}\mathrm{e}^{-\frac{x}{2}}W(x), \\ x &= \Pr\bar{\tau}, \end{aligned} \tag{7.95}$$

the equation (7.94) can be transformed into the Whittaker equation

$$\frac{\mathrm{d}^2W}{\mathrm{d}x^2} + \left(-\frac{1}{4} + \frac{\kappa}{x} + \frac{\frac{1}{4}-\mu}{x^2}\right)W = 0, \tag{7.96}$$

where

$$\kappa = n + \frac{1}{2}, \quad \mu = 0. \tag{7.97}$$

The fundamental solutions of (7.96) are

$$W(x) = \begin{cases} x^{\frac{1}{2}}\mathrm{e}^{-\frac{x}{2}}M(n+1,1,x), \\ x^{\frac{1}{2}}\mathrm{e}^{-\frac{x}{2}}U(n+1,1,x). \end{cases} \tag{7.98}$$

Thus, the fundamental solutions of (7.94) are

$$\overline{W}_{n,1}(\Pr\bar{\tau}) = \begin{cases} P_n(\Pr\bar{\tau}) = \mathrm{e}^{-\frac{\Pr\bar{\tau}}{2}}M(n+1,1,\Pr\bar{\tau}), \\ \widehat{W}_{n,1}(\Pr\bar{\tau}) = \mathrm{e}^{-\frac{\Pr\bar{\tau}}{2}}U(n+1,1,\Pr\bar{\tau}). \end{cases} \tag{7.99}$$

Note that the solution

$$P_n(x) = \mathrm{e}^{-\frac{x}{2}}M(n+1,1,x), \tag{7.100}$$

is a polynomial of degree n, subject to the recurrence formula

$$\begin{aligned} P_0(x) &= 1, \\ P_{n+1}(x) &= P_n(x) + \tfrac{x}{n+1}\Big[P_n(x) + P_n'(x)\Big]. \end{aligned} \tag{7.101}$$

Hence, we have

$$\begin{aligned} P_1(x) &= 1 + x, \\ P_2(x) &= \tfrac{1}{2!}(2 + 4x + x^2), \\ P_3(x) &= \tfrac{1}{3!}(6 + 18x + 9x^2 + x^3), \\ P_4(x) &= \tfrac{1}{4!}(24 + 96x + 72x^2 + 16x^3 + x^4), \\ &\vdots \end{aligned} \tag{7.102}$$

Taken into account the far-field condition, we have

$$\overline{W}_{n,1} = \bar{d}_{n,1}\widehat{W}_{n,1}(\mathrm{Pr}\,\bar{\tau}) = \bar{d}_{n,1}\mathrm{e}^{-\frac{\mathrm{Pr}\,\bar{\tau}}{2}}U(n+1,1,\mathrm{Pr}\,\bar{\tau}). \qquad (7.103)$$

Especially, we have

$$\begin{aligned}
&\widehat{W}_{0,1}(\mathrm{Pr}\,\bar{\tau}) = E_1(\mathrm{Pr}\,\bar{\tau}),\\
&\widehat{W}_{1,1}(\mathrm{Pr}\,\bar{\tau}) = \Big[E_1(\mathrm{Pr}\,\bar{\tau}) - E_2(\mathrm{Pr}\,\bar{\tau})\Big],\\
&\vdots\\
&\widehat{W}_{n,1}(\mathrm{Pr}\,\bar{\tau}) = (-1)^n\sum_{k=0}^{n}\frac{(-1)^k}{(n-k)!k!}E_{k+1}(\mathrm{Pr}\,\bar{\tau}).
\end{aligned} \qquad (7.104)$$

For the second part of the solution, $\overline{P}_{n,1}$, we need to find the particular solution of the equations

$$\bar{\tau}\frac{\mathrm{d}^2\overline{P}_{n,1}}{\mathrm{d}\bar{\tau}^2} + (1+\mathrm{Pr}\,\bar{\tau})\frac{\mathrm{d}\overline{P}_{n,1}}{\mathrm{d}\bar{\tau}} - n\mathrm{Pr}\overline{P}_{n,1}(\bar{\tau}) = H_n(\bar{\tau}) \qquad (7.105)$$

for $(n = 0, 1, 2, \ldots,)$, where

$$H_n(\bar{\tau}) = \Big[(n+1) - (n+2)\mathrm{Pr}\Big]\overline{P}_{n+1,1} - \sum_{m=n+2}^{\infty}\overline{P}_{m,1} + \frac{1}{2\mathrm{Pr}}\mathcal{N}_T\Big\{D_0, \bar{\Psi}_0^{(II)}\Big\}. \qquad (7.106)$$

Note that as $n \to \infty$, $\bar{D}_{n,1} \to 0$. So, we can truncate the system at $n = N$, by approximately setting

$$\overline{P}_{n,1} = 0, \qquad (n \geq N+1). \qquad (7.107)$$

By using the method of variation parameter, we derive

$$\overline{P}_{n,1}(\bar{\tau}) = \overline{\mathcal{L}}_n\Big\{H_n(\bar{\tau})\Big\}\widehat{W}_{n,1}(\mathrm{Pr}\,\bar{\tau}) + \widehat{\mathcal{L}}_n\Big\{H_n(\bar{\tau})\Big\}P_n(\mathrm{Pr}\,\bar{\tau}) \quad (7.108)$$

with the linear operators

$$\begin{aligned}
\overline{\mathcal{L}}_n\Big\{H_n(\bar{\tau})\Big\} &= \int_{\bar{\tau}}^{\infty}\frac{P_n(\mathrm{Pr}\,\tilde{\tau})H_n(\tilde{\tau})}{\tilde{\tau}\Delta(\widehat{W}_{n,1}, P_n)}\mathrm{d}\tilde{\tau},\\
\widehat{\mathcal{L}}_n\Big\{H_n(\bar{\tau})\Big\} &= -\int_{\bar{\tau}}^{\infty}\frac{\widehat{W}_{n,1}(\mathrm{Pr}\,\tilde{\tau})H_n(\tilde{\tau})}{\tilde{\tau}\Delta(\widehat{W}_{n,1}, P_n)}\mathrm{d}\tilde{\tau},
\end{aligned} \qquad (7.109)$$

where $\Delta(\widehat{W}_{n,1}, P_n)$ is the Wronskian of the solutions $\{\widehat{W}_{n,1}(\Pr\bar{\tau}), P_n(\Pr\bar{\tau})\}$. One can find $\overline{P}_{n,1}(\bar{\tau})$ first, then consecutively find $\overline{P}_{N-1,1}(\bar{\tau})$, $\overline{P}_{N-2,1}(\bar{\tau})$, ..., etc.

For the third part of the solution, $\overline{Q}_{n,1}$, we have

$$\overline{Q}_{n,1} = 0 \quad (n = 1, 2, 3, \ldots) \tag{7.110}$$

and

$$\bar{\tau}\overline{Q}''_{0,1}(\bar{\tau}) + (1 + \Pr\bar{\tau})\overline{Q}'_{0,1}(\bar{\tau}) = H_{0,1}(\bar{\tau}), \tag{7.111}$$

where

$$\begin{aligned} H_{0,1}(\bar{\tau}) &= -\frac{1}{2\Pr}\mathcal{N}_T\Big\{\bar{D}_{0,0}, \bar{\Psi}_0^{(I)}\Big\} \\ &= \frac{I_0}{\Pr}\frac{\mathrm{e}^{-\Pr\bar{\tau}}}{\bar{\tau}}\bar{\sigma}\Big[\bar{b}_{0,0}E_2(\bar{\tau}) + I_0\widehat{B}_0 + \sum_{k=0}^{\infty} C_{0,k}\mathcal{A}_{0,k}(\bar{\tau})\Big]. \end{aligned} \tag{7.112}$$

We derive that

$$\overline{Q}_{0,1}(\bar{\tau}) = \overline{\mathcal{L}}_n\Big\{H_{0,1}(\bar{\tau})\Big\}\widehat{W}_{n,1}(\Pr\bar{\tau}) + \hat{\mathcal{L}}_n\Big\{H_{0,1}(\bar{\tau})\Big\}P_n(\Pr\bar{\tau}). \tag{7.113}$$

Thus, one can easily see that all three parts of the solution $\bar{D}_{n,1}$ that we obtained above,

$$\Big\{\overline{W}_{n,1}; \overline{P}_{n,1}; \overline{Q}_{n,1}\Big\} \to 0 \qquad \text{(exponentially)},$$

as $\bar{\tau} \to \infty$.

We now turn to consider the interface conditions. The procedure will be very much similar to Chap. 5. In the first order approximation, the interface conditions for the temperature field can be written in the following forms: at $\bar{\tau} = \bar{\tau}_{0,0}$,

$$\begin{aligned} &\bar{T}_1 + \bar{T}_0'(\bar{\tau}_{0,1} + \bar{h}_0) = 0, \\ &\bar{\tau}_{0,0}\left[\frac{\partial \bar{T}_1}{\partial \bar{\tau}} + \bar{T}_0''(\bar{\tau}_{0,1} + \bar{h}_0)\right] + \bar{T}_0'(\bar{\tau}_{0,1} + \bar{h}_0) \\ &\qquad + \Big[\bar{\tau}_{0,1} + \bar{h}_0 + \bar{\sigma}\bar{h}_0'(\bar{\sigma})\Big] = 0. \end{aligned} \tag{7.114}$$

In terms of the formulations,

$$\bar{D}'_{0,0}(\bar{\tau}_{0,0}) = -1, \quad \bar{\tau}_{0,0}\bar{D}''_{0,0}(\bar{\tau}_{0,0}) = (1 + \Pr)(\bar{\tau}_{0,1} + \bar{h}_0), \tag{7.115}$$

we derive that

$$\begin{aligned} &\bar{D}_{0,1}(\bar{\tau}_{0,0}) + \bar{D}'_{0,0}(\bar{\tau}_{0,0})(\bar{\tau}_{0,1} + \bar{h}_{0,0}) = 0, \\ &\bar{\tau}_{0,0}\Big[\bar{D}'_{0,1}(\bar{\tau}_{0,0}) + \bar{D}''_{0,0}(\tau_{0,0})(\bar{\tau}_{0,1} + \bar{h}_{0,0})\Big] - 2\bar{h}_{1,0} = 0, \end{aligned} \tag{7.116}$$

and

$$\begin{aligned} &\bar{D}_{n,1}(\bar{\tau}_{0,0}) + \bar{D}'_{0,0}(\bar{\tau}_{0,0})\bar{h}_{n,0} = 0, \\ &\bar{\tau}_{0,0}\Big[\bar{D}'_{n,1}(\bar{\tau}_{0,0}) + \bar{D}''_{0,0}(\bar{\tau}_{0,0})\bar{h}_{n,0}\Big] + nh_{n,0} - (n+2)h_{n+1,0} = 0, \end{aligned} \tag{7.117}$$

for $(n = 1, 2, \cdots, N)$. Now (7.116) and (7.117) may be re-written as

$$\begin{aligned} &\bar{d}_{0,1}\widehat{W}_{0,1}(\mathrm{Pr}\bar{\tau}_{0,0}) - (\bar{\tau}_{0,1} + \bar{h}_{0,0}) = -\Big[\overline{Q}_{0,1}(\bar{\tau}_{0,0}) + \overline{P}_{0,1}(\bar{\tau}_{0,0})\Big], \\ &\mathrm{Pr}\bar{\tau}_{0,0}\bar{d}_{0,1}\widehat{W}'_{0,1}(\mathrm{Pr}\bar{\tau}_{0,0}) + (1 + \mathrm{Pr}\bar{\tau}_{0,0})(\bar{\tau}_{0,1} + \bar{h}_{0,0}) - 2h_{1,0} \\ &\qquad = -\bar{\tau}_{0,0}\Big[\overline{Q}'_{0,1}(\bar{\tau}_{0,0}) + \overline{P}'_{0,1}(\bar{\tau}_{0,0})\Big], \end{aligned} \tag{7.118}$$

and

$$\begin{aligned} &\bar{d}_{n,1}\widehat{W}_{n,1}(\mathrm{Pr}\bar{\tau}_{0,0}) - \bar{h}_{n,0} = -\overline{P}_{n,1}(\bar{\tau}_{0,0}), \\ &\mathrm{Pr}\bar{\tau}_{0,0}\bar{d}_{n,1}\widehat{W}'_{n,1}(\mathrm{Pr}\bar{\tau}_{0,0}) + (n + 1 + \mathrm{Pr}\bar{\tau}_{0,0})\bar{h}_{n,0} \\ &\qquad -(n+2)h_{n+1,0} = -\bar{\tau}_{0,0}\overline{P}'_{n,1}(\bar{\tau}_{0,0}), \end{aligned} \tag{7.119}$$

for $(n = 1, 2, \ldots, N)$. We truncate the series at $n = N$, by setting $\bar{h}_{n,0} = \bar{d}_{n,1} = 0$, $(n > N)$. Then the formulas (7.118) and (7.119), combining with the tip condition,

$$\bar{h}_{\mathrm{s}}(0, \epsilon_0) = \sum_{n=0}^{N} \bar{h}_{n,0} = 0. \tag{7.120}$$

uniquely determine the $(2N+3)$ unknown constants: $\{\bar{d}_{n,1}, \bar{h}_{n,0}\}, (n = 0, 1, 2, \ldots, N)$ and $\bar{\tau}_{0,1}$.

With the above results, we may finally determine the interface shape function in the form

$$\bar{\tau} = \bar{\tau}_{\mathrm{s}}(\bar{\sigma}) = \bar{\tau}_0 + \epsilon_0 \sum_{n=0}^{N} h_{n,0} L_n^{(1)}(\bar{\sigma}) + \cdots, \tag{7.121}$$

and

$$\bar{\tau}_0 = \bar{\tau}_{0,0} + \epsilon_0 \bar{\tau}_{0,1} + \cdots. \tag{7.122}$$

In the coordinate system (ξ, η), we have

$$\eta = \eta_{\mathrm{s}}(\xi) = \sqrt{\frac{\bar{\tau}_{\mathrm{s}}}{\bar{\tau}_0}} = 1 + \epsilon_0 \frac{\mathrm{Pr}}{\eta_0^2} \sum_{n=0}^{N} \bar{h}_{n,0} L_n^{(1)}(\bar{\sigma}) + \cdots. \tag{7.123}$$

4. Summary

In this chapter, we study the steady dendritic growth with small buoyancy effect, $\overline{\mathrm{G}} = \epsilon_0 \ll 1$. The solution is expressed in the Laguerre series representation, and we find a uniformly valid asymptotic expansion of the solution in the limit $\epsilon_0 \to 0$ in the entire flow field. From the structure of the solution derived in this chapter, at least one can conclude that, with any small buoyancy effect ($\epsilon_0 \neq 0$), the solution of dendritic growth is no longer an exact similar solution. Accordingly, in the first-order approximation, the shape of the dendrite is no longer a pure paraboloid. The small buoyancy effect introduces many small non-similar ingredients to the solution, which are represented, in the Laguerre series representation, by the infinitely many components $L_n^{(1)}(\bar{\sigma})$ $(n = 0, 1, 2, \ldots)$. So, it is proper to call our solution a 'nearly' similar solution.

From (7.121), we derive the mean curvature of the tip

$$\mathcal{K}(0) = \frac{2}{\eta_0^2}\left\{1 - \epsilon_0\left[\frac{\eta_0^2}{\mathrm{Pr}}\frac{\mathrm{d}}{\mathrm{d}\bar{\sigma}}\bar{\eta}_\mathrm{s}(0) + \bar{\eta}_\mathrm{s}(0)\right]\right\}. \tag{7.124}$$

Thus, the Peclet number of dendritic growth with the buoyancy effect is obtained as

$$\begin{aligned}\mathrm{Pe} = \frac{\ell_\mathrm{t}}{\ell_\mathrm{T}} = \frac{2}{\mathcal{K}(0)} &= \eta_0^2\left\{1 + \epsilon_0\left[\frac{\eta_0^2}{\mathrm{Pr}}\frac{\mathrm{d}}{\mathrm{d}\bar{\sigma}}\bar{\eta}_\mathrm{s}(0) + \bar{\eta}_\mathrm{s}(0)\right]\right\} \\ &= \eta_0^2\left\{1 + \epsilon_0\sum_{n=0}^{\infty}\bar{h}_{n,0}\left[L_n^{(1)\prime}(0) + \frac{\mathrm{Pr}}{\eta_0^2}L_n^{(1)}(0)\right]\right\},\end{aligned} \tag{7.125}$$

where

$$\eta_0^2 = 2\mathrm{Pr}(\bar{\tau}_{0,0} + \epsilon_0\bar{\tau}_{0,1} + \cdots) = \mathrm{Pe}_0\left(1 + \epsilon_0\frac{\bar{\tau}_{0,1}}{\bar{\tau}_{0,0}} + \cdots\right) \tag{7.126}$$

and Pe_0 is the Peclet number of dendrite growth in zero gravity. Combining (7.125) and (7.126), we derive

$$\frac{\mathrm{Pe}}{\mathrm{Pe}_0} = 1 + \epsilon_0\left\{\frac{\bar{\tau}_{0,1}}{\bar{\tau}_{0,0}} + \sum_{n=0}^{\infty}\bar{h}_{n,0}\left[L_n^{(1)\prime}(0) + \frac{\mathrm{Pr}}{\eta_0^2}L_n^{(1)}(0)\right]\right\}. \tag{7.127}$$

Chapter 8

STEADY DENDRITIC GROWTH WITH NATURAL CONVECTION (II): THE CASE OF PR $\gg$ 1 AND $\overline{\mathrm{G}} = O(1)$

In this chapter, we study the case of $\mathrm{Pr} \gg 1$, but $\overline{\mathrm{G}} = O(1)$. For this case, we shall use $\epsilon_2 = 1/\mathrm{Pr} \ll 1$ as the basic small parameter, and look for the asymptotic expansion solution in the limit $\epsilon_2 \to 0$.

Note that in the coordinate system (ξ, η), the interface shape function $\eta_{\mathrm{s}} = O(1)$. However, in the new coordinate system $(\bar{\sigma}, \bar{\tau})$, the interface shape function becomes

$$\bar{\tau}_{\mathrm{s}}(\bar{\sigma}, \epsilon_2) = \frac{\eta_0^2}{2\mathrm{Pr}}\eta_{\mathrm{s}}^2 = \epsilon_2 \frac{\eta_0^2 \eta_{\mathrm{s}}^2}{2}. \tag{8.1}$$

At the dendrite's tip, $\xi = \bar{\sigma} = 0$, from the normalization condition we derive

$$\bar{\tau}_{\mathrm{s}}(0) = \bar{\tau}_0 = \epsilon_2 \frac{\eta_0^2}{2}. \tag{8.2}$$

We further assume that $\frac{\eta_0^2}{2} = O(1)$. Consequently, we have $\bar{\tau}_0 = O(\epsilon_2)$. Hence, very similar to the case studied in Chap. 6, the preset problem is also a *singular boundary problem*, which can be solved by following almost the same approach carried out in Chap. 6. Thus, the following presentation will largely repeat Chap. 6. We first divide the whole physical region into two sub-regions:

- the outer region away from the interface, $\bar{\tau} = O(1)$,
- the inner region near the interface, $\bar{\tau} \ll 1$,

and derive the outer and inner expansion solutions in each sub-region, then match them in the intermediate region.

1. Laguerre Series Representation and Asymptotic Forms of Solutions

1.1 Laguerre Series Representation of the Solution

As in the previous chapters, we expand the solution in the Laguerre series

$$\begin{cases} \bar{\zeta}(\bar{\sigma},\bar{\tau}) = \bar{\sigma} \sum\limits_{n=0}^{\infty} \bar{A}_n(\bar{\tau},\epsilon_2) L_n^{(1)}(\bar{\sigma}), \\ \bar{\Psi}(\bar{\sigma},\bar{\tau}) = \bar{\sigma} \sum\limits_{n=0}^{\infty} \bar{B}_n(\bar{\tau},\epsilon_2) L_n^{(1)}(\bar{\sigma}), \\ \bar{T}(\bar{\sigma},\bar{\tau}) = \sum\limits_{n=0}^{\infty} \bar{D}_n(\bar{\tau},\epsilon_2) L_n^{(1)}(\bar{\sigma}). \end{cases} \tag{8.3}$$

We then derive that for the flow field, we have

$$\begin{aligned} \frac{\mathrm{d}^2 \bar{B}_n}{\mathrm{d}\bar{\tau}^2} &= \frac{n+2}{\bar{\tau}} \bar{B}_{n+1} - \left[1 + \frac{2(n+1)}{\bar{\tau}}\right] \bar{A}_n \\ &\quad + \tfrac{1}{\bar{\tau}}\Big[(n+2)\bar{A}_{n+1} + n\bar{A}_{n-1}\Big], \\ \frac{\mathrm{d}^2 \bar{A}_n}{\mathrm{d}\bar{\tau}^2} + \frac{\mathrm{d}\bar{A}_n}{\mathrm{d}\bar{\tau}} - \frac{n+1}{\bar{\tau}}\bar{A}_n &= \overline{\mathrm{G}}\Big(\frac{\mathrm{d}\bar{D}_n}{\mathrm{d}\bar{\tau}} - \sum_{m=n+1}^{\infty} \bar{D}_m\Big) \\ &\quad - \tfrac{\epsilon_2^2}{2}\mathcal{N}\Big\{\bar{A}_n; \bar{B}_n\Big\} \\ &\quad (n = 0, 1, 2, \ldots;) \end{aligned} \tag{8.4}$$

for the temperature field, we have

$$\epsilon_2 \bar{\tau} \frac{\mathrm{d}^2 \bar{D}_n}{\mathrm{d}\bar{\tau}^2} + (\epsilon_2 + \bar{\tau}) \frac{\mathrm{d}\bar{D}_n}{\mathrm{d}\bar{\tau}} - n\bar{D}_n = \Big[\epsilon_2(n+1) - (n+2)\Big]\bar{D}_{n+1}$$

$$-\epsilon_2 \sum_{m=n+2}^{\infty} \bar{D}_n + \frac{\epsilon_2^2}{2}\mathcal{N}_T\Big\{\bar{D}_n, \bar{B}_n\Big\}. \quad (n = 0, 1, 2, \ldots). \tag{8.5}$$

1.2 Outer Expansion Form of the Solution

In the outer region $\tau = O(1)$, we can make the following outer asymptotic expansion in the limit $\epsilon_2 \to 0$ for the solution of the flow field:

$$\begin{aligned} \bar{A}_n(\bar{\tau},\epsilon_2) &= \bar{\nu}_0(\epsilon_2)\bar{A}_{0,n}(\bar{\tau}) + \bar{\nu}_1(\epsilon_2)\bar{A}_{1,n}(\bar{\tau}) + \cdots, \\ \bar{B}_n(\bar{\tau},\epsilon_2) &= \bar{\nu}_0(\epsilon_2)\bar{B}_{0,n}(\bar{\tau}) + \bar{\nu}_1(\epsilon_2)\bar{B}_{1,n}(\bar{\tau}) + \cdots, \end{aligned} \tag{8.6}$$

for $(n = 0, 1, \ldots)$. The asymptotic factors, $\bar{\nu}_n(\epsilon_2)$,

$$\bar{\nu}_0(\epsilon_2) \gg \bar{\nu}_1(\epsilon_2) \gg \bar{\nu}_2(\epsilon_2) \gg \cdots$$

are to be determined later. Thus, the outer solution will have the following general form of asymptotic expansion:

$$\begin{aligned}\zeta &= \nu_0(\epsilon_2)\zeta_0 + \bar{\nu}_1(\epsilon_2)\bar{\zeta}_1 + \bar{\nu}_2(\epsilon_2)\bar{\zeta}_2 + \cdots, \\ \bar{\Psi} &= \bar{\nu}_0(\epsilon_2)\bar{\Psi}_0 + \bar{\nu}_1(\epsilon_2)\bar{\Psi}_1 + \bar{\nu}_2(\epsilon_2)\bar{\Psi}_2 + \cdots .\end{aligned} \tag{8.7}$$

With the factor $\bar{\nu}_m(\epsilon_2)$, the solution $\bar{\Psi}_m$ will have the asymptotic structure

$$\bar{\Psi}_m = \bar{B}_{m,0}L_0^{(1)}(\bar{\sigma}) + \bar{B}_{m,1}L_1^{(1)}(\bar{\sigma}) + \bar{B}_{m,2}L_2^{(1)}(\bar{\sigma}) + \cdots . \tag{8.8}$$

The solution of vorticity function has similar asymptotic structure.

1.3 Inner Expansion Form of the Solution

Noting that the parameter $\bar{\tau}_0$ can be expressed as:

$$\bar{\tau}_0 = \epsilon_2\hat{\tau}_0, \quad \hat{\tau}_0 = \frac{\eta_0^2}{2} = O(1), \tag{8.9}$$

in the inner region near the interface, $\bar{\tau} = O(\epsilon_2)$, we introduce the inner variables $(\bar{\sigma}, \hat{\tau})$, where

$$\hat{\tau} = \frac{\bar{\tau}}{\epsilon_2} . \tag{8.10}$$

With the inner variables, we have

$$\begin{aligned}\Psi &= \bar{\Psi}_*(\bar{\sigma}, \hat{\tau}) + \hat{\Psi}(\bar{\sigma}, \hat{\tau}, \epsilon_2), \\ \zeta &= \frac{1}{\mathrm{Pr}^2}\hat{\zeta}(\bar{\sigma}, \hat{\tau}, \epsilon_2), \\ T &= T_\infty + \mathrm{Pr}\hat{T}(\bar{\sigma}, \hat{\tau}, \epsilon_2),\end{aligned} \tag{8.11}$$

where

$$\bar{\Psi}_* = \frac{2}{\epsilon_2}\bar{\sigma}\hat{\tau} \tag{8.12}$$

and

$$\begin{cases}\hat{\Psi}(\bar{\sigma}, \hat{\tau}, \epsilon_2) = \bar{\Psi}(\bar{\sigma}, \epsilon_2\hat{\tau}, \epsilon_2), \\ \hat{\zeta}(\bar{\sigma}, \hat{\tau}, \epsilon_2) = \bar{\zeta}(\bar{\sigma}, \epsilon_2\hat{\tau}, \epsilon_2), \\ \hat{T}(\bar{\sigma}, \hat{\tau}, \epsilon_2) = \bar{T}(\bar{\sigma}, \epsilon_2\hat{\tau}, \epsilon_2).\end{cases} \tag{8.13}$$

We expand the inner solution in the Laguerre series

$$\begin{cases} \hat{\Psi}(\bar{\sigma},\hat{\tau}) = \bar{\sigma}\sum\limits_{n=0}^{\infty}\hat{B}_n(\hat{\tau},\epsilon_2)L_n^{(1)}(\bar{\sigma}), \\ \hat{\zeta}(\bar{\sigma},\hat{\tau}) = \bar{\sigma}\sum\limits_{n=0}^{\infty}\hat{A}_n(\hat{\tau},\epsilon_2)L_n^{(1)}(\bar{\sigma}), \\ \hat{T}(\bar{\sigma},\hat{\tau}) = \sum\limits_{n=0}^{\infty}\hat{D}_n(\hat{\tau},\epsilon_2)L_n^{(1)}(\bar{\sigma}). \end{cases} \tag{8.14}$$

Then, the inner system can be obtained from (8.4) as follows:

$$\begin{aligned} \frac{d^2\hat{B}_n}{d\hat{\tau}^2} &= \epsilon_2\frac{n+2}{\hat{\tau}}\hat{B}_{n+1} - \left[\epsilon_2^2 + \epsilon_2\frac{2(n+1)}{\hat{\tau}}\right]\hat{A}_n \\ &+\frac{\epsilon_2}{\hat{\tau}}\Big[(n+2)\hat{A}_{n+1} + n\hat{A}_{n-1}\Big], \\ \frac{d^2\hat{A}_n}{d\hat{\tau}^2} &+ \epsilon_2\frac{d\hat{A}_n}{d\hat{\tau}} - \epsilon_2\frac{n+1}{\hat{\tau}}\hat{A}_n = \epsilon_2\overline{\mathrm{G}}\Big(\frac{d\hat{D}_n}{d\hat{\tau}} - \epsilon_2\sum_{m=n+1}^{\infty}\hat{D}_m\Big) \\ &-\frac{\epsilon_2^3}{2}\hat{\mathcal{N}}\Big\{\hat{A}_n;\hat{B}_n\Big\}, \quad (n=0,1,2,\cdots), \end{aligned} \tag{8.15}$$

and

$$\begin{aligned} \hat{\tau}\frac{d^2\hat{D}_n}{d\hat{\tau}^2} + (1+\hat{\tau})\frac{d\hat{D}_n}{d\hat{\tau}} - n\hat{D}_n &= \Big[\epsilon_2(n+1) - (n+2)\Big]\hat{D}_{n+1} \\ -\epsilon_2\sum_{m=n+2}^{\infty}\hat{D}_n &+ \frac{\epsilon_2}{2}\hat{\mathcal{N}}_T\Big\{\hat{D}_n,\hat{B}_n\Big\}, \quad (n=0,1,2,\cdots). \end{aligned} \tag{8.16}$$

We make the following inner asymptotic expansion in the limit $\epsilon_2 \to 0$:

$$\begin{aligned} \hat{A}_n(\hat{\tau},\epsilon_2) &= \hat{\nu}_0(\epsilon_2)\hat{A}_{0,n}(\hat{\tau}) + \hat{\nu}_1(\epsilon_2)\hat{A}_{1,n}(\hat{\tau}) + \cdots, \\ \hat{B}_n(\hat{\tau},\epsilon_2) &= \epsilon_2\Big\{\hat{\nu}_0(\epsilon_2)\hat{B}_{0,n}(\hat{\tau}) + \hat{\nu}_1(\epsilon_2)\hat{B}_{1,n}(\hat{\tau}) + \cdots\Big\}, \\ \hat{D}_n(\hat{\tau},\epsilon_2) &= \hat{\gamma}_0(\epsilon_2)\hat{D}_{0,n}(\hat{\tau}) + \hat{\gamma}_1(\epsilon_2)\hat{D}_{0,n}(\hat{\tau}) + \hat{\gamma}_2(\epsilon_2)\hat{D}_{1,n}(\hat{\tau}) + \cdots \end{aligned} \tag{8.17}$$

for $(n = 0, 1, \ldots)$. The asymptotic factors, $\hat{\nu}_n(\epsilon_2)$ and $\hat{\gamma}_n(\epsilon_2)$,

$$\hat{\nu}_0(\epsilon_2) \gg \hat{\nu}_1(\epsilon_2) \gg \hat{\nu}_2(\epsilon_2) \gg \cdots,$$

$$\hat{\gamma}_0(\epsilon_2) \gg \hat{\gamma}_1(\epsilon_2) \gg \hat{\gamma}_2(\epsilon_2) \gg \cdots,$$

are to be determined later. So, the inner solutions have the following general form of expansion:

$$\begin{aligned} \hat{\zeta} &= \hat{\nu}_0(\epsilon_2)\hat{\zeta}_0 + \hat{\nu}_1(\epsilon_2)\hat{\zeta}_1 + \hat{\nu}_2(\epsilon_2)\hat{\zeta}_2 + \cdots, \\ \hat{\Psi} &= \epsilon_2\left[\hat{\nu}_0(\epsilon_2)\hat{\Psi}_0 + \hat{\nu}_1(\epsilon_2)\hat{\Psi}_1 + \hat{\nu}_2(\epsilon_2)\hat{\Psi}_2 + \cdots\right], \\ \hat{T} &= \hat{\gamma}_0(\epsilon_2)\hat{T}_0 + \hat{\gamma}_1(\epsilon_2)\hat{T}_1 + \hat{\gamma}_2(\epsilon_2)\hat{T}_2 + \cdots. \end{aligned} \tag{8.18}$$

With the factor $\hat{\nu}_m(\epsilon_2)$, the solution $\hat{\Psi}_m$ will have the structure

$$\hat{\Psi}_m = \hat{B}_{m,0}L_0^{(1)}(\hat{\sigma}) + \hat{B}_{m,1}L_1^{(1)}(\hat{\sigma}) + \hat{B}_{m,2}L_2^{(1)}(\hat{\sigma}) + \cdots. \quad (8.19)$$

The solutions of the vorticity function and temperature have similar asymptotic structure.

With the inner variable, the interface shape can be written in the form $\hat{\tau} = \hat{\tau}_s(\bar{\sigma}) = \hat{\tau}_0 + \hat{h}_s(\bar{\sigma})$, where $|\hat{h}_s| \ll \hat{\tau}_0$. From the normalization condition (8.2), it follows that

$$\hat{h}_s(0) = 0. \quad (8.20)$$

Similarly, we make the Laguerre expansion for the function $\hat{h}_s(\bar{\sigma})$:

$$\hat{h}_s(\bar{\sigma}) = \sum_{n=0}^{\infty} h_n(\epsilon_2)L_n^{(1)}(\bar{\sigma}). \quad (8.21)$$

As $\epsilon_2 \to 0$, the coefficients $\hat{h}_n (n = 0, 1, 2, \ldots)$ have the asymptotic structure

$$h_n(\epsilon_2) = \hat{\delta}_0(\epsilon_2)\hat{h}_{0,n} + \hat{\delta}_1(\epsilon_2)\hat{h}_{1,n} + \cdots. \quad (8.22)$$

Alternatively, we may make the asymptotic expansion

$$\hat{h}_s = \hat{\delta}_0(\epsilon_2)\hat{h}_0(\bar{\sigma}) + \hat{\delta}_1(\epsilon_2)\hat{h}_1(\bar{\sigma}) + \hat{\delta}_2(\epsilon_2)\hat{h}_2(\bar{\sigma}) + \cdots \quad (8.23)$$

as $\epsilon_2 \to 0$. Then make a Laguerre expansion for each solution $\hat{h}_m(\bar{\sigma})$, $(m = 0, 1, 2, \ldots)$:

$$\hat{h}_m(\bar{\sigma}) = \hat{h}_{m,0}L_0^{(1)}(\hat{\sigma}) + \hat{h}_{m,1}L_1^{(1)}(\hat{\sigma}) + \hat{h}_{m,2}L_2^{(1)}(\hat{\sigma}) + \cdots. \quad (8.24)$$

Moreover, note that $\hat{\tau}_0$, like the parameter η_0^2, depends on ϵ_2 and other physical parameters. We assume that as $\epsilon_2 = 0$,

$$\eta_0^2 = \eta_{0,0}, \qquad \hat{\tau}_0 = \hat{\tau}_{0,0}. \quad (8.25)$$

Letting

$$\begin{aligned} \eta_0^2(\epsilon_2) &= \eta_{0,0}^2 + \tilde{\eta}_0(\epsilon_2), \\ \hat{\tau}_0(\epsilon_2) &= \hat{\tau}_{0,0} + \tilde{\tau}_0(\epsilon_2), \end{aligned} \quad (8.26)$$

we may make the expansion

$$\begin{aligned} \tilde{\eta}_0(\epsilon_2) &= \hat{\delta}_0(\epsilon_2)\eta_{0,1} + \hat{\delta}_1(\epsilon_2)\eta_{0,2} + \cdots, \\ \tilde{\tau}_0(\epsilon_2) &= \hat{\delta}_0(\epsilon_2)\hat{\tau}_{0,1} + \hat{\delta}_1(\epsilon_2)\hat{\tau}_{0,2} + \cdots. \end{aligned} \quad (8.27)$$

Since $(\tilde{\tau}_0+\hat{h}_{\rm s})\ll 1$, we can make a Taylor expansion for the interface conditions (7.26)–(7.29) around $\hat{\tau}=\hat{\tau}_{0,0}$ as follows: at $\hat{\tau}=\hat{\tau}_{0,0}$,

$$\frac{\partial\hat{\Psi}}{\partial\bar{\sigma}}+\frac{\partial^2\hat{\Psi}}{\partial\bar{\sigma}\partial\hat{\tau}}(\tilde{\tau}_0+\hat{h}_{\rm s})+\frac{1}{2!}\frac{\partial^3\hat{\Psi}}{\partial\bar{\sigma}\partial\hat{\tau}^2}(\tilde{\tau}_0+\hat{h}_{\rm s})^2+\cdots$$
$$+\hat{h}_{\rm s}'\left(\frac{\partial\hat{\Psi}}{\partial\hat{\tau}}+\frac{\partial^2\hat{\Psi}}{\partial^2\hat{\tau}}(\tilde{\tau}_0+\hat{h}_{\rm s})+\frac{1}{2!}\frac{\partial^3\hat{\Psi}}{\partial\hat{\tau}^3}(\tilde{\tau}_0+\hat{h}_{\rm s})^2+\cdots\right)=0, \quad (8.28)$$

$$(\hat{\tau}_{0,0}+\tilde{\tau}_0+\hat{h}_{\rm s})\left(\frac{\partial\hat{\Psi}}{\partial\hat{\tau}}+\frac{\partial^2\hat{\Psi}}{\partial\hat{\tau}^2}(\tilde{\tau}_0+\hat{h}_{\rm s})+\frac{1}{2!}\frac{\partial^3\hat{\Psi}}{\partial\hat{\tau}^3}(\tilde{\tau}_0+\hat{h}_{\rm s})^2+\cdots\right)$$
$$-\epsilon_2\bar{\sigma}\hat{h}_{\rm s}'\left(\frac{\partial\hat{\Psi}}{\partial\bar{\sigma}}+\frac{\partial^2\hat{\Psi}}{\partial\bar{\sigma}\partial\hat{\tau}}(\tilde{\tau}_0+\hat{h}_{\rm s})+\frac{1}{2!}\frac{\partial^3\hat{\Psi}}{\partial\bar{\sigma}\partial\hat{\tau}^2}(\tilde{\tau}_0+\hat{h}_{\rm s})^2+\cdots\right)=0, (8.29)$$

$$\epsilon_2T_\infty+\left(\hat{T}+\frac{\partial\hat{T}}{\partial\hat{\tau}}(\tilde{\tau}_0+\hat{h}_{\rm s})+\frac{1}{2!}\frac{\partial^2\hat{T}}{\partial\hat{\tau}^2}(\tilde{\tau}_0+\hat{h}_{\rm s})^2+\cdots\right)=0, \quad (8.30)$$

$$(\hat{\tau}_{0,0}+\tilde{\tau}_0+\hat{h}_{\rm s})\left(\frac{\partial\hat{T}}{\partial\hat{\tau}}+\frac{\partial^2\hat{T}}{\partial\hat{\tau}^2}(\tilde{\tau}_0+\hat{h}_{\rm s})+\frac{1}{2!}\frac{\partial^3\hat{T}}{\partial\hat{\tau}^3}(\tilde{\tau}_0+\hat{h}_{\rm s})^2+\cdots\right)$$
$$-\epsilon_2\bar{\sigma}\hat{h}_{\rm s}'\left(\frac{\partial\hat{T}}{\partial\bar{\sigma}}+\frac{\partial^2\hat{T}}{\partial\bar{\sigma}\partial\hat{\tau}}(\tilde{\tau}_0+\hat{h}_{\rm s})+\frac{1}{2!}\frac{\partial^3\hat{T}}{\partial\bar{\sigma}\partial\hat{\tau}^2}(\tilde{\tau}_0+\hat{h}_{\rm s})^2+\cdots\right)$$
$$+\epsilon_2(\hat{\tau}_{0,0}+\tilde{\tau}_0+\hat{h}_{\rm s}+\bar{\sigma}\hat{h}_{\rm s}')=0. \quad (8.31)$$

2. Leading-Order Asymptotic Expansion Solutions

2.1 Leading-Order Asymptotic Expansion Solution of the Temperature Field

It is evident that for the problem under study, the temperature perturbation plays an active role. One has the non-uniform, perturbed temperature field first. Then, due to the buoyancy effect, it induces the perturbed flow field. Hence, to proceed, one needs to find the solution for the perturbed temperature field first. Moreover, as we have pointed out in the last chapter, as a singular perturbation problem, the temperature perturbation is restricted in the inner region and governed by the system (8.16); its outer solution can be set to be zero. Therefore, the interaction between the temperature field and flow field is only through the inner solutions of the perturbed temperature and stream function.

The first sequence of solutions $\hat{\gamma}_0(\epsilon_2)\hat{T}_0$ with the leading the factor $\hat{\gamma}_0(\epsilon_2)$ is determined by the latent heat release without involving flow field. Hence, it can be assumed to be similarity solutions only depending on the variable $\hat{\tau}$. Namely, we have

$$\hat{D}_{0,n} = 0 \quad \text{for} \quad n = 1, 2, \ldots, \tag{8.32}$$

and

$$\hat{T}_0 = \hat{\gamma}_0(\epsilon_2)\hat{D}_{0,0}(\hat{\tau})L_0^{(1)}(\bar{\sigma}). \tag{8.33}$$

The governing equation of $\hat{D}_{0,0}(\hat{\tau})$ is

$$\frac{\mathrm{d}^2\hat{D}_{0,0}}{\mathrm{d}\hat{\tau}^2} + \left(1 + \frac{1}{\hat{\tau}}\right)\frac{\mathrm{d}\hat{D}_{0,0}}{\mathrm{d}\hat{\tau}} = 0. \tag{8.34}$$

From the leading-order approximation of the interface condition (8.31), we derive

$$\hat{\gamma}_0(\epsilon_2) = \epsilon_2, \tag{8.35}$$

and from (8.30)–(8.31), we derive that, at $\hat{\tau} = \hat{\tau}_{0,0}$,

$$\hat{D}_{0,0}(\hat{\tau}_{0,0}) + T_\infty = 0, \tag{8.36}$$

$$\hat{D}'_{0,0}(\hat{\tau}_{0,0}) + 1 = 0\,. \tag{8.37}$$

The solution is obtained as

$$\hat{D}_{0,0} = I_* + I_{0,0}E_1(\hat{\tau}). \tag{8.38}$$

From the far-field condition,

$$\hat{D}_{0,0} \to 0, \qquad \text{as} \quad \hat{\tau} \to \infty, \tag{8.39}$$

we derive

$$I_* = 0. \tag{8.40}$$

Furthermore, from interface condition (8.37), we derive

$$I_{0,0} = \hat{\tau}_{0,0}\mathrm{e}^{\hat{\tau}_{0,0}}, \tag{8.41}$$

so that we have the solution

$$\hat{D}_{0,0} = \hat{\tau}_{0,0}\mathrm{e}^{\hat{\tau}_{0,0}}E_1(\hat{\tau}). \tag{8.42}$$

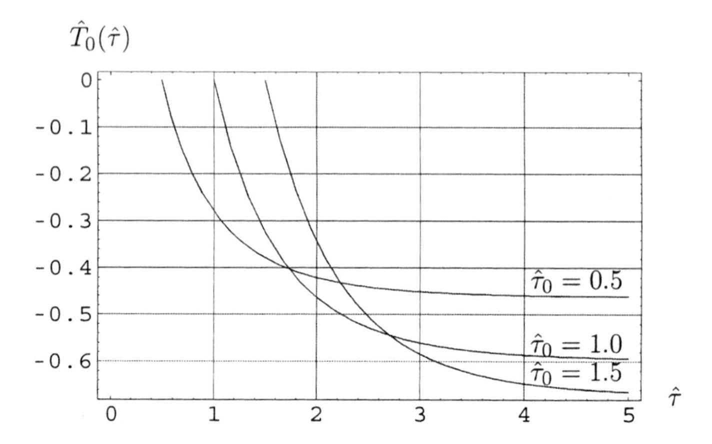

Figure 8.1. The graphs of the solution $\hat{T}_0(\hat{\tau})$ versus $\hat{\tau}$ for $\hat{\tau}_{0,0} = 0.5,\ 1.0,\ 1.5$.

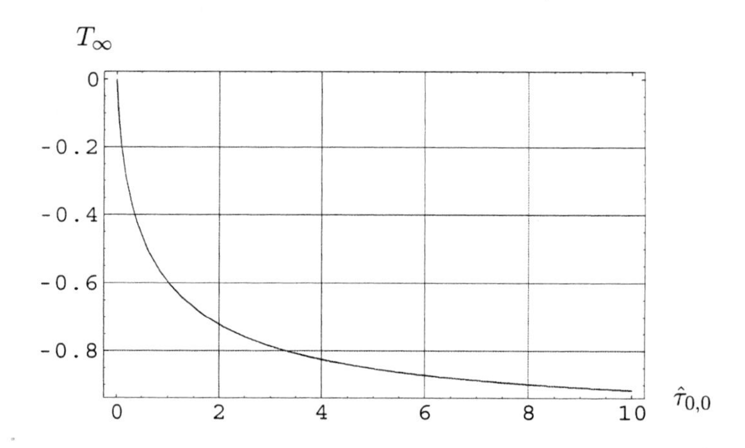

Figure 8.2. The graph of T_∞ versus $\hat{\tau}_{0,0}$.

Finally, from interface condition (8.36), we derive

$$T_\infty = -\hat{\tau}_{0,0} e^{\hat{\tau}_{0,0}} E_1(\hat{\tau}_{0,0}). \tag{8.43}$$

This solution is just the Ivantsov solution, as expected.

In Fig 8.1, we show the distributions of the temperature $\hat{T}_0(\hat{\tau})$ along $\hat{\tau}$-axis for the cases, $\hat{\tau}_{0,0} = 0.5, 1.0, 1.5$, while in Fig 8.2, we show the variation of undercooling temperature T_∞ with the parameter $\hat{\tau}_{0,0}$.

2.2 Leading-Order Inner Solutions of the Velocity Field $O(\epsilon_2^2)$

Due to the buoyancy effect, the non-uniform temperature distribution induces a perturbed flow field in the inner region, which will also be of $O(\epsilon_2)$. Hence, with the factor $\hat{\nu}_0(\epsilon_2) = \epsilon_2\hat{\gamma}_0(\epsilon_2) = \epsilon_2^2$, the first sequence of inner solutions of flow, $\{\hat{A}_{0,n}, \hat{B}_{0,n}\}$ are governed by the following system:

(1) For $n = 0$,

$$\frac{\mathrm{d}^2\hat{\mathrm{B}}_{0,0}}{\mathrm{d}\hat{\tau}^2} = -\frac{2}{\hat{\tau}}\hat{A}_{0,0} + \frac{2}{\hat{\tau}}\hat{A}_{0,1}\,, \tag{8.44}$$

$$\frac{\mathrm{d}^2\hat{\mathrm{A}}_{0,0}}{\mathrm{d}\hat{\tau}^2} = \overline{\mathrm{G}}\hat{D}'_{0,0}(\hat{\tau}) = -\overline{\mathrm{G}}I_{0,0}\frac{\mathrm{e}^{-\hat{\tau}}}{\hat{\tau}}\,. \tag{8.45}$$

(2) For $n = 1$,

$$\frac{\mathrm{d}^2\hat{\mathrm{B}}_{0,1}}{\mathrm{d}\hat{\tau}^2} = -\frac{4}{\hat{\tau}}\hat{A}_{0,1} + \frac{3}{\hat{\tau}}\hat{A}_{0,2} + \frac{1}{\hat{\tau}}\bar{A}_{0,0}\,, \tag{8.46}$$

$$\frac{\mathrm{d}^2\hat{\mathrm{A}}_{0,1}}{\mathrm{d}\hat{\tau}^2} = 0. \tag{8.47}$$

(3) For $n = 2, 3, \ldots$,

$$\frac{\mathrm{d}^2\hat{\mathrm{B}}_{0,\mathrm{n}}}{\mathrm{d}\hat{\tau}^2} = -\frac{2n+2}{\hat{\tau}}\hat{A}_{0,n} + \frac{n+2}{\hat{\tau}}\hat{A}_{0,n+1} + \frac{n}{\hat{\tau}}\bar{A}_{0,n-1}\,, \tag{8.48}$$

$$\frac{\mathrm{d}^2\hat{\mathrm{A}}_{0,\mathrm{n}}}{\mathrm{d}\hat{\tau}^2} = 0, \tag{8.49}$$

with the boundary condition,

$$\hat{B}_{0,n}(\hat{\tau}_{0,0}) = \hat{B}'_{0,n}(\hat{\tau}_{0,0}) = 0. \tag{8.50}$$

From the above system, one derives that

$$\hat{A}_{0,n} = 0 \quad (n = 1, 2, 3, \ldots), \tag{8.51}$$

$$\hat{B}_{0,n} = 0 \quad (n = 2, 3, 4, \ldots), \tag{8.52}$$

and

$$\hat{A}_{0,0} = \overline{\mathrm{G}}I_{0,0}\hat{P}_0(\hat{\tau}), \tag{8.53}$$

where

$$\hat{P}_0(\hat{\tau}) = \hat{\tau}E_1(\hat{\tau}) - \mathrm{e}^{-\hat{\tau}}. \tag{8.54}$$

Furthermore, one derives that

$$\begin{aligned} \hat{B}_{0,0} &= -2\overline{\mathrm{G}}I_{0,0}\left[\widehat{Q}_0(\hat{\tau}) - \widehat{Q}_0(\hat{\tau}_{0,0}) - \widehat{Q}_0'(\hat{\tau}_{0,0})(\hat{\tau} - \hat{\tau}_{0,0})\right], \\ \hat{B}_{0,1} &= \overline{\mathrm{G}}I_{0,0}\left[\widehat{Q}_0(\hat{\tau}) - \widehat{Q}_0(\hat{\tau}_{0,0}) - \widehat{Q}_0'(\hat{\tau}_{0,0})(\hat{\tau} - \hat{\tau}_{0,0})\right], \end{aligned} \tag{8.55}$$

where

$$\begin{aligned} \widehat{Q}_0''(\hat{\tau}) &= \tfrac{\hat{P}_0(\hat{\tau})}{\hat{\tau}} = E_1(\hat{\tau}) - \tfrac{\mathrm{e}^{-\hat{\tau}}}{\hat{\tau}}, \\ \widehat{Q}_0'(\hat{\tau}) &= E_1(\hat{\tau})(1+\hat{\tau}) - \mathrm{e}^{-\hat{\tau}}, \\ \widehat{Q}_0(\hat{\tau}) &= \frac{\hat{\tau}}{2}E_1(\hat{\tau})(2+\hat{\tau}) - \frac{1}{2}\mathrm{e}^{-\hat{\tau}}(1+\hat{\tau}). \end{aligned} \tag{8.56}$$

Thus, we may write the first sequence of the inner solutions of the velocity field as follows:

$$\begin{cases} \epsilon_2\hat{\nu}_0(\epsilon_2)\hat{\Psi}_0(\bar{\sigma},\hat{\tau},\epsilon_2) = -\epsilon_2\hat{\nu}_0(\epsilon_2)\overline{\mathrm{G}}I_{0,0}\;\bar{\sigma}\left[\widehat{Q}_0(\hat{\tau}) - \widehat{Q}_0(\hat{\tau}_{0,0})\right. \\ \qquad \left. -\widehat{Q}_0'(\hat{\tau}_{0,0})(\hat{\tau} - \hat{\tau}_{0,0})\right]\left[2L_0^{(1)}(\bar{\sigma}) - L_1^{(1)}(\bar{\sigma})\right] \\ \qquad +O(\epsilon_2^2\hat{\nu}_0(\epsilon_2)), \\ \hat{\nu}_0(\epsilon_2)\;\hat{\zeta}_0(\bar{\sigma},\hat{\tau},\epsilon_2) = -\hat{\nu}_0(\epsilon_2)\overline{\mathrm{G}}I_{0,0}\;\bar{\sigma}\hat{P}_0(\hat{\tau}) + O(\epsilon_2\hat{\nu}_0(\epsilon_2)). \end{cases} \tag{8.57}$$

The above inner solution is fully determined without any unknown constant. In Fig 8.3, we show the distributions of the stream function $\hat{\Psi}_0(\bar{\sigma},\hat{\tau})$ along the $\hat{\tau}$-axis for different $\bar{\sigma} = 0.5, 1.0, 5.0$. It is seen that this solution is not uniformly valid in the far field. It diverges when $\hat{\tau} \to \infty$.

In order to derive uniformly valid asymptotic solution for the flow field, one must find its outer solution and match it with the inner solution just obtained, and satisfy the far-field conditions. This outer solution can be uniquely determined. It is interesting to see that for the problem under investigation, the outer solution of the flow field is totally passive, which does not react on the temperature field, nor the inner solution of the flow field.

2.3 Leading-Order Outer Solutions of the Velocity Field $O(\bar{\nu}_0(\epsilon_2))$

Since in the outer region the solution of the perturbed temperature field is zero, the leading-order approximation of the system of flow can

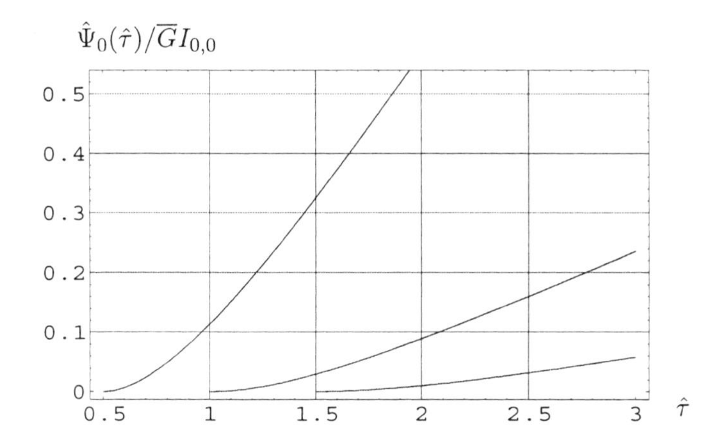

Figure 8.3. The variations of the inner solution of flow field, $\hat{\Psi}_0(\hat{\tau})$ along $\hat{\tau}$ axis for $\bar{\sigma} = 1.5$ and $\hat{\tau}_{0,0} = 0.5, 1.0, 1.5$ from bottom to top

be written as follows:

$$\begin{aligned} \frac{\mathrm{d}^2 \bar{B}_{0,n}}{\mathrm{d}\bar{\tau}^2} &= \frac{n+2}{\bar{\tau}} \bar{B}_{0,n+1} - \left[1 + \frac{2(n+1)}{\bar{\tau}}\right] \bar{A}_{0,n} \\ &+ \frac{1}{\bar{\tau}}\left[(n+2)\bar{A}_{0,n+1} + n\bar{A}_{0,n-1}\right], \\ \frac{\mathrm{d}^2 \bar{A}_{0,n}}{\mathrm{d}\bar{\tau}^2} &+ \frac{\mathrm{d}\bar{A}_{0,n}}{\mathrm{d}\bar{\tau}} - \frac{n+1}{\bar{\tau}} \bar{A}_{0,n} = 0 \\ &(n = 0, 1, 2, \ldots). \end{aligned} \tag{8.58}$$

This system is just the Oseen model of fluid dynamics, which has been solved in the previous chapters.

The general solution of the stream function is obtained as

$$\bar{\nu}_0(\epsilon_2)\bar{\Psi}_0 = \bar{\nu}_0(\epsilon_2)\bar{\sigma} \sum_{n=0}^{\infty} \left\{ \sum_{k=0}^{\infty} \mathcal{C}_{0,k} \mathcal{A}_{n,k}(\bar{\tau}) + \bar{b}_{0,n} \mathcal{F}_n(\bar{\tau}) \right\} L_n^{(1)}(\bar{\sigma}), \tag{8.59}$$

while the general solution for the vorticity function is

$$\bar{\nu}_0(\epsilon_2)\bar{\zeta}_0 = \bar{\nu}_0(\epsilon_2)\bar{\sigma} \sum_{n=0}^{\infty} \bar{a}_{0,n} \mathcal{F}_n(\bar{\tau}) L_n^{(1)}(\bar{\sigma}). \tag{8.60}$$

The coefficients $\bar{a}_{n,0}$ and $\bar{b}_{n,0}$ are related to each other with the formula

$$\bar{b}_{0,n} = -\bar{a}_{0,n} + n\bar{a}_{0,n-1}, \quad (n = 1, 2, \ldots). \tag{8.61}$$

Solution (8.59) involves two sequences of unknowns, $\{\bar{b}_{0,n}\ \mathcal{C}_{0,k}(n, k = 0, 1, 2, \ldots)\}$ to be determined by matching conditions with the inner solution.

2.4 Matching Conditions for the Leading Order Solutions of the Flow Field

We now turn to matching the outer solution

$$\bar{\nu}_0(\epsilon_2)\bar{\Psi}_0 = \bar{\nu}_{0,0}(\epsilon_2)\sum_{m=0}^{\infty}\gamma_{0,m}\Psi_{*m}(\bar{\sigma},\bar{\tau})$$

$$+\ \bar{\nu}_0(\epsilon_2)\bar{\sigma}\sum_{n=0}^{\infty}\bar{b}_{0,n,0}\mathcal{F}_n(\bar{\tau})L_n^{(1)}(\bar{\sigma}), \tag{8.62}$$

$$\bar{\nu}_0(\epsilon_2)\bar{\zeta}_0 = \bar{\nu}_{0,0}(\epsilon_2)\sum_{m=0}^{\infty}\gamma_{0,m}\zeta_{*m}(\bar{\sigma},\bar{\tau})$$

$$+\ \bar{\nu}_0(\epsilon_2)\bar{\sigma}\left\{\sum_{n=0}^{\infty}\bar{a}_{0,n,0}\mathcal{F}_n(\bar{\tau})L_n^{(1)}(\bar{\sigma})\right\}, \tag{8.63}$$

with the inner solution

$$\begin{aligned}\epsilon_2\hat{\nu}_0(\epsilon_2)\hat{\Psi}_0(\bar{\sigma},\hat{\tau},\epsilon_2) &= -\epsilon_2^3\overline{\mathrm{G}}I_{0,0}\ \bar{\sigma}\Big[\widehat{Q}_0(\hat{\tau})-\widehat{Q}_0(\hat{\tau}_{0,0})\\ &-\widehat{Q}_0'(\hat{\tau}_{0,0})(\hat{\tau}-\hat{\tau}_{0,0})\Big]\Big[2-L_1^{(1)}(\bar{\sigma})\Big]\\ &+O(\epsilon_2^3).\end{aligned} \tag{8.64}$$

We first change the outer variable $\bar{\tau}$ in the outer solution (8.62) to the inner variable $\hat{\tau}$, then match it with the inner solution (8.64).

(1) By matching the terms of $\{\bar{\sigma}\hat{\tau}\}$ in the component $L_0^{(1)}(\bar{\sigma})$, we obtain

$$\begin{aligned}\bar{\nu}_0(\epsilon_2) &= \frac{\epsilon_2^2}{\ln\epsilon_2},\\ \bar{b}_{0,0} &= 2\overline{\mathrm{G}}I_{0,0}\widehat{Q}_0'(\hat{\tau}_{0,0}).\end{aligned} \tag{8.65}$$

(2) By matching the terms of $\{\bar{\sigma}\}$ in the component $L_0^{(1)}(\bar{\sigma})$, we obtain

$$\begin{aligned}\bar{\nu}_{0,0}(\epsilon_2) &= \bar{\nu}_0(\epsilon_2) = \frac{\epsilon_2^2}{\ln\epsilon_2},\\ \gamma_{0,0}+\bar{b}_{0,0} &= 0.\end{aligned} \tag{8.66}$$

(3) By matching the terms of $\{\bar{\sigma}\hat{\tau}\}$ in the component $L_1^{(1)}(\bar{\sigma})$,

$$\begin{aligned}\bar{b}_{0,1} &= -\overline{\mathrm{G}}I_{0,0}\hat{Q}_0'(\hat{\tau}_{0,0}),\\ \frac{\bar{b}_{0,n}}{(n+1)!} &= 0 \qquad (n=2,3,4,\ldots).\end{aligned} \tag{8.67}$$

(4) By matching the terms of $\{\bar{\sigma}\}$ of $O(\epsilon_2/\ln\epsilon_2)$ in the component $L_1^{(1)}(\bar{\sigma})$,

$$\begin{aligned} &\gamma_{0,1} + \frac{\bar{b}_{0,1}}{2} = 0, \\ &\gamma_{0,n} = \frac{\bar{b}_{0,n}}{(n+1)!} = 0 \qquad (n = 2, 3, 4, \ldots). \end{aligned} \tag{8.68}$$

The terms of $\{\bar{\sigma}\}$ of $O(\epsilon_2^2)$ in the component $L_1^{(1)}(\bar{\sigma})$,

$$\left\{\epsilon_2^2\bar{\sigma}L_1^{(1)}(\bar{\sigma})\overline{\mathrm{G}}I_{0,0}\left[\widehat{Q}_0'(\hat{\tau}_{0,0})\hat{\tau}_{0,0} - \widehat{Q}_0(\hat{\tau}_{0,0})\right]\right\},$$

which is contained in the inner solution, will be balanced by the higher-order terms of the outer solution: $\left\{\epsilon_2^2\Psi_{*1}(\bar{\sigma},\bar{\tau})\right\}$.

In summary, we derive that

$$\bar{\nu}_{0,0}(\epsilon_2) = \bar{\nu}_0(\epsilon_2) = \frac{\epsilon_2^2}{\ln\epsilon_2} \tag{8.69}$$

and the outer solutions of the stream function

$$\begin{aligned} \tfrac{\epsilon_2^2}{\ln\epsilon_2}\bar{\Psi}_0(\bar{\sigma},\bar{\tau},\epsilon_2) = \tfrac{\epsilon_2^2}{\ln\epsilon_2}\bar{\sigma}\Big\{&\bar{b}_{0,0}\left[E_2(\bar{\tau}) - \Psi_{*0}(\bar{\sigma},\bar{\tau})\right] \\ &+\bar{b}_{0,1}\left[(3E_4(\bar{\tau}) - 2E_3(\bar{\tau}))L_1^{(1)}(\bar{\sigma}) - \tfrac{1}{2}\Psi_{*1}(\bar{\sigma},\bar{\tau})\right]\Big\} \\ &+\cdots. \end{aligned} \tag{8.70}$$

3. First-Order Asymptotic Expansion Solutions

3.1 First-Order Asymptotic Solution for the Temperature Field

The zeroth-order inner solutions of induced flow field $\epsilon_2^3\hat{\Psi}_0$ can affect the temperature field and, subsequently, the interface shape. It yields the first-order correction of temperature field, $\hat{\gamma}_1(\epsilon_2)\hat{T}_1$ and leading-order correction of the interface shape function, $\hat{\delta}_0(\epsilon_2)\hat{h}_0$. Since the solution $\hat{\Psi}_0$ only involves the components $L_0^{(1)}(\hat{\sigma})$ and $L_1^{(1)}(\hat{\sigma})$, one can assume that $\hat{T}_1$ will only involve the components $L_0^{(1)}(\hat{\sigma})$ and $L_1^{(1)}(\hat{\sigma})$, and has the asymptotic form

$$\hat{\gamma}_1(\epsilon_2)\hat{T}_1 = \hat{\gamma}_1(\epsilon_2)\left[\hat{D}_{1,0}(\bar{\tau})L_0^{(1)}(\bar{\sigma}) + \hat{D}_{1,1}(\bar{\tau})L_1^{(1)}(\bar{\sigma})\right]. \tag{8.71}$$

Furthermore, due to $\epsilon_2\hat{\nu}_0(\epsilon_2)\hat{\Psi}_0 = O(\epsilon_2^3)$, one has the interaction term

$$\frac{\epsilon_2}{2}\hat{\mathcal{N}}_T\Big\{\hat{T}_0, \hat{\Psi}_0\Big\} = O(\epsilon_2^4).$$

Hence, we derive that

$$\hat{\gamma}_1(\epsilon_2) = \epsilon_2^5. \tag{8.72}$$

The governing equation for $\hat{D}_{1,0}(\hat{\tau})$ is

$$\frac{\mathrm{d}^2\hat{\mathrm{D}}_{1,0}}{\mathrm{d}\hat{\tau}^2} + \left(1 + \frac{1}{\hat{\tau}}\right)\frac{\mathrm{d}\hat{\mathrm{D}}_{1,0}}{\mathrm{d}\hat{\tau}} = -2\overline{\mathrm{G}}I_{0,0}^2\hat{H}_0, \tag{8.73}$$

where

$$\begin{aligned}\hat{H}_0 &= -\Big[\widehat{Q}_0(\hat{\tau}) - \widehat{Q}_0(\hat{\tau}_{0,0}) - \widehat{Q}_0'(\hat{\tau}_{0,0})(\hat{\tau} - \hat{\tau}_{0,0})\Big]\hat{D}_{0,0}' \\ &= \Big[\widehat{Q}_0(\hat{\tau}) - \widehat{Q}_0(\hat{\tau}_{0,0}) - \widehat{Q}_0'(\hat{\tau}_{0,0})(\hat{\tau} - \hat{\tau}_{0,0})\Big]\frac{\mathrm{e}^{-\hat{\tau}}}{\hat{\tau}}.\end{aligned} \tag{8.74}$$

Furthermore, the governing equation for $\hat{D}_{1,1}(\bar{\tau})$ is

$$\frac{\mathrm{d}^2\hat{\mathrm{D}}_{1,1}}{\mathrm{d}\hat{\tau}^2} + \left(1 + \frac{1}{\hat{\tau}}\right)\frac{\mathrm{d}\hat{\mathrm{D}}_{1,1}}{\mathrm{d}\hat{\tau}} = \overline{\mathrm{G}}I_{0,0}^2\hat{H}_0. \tag{8.75}$$

In the above, in deriving the inhomogeneous terms, we have applied the formula

$$\frac{\partial}{\partial\bar{\sigma}}\Big[\bar{\sigma}(2L_0^{(1)}(\bar{\sigma}) - L_1^{(1)}(\bar{\sigma}))\Big] = 4L_0^{(1)}(\bar{\sigma}) - 2L_1^{(1)}(\bar{\sigma}). \tag{8.76}$$

The general solutions are obtained as

$$\begin{aligned}\hat{D}_{1,0}(\hat{\tau}) &= I_{1,0}E_1(\hat{\tau}) - 2\overline{\mathrm{G}}I_{0,0}^2\widehat{R}_1(\hat{\tau}), \\ \hat{D}_{1,1}(\hat{\tau}) &= I_{1,1}E_1(\hat{\tau}) + \overline{\mathrm{G}}I_{0,0}^2\widehat{R}_1(\hat{\tau}).\end{aligned} \tag{8.77}$$

We derive that

$$\begin{aligned}\widehat{R}_1'(\hat{\tau}) = \mathrm{e}^{-\hat{\tau}}E_1(\hat{\tau})\Big(\frac{\hat{\tau}}{2} + \frac{\hat{\tau}^2}{6}\Big) + \frac{\mathrm{e}^{-2\hat{\tau}}}{\hat{\tau}}\Big(\frac{1}{6} - \frac{\hat{\tau}}{3} - \frac{\hat{\tau}^2}{6}\Big) \\ + \frac{\mathrm{e}^{-\hat{\tau}}}{\hat{\tau}}\Big[\widehat{Q}_0'(\hat{\tau}_{0,0})\hat{\tau}_{0,0} - \widehat{Q}_0(\hat{\tau}_{0,0})\Big](\hat{\tau} - \hat{\tau}_{0,0}) \\ - \frac{\mathrm{e}^{-\hat{\tau}}}{2\hat{\tau}}\widehat{Q}_0'(\hat{\tau}_{0,0})(\hat{\tau}^2 - \hat{\tau}_{0,0}^2)\end{aligned} \tag{8.78}$$

and subsequently,

$$\begin{aligned}\widehat{R}_1(\hat{\tau}) &= \frac{2}{3}E_1(2\hat{\tau}) + \mathrm{e}^{-2\hat{\tau}}\left(\frac{2}{3}+\frac{\hat{\tau}}{6}\right) \\ &+ \hat{\tau}_{0,0}E_1(\hat{\tau})\left[\widehat{Q}_0'(\hat{\tau}_{0,0})\hat{\tau}_{0,0} - \widehat{Q}_0(\hat{\tau}_{0,0}) - \frac{1}{2}\widehat{Q}_0'(\hat{\tau}_{0,0})\hat{\tau}_{0,0}\right] \\ &+ \mathrm{e}^{-\hat{\tau}}\left[\frac{1}{2}\widehat{Q}_0'(\hat{\tau}_{0,0})(1+\hat{\tau}) - \widehat{Q}_0'(\hat{\tau}_{0,0})\hat{\tau}_{0,0} + \widehat{Q}_0(\hat{\tau}_{0,0})\right] \\ &- \mathrm{e}^{-\hat{\tau}}E_1(\hat{\tau})\left[\frac{5}{6}(1+\hat{\tau}) + \frac{\hat{\tau}^2}{6}\right] \end{aligned} \tag{8.79}$$

with

$$\widehat{R}_1(\infty) = 0. \tag{8.80}$$

On the other hand, from the interface condition (8.30), we derive

$$\hat{\delta}_0(\epsilon_2) = \epsilon_2^4. \tag{8.81}$$

Hence, we may write

$$\hat{h}_{\mathrm{s}}(\bar{\sigma}, \epsilon_2) = \epsilon_2^4\left[\hat{h}_{0,0} + \hat{h}_{0,1}L_1^{(1)}(\bar{\sigma})\right] + \cdots. \tag{8.82}$$

Noting that

$$\hat{D}_{1,0}'(\hat{\tau}_{0,0}) = -1, \quad \hat{D}_{1,0}''(\hat{\tau}_{0,0}) = \left(1 + \frac{1}{\hat{\tau}_{0,0}}\right)$$

and

$$\bar{\sigma}\hat{h}_{\mathrm{s}}'(\bar{\sigma}, \epsilon_2) = \epsilon_2^4\hat{h}_{0,1}\left[-2L_1^{(0)}(\bar{\sigma}) + L_1^{(1)}(\bar{\sigma})\right] + \cdots, \tag{8.83}$$

we derive the interface conditions for the first-order solution as follows: at $\hat{\tau} = \hat{\tau}_{0,0}$,

$$\begin{cases} \hat{D}_{1,0}(\hat{\tau}_{0,0}) - (\hat{\tau}_{1,0} + \hat{h}_{0,0}) = 0, \\ \hat{\tau}_{0,0}\hat{D}_{1,0}'(\hat{\tau}_{0,0}) + (1+\hat{\tau}_{0,0})\,(\hat{\tau}_{1,0} + \hat{h}_{0,0}) - 2\hat{h}_{0,1} = 0 \end{cases} \tag{8.84}$$

and

$$\begin{cases} \hat{D}_{1,1}(\hat{\tau}_{0,0}) - \hat{h}_{0,1} = 0, \\ \hat{\tau}_{0,0}\hat{D}_{1,1}'(\hat{\tau}_{0,0}) + (1+\hat{\tau}_{0,0})\,\hat{h}_{0,1} + \hat{h}_{0,1} = 0. \end{cases} \tag{8.85}$$

From these interface conditions, we obtain the two sets of constants $\{I_{1,0}, (\hat{\tau}_{1,0} + \hat{h}_{0,0})\}$ and $\{I_{1,1}, \hat{h}_{0,1}\}$ as follows:

$$I_{1,0} = \overline{\mathrm{G}} I_{0,0}^2 \tilde{I}_0, \qquad I_{1,1} = \overline{\mathrm{G}} I_{0,0}^2 \tilde{I}_1, \tag{8.86}$$

and

$$(\hat{\tau}_{1,0} + \hat{h}_{0,0}) = \overline{\mathrm{G}} I_{0,0}^2 \left[E_1(\hat{\tau}_{0,0}) \tilde{I}_0 - 2\widehat{R}_1(\hat{\tau}_{0,0}) \right], \qquad \hat{h}_{0,1} = \overline{\mathrm{G}} I_{0,0}^2 \tilde{h}_1, \tag{8.87}$$

where

$$\tilde{I}_0 = 2\frac{\tilde{h}_1 + \hat{\tau}_{0,0}\widehat{R}_1'(\hat{\tau}_{0,0}) + (1+\hat{\tau}_{0,0})\widehat{R}_1(\hat{\tau}_{0,0})}{(1+\hat{\tau}_{0,0})E_1(\hat{\tau}_{0,0}) - \mathrm{e}^{-\hat{\tau}_{0,0}}},$$
$$\tilde{I}_1 = \frac{\widehat{R}_1(\hat{\tau}_{0,0})(2+\hat{\tau}_{0,0}) + \hat{\tau}_{0,0}\widehat{R}_1'(\hat{\tau}_{0,0})}{\mathrm{e}^{-\hat{\tau}_{0,0}} - (2+\hat{\tau}_{0,0})E_1(\hat{\tau}_{0,0})}, \tag{8.88}$$
$$\tilde{h}_1 = \widehat{R}_1(\hat{\tau}_{0,0}) + \tilde{I}_1 E_1(\hat{\tau}_{0,0}).$$

Due to the normalization condition

$$\hat{h}_0(0) = \hat{h}_{0,0} + \hat{h}_{0,1} L_1^{(1)}(0) = 0, \tag{8.89}$$

we derive

$$\hat{h}_{0,0} = -2\hat{h}_{0,1}, \tag{8.90}$$

and

$$\hat{h}_0(\bar{\sigma}) = \hat{h}_{0,0} + \hat{h}_{0,1} L_1^{(1)}(\bar{\sigma}) = -\hat{h}_{0,1}\bar{\sigma}. \tag{8.91}$$

Subsequently, we obtain

$$\hat{\tau}_{1,0} = \overline{\mathrm{G}} I_{0,0}^2 \Big[2\tilde{h}_1 + E_1(\hat{\tau}_{0,0})\tilde{I}_0 - 2\widehat{R}_1(\hat{\tau}_{0,0}) \Big] = 0. \tag{8.92}$$

We finally obtain the solution of the interface shape function:

$$\hat{\tau}_{\mathrm{s}} = \hat{\tau}_0 + \epsilon_2^4 \hat{h}_0(\bar{\sigma}) + O\,(\mathrm{h.r.t}) = \hat{\tau}_{0,0} - \epsilon_2^4 \overline{\mathrm{G}}\ I_{0,0}^2 \tilde{h}_1 \bar{\sigma} + O\,(\mathrm{h.r.t})$$
$$= \hat{\tau}_{0,0} - \epsilon_2^4 \overline{\mathrm{G}}\ \hat{\tau}_{0,0}^2 \mathrm{e}^{2\hat{\tau}_{0,0}} \tilde{h}_1(\hat{\tau}_{0,0})\bar{\sigma} + O\,(\mathrm{h.r.t})\,. \tag{8.93}$$

Accordingly, we have

$$\eta_0^2 = 2\hat{\tau}_0 = 2\hat{\tau}_{0,0}\Big[1 + O\,(\mathrm{h.r.t})\Big], \tag{8.94}$$

and

$$\eta_s^2 = \frac{\hat{\tau}_s}{\hat{\tau}_0} = 1 + \bar{\eta}_s = 1 - \epsilon_2^4 \overline{G}\ \hat{\tau}_{0,0} e^{2\hat{\tau}_{0,0}} \tilde{h}_1(\hat{\tau}_{0,0})\bar{\sigma} + O\,(\text{h.r.t})$$

$$= 1 - \frac{\hat{\tau}_{0,0} e^{\hat{\tau}_{0,0}} \tilde{h}_1(\hat{\tau}_{0,0})}{E_1(\hat{\tau}_{0,0})} \frac{\text{Gr}}{\text{Pr}^2} \xi^2 + O\,(\text{h.r.t})\,. \tag{8.95}$$

The twice mean curvature of the tip can be calculated as

$$\mathcal{K}(0) = \frac{2}{\eta_0^2} \left\{ 1 - \left[\frac{\eta_0^2}{\text{Pr}} \frac{\text{d}}{\text{d}\bar{\sigma}} \bar{\eta}_s(0) + 2\bar{\eta}_s(0) \right] \right\}, \tag{8.96}$$

as $\bar{\eta}_s \ll 1$. Then the Peclet number is obtained as follows:

$$\text{Pe} = \frac{\ell_t}{\ell_T} = \frac{2}{\mathcal{K}(0)} \approx \text{Pe}_0 \left[1 - \frac{\hat{\tau}_{0,0} e^{\hat{\tau}_{0,0}} \tilde{h}_1(\hat{\tau}_{0,0})}{E_1(\hat{\tau}_{0,0})} \frac{\text{Gr}}{\text{Pr}^2} \right], \tag{8.97}$$

where $\text{Pe}_0 = \eta_{0,0}^2 = 2\hat{\tau}_{0,0}$ is the Peclet number of dendritic growth in zero gravity.

From the above results, one sees that for the limiting case: $\text{Pr} = \infty$, with the fixed parameter $\widehat{\text{Gr}} = \frac{\text{Gr}}{\text{Pr}^2} = O(1)$, the shape of dendrite growing in the system with buoyancy-induced convection is described by the function:

$$\eta = \eta_s(\xi) = \sqrt{1 - \frac{\hat{\tau}_{0,0} e^{\hat{\tau}_{0,0}} \tilde{h}_1(\hat{\tau}_{0,0})}{E_1(\hat{\tau}_{0,0})} \widehat{\text{Gr}}\ \xi^2}, \tag{8.98}$$

while the normalization parameter η_0^2 of the system remains unaffected.

In Fig 8.4, we show the graphs of the function $\tilde{h}_1(\hat{\tau}_{0,0})$. It is seen that $\tilde{h}_1(\hat{\tau}_{0,0})$ is always negative, its magnitude increases with increasing $\hat{\tau}_{0,0}$. Consequently, with positive Grashof number $\text{Gr} > 0$, the Peclet number of growth becomes larger. In Fig 8.5, we show the shapes of dendrites with $\text{Pr} = 22$, $\hat{\tau}_{0,0} = 0.01, 0.1$ and $\text{Gr} = 0,\ 0.02,\ 0.04,\ 0.06$. It is seen that due to the buoyancy effect, the shape of dendrite becomes fatter. Finally, in Fig 8.6, we show the graphs of the solution $T_1(\sigma, \hat{\tau})$ versus $\hat{\tau}$ for the case $\hat{\tau}_{0,0} = 0.2; 2.0$, $\overline{G} = 1.0$ at $\bar{\sigma} = 0, 0.5, 1, 2, 2.5$.

4. Summary of the Results

In the above, we studied steady dendritic growth with natural convection in the case of large Prandtl number and a moderate modified Grashof number $\overline{\text{G}} = O(1)$. We find the uniformly valid asymptotic expansion solution to the problem as $\text{Pr} \to \infty$, in the physical domain, from the interface extended to the up-stream far field:

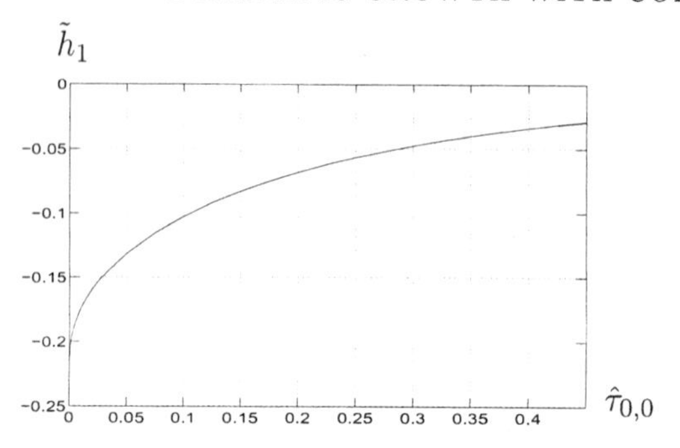

Figure 8.4. The graph of function $\tilde{h}_1(\hat{\tau}_{0,0})$

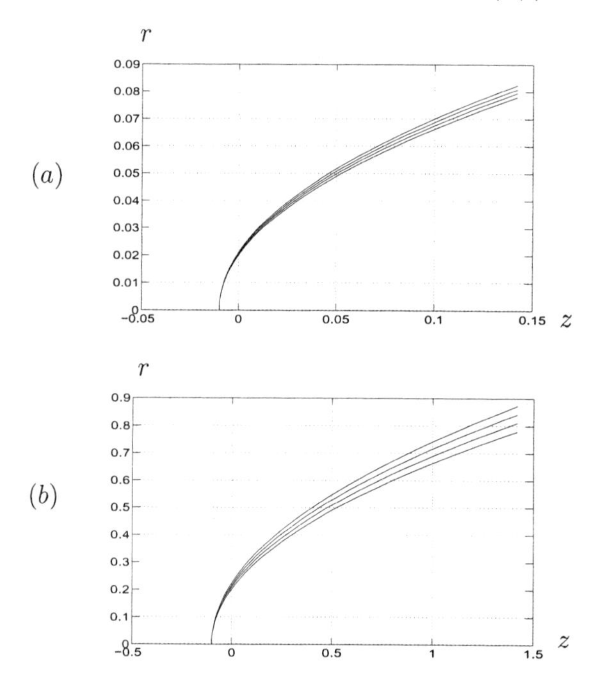

Figure 8.5. The Shapes of dendrite's interface with Pr = 22 and Gr = 0, 0.02, 0.04, 0.06, from bottom to top for the cases: (a) $\hat{\tau}_{0,0} = 0.01$, $|T|_\infty = 0.04079$; (b) $\hat{\tau}_{0,0} = 0.1$, $|T|_\infty = 0.2015$

$\{0 < \bar{\eta} < \infty,\ 0 \leq \bar{\sigma} < \bar{\sigma}_{\max} < \infty\}$. The perturbation of temperature plays an active role, which has a boundary layer near the interface with the thickness of $O(\epsilon_2)$, and restricted only inside this layer. The perturbation of temperature vanishes outside of the temperature boundary layer. The perturbed flow field is passive, driven by the perturbed temperature field via the buoyancy effect. The perturbed flow field also

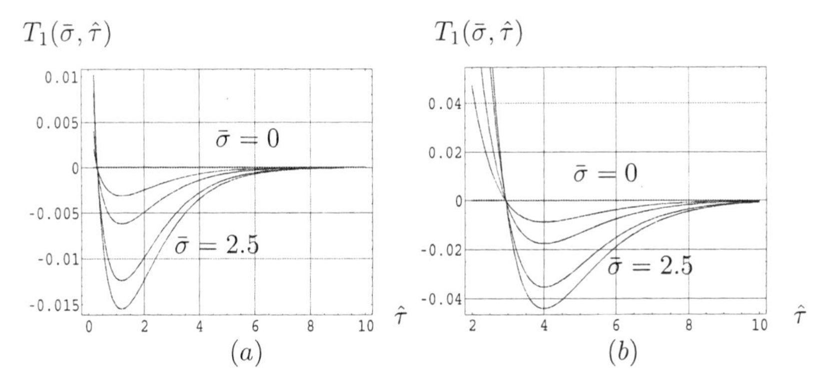

Figure 8.6. The graphs of the solution $T_1(\bar{\sigma}, \hat{\tau})$ versus $\hat{\tau}$ at $\bar{\sigma} = 0, 0.5, 1, 2, 2.5$, for the case, $\overline{G} = 1.0$, (a) $\hat{\tau}_{0,0} = 0.2$; (b) $\hat{\tau}_{0,0} = 2.0$

has a boundary layer with the same thickness of $O(\epsilon_2)$, as the boundary layer of temperature field. Both boundary layers are called the inner region near the interface. The interaction between the temperature and flow field is only through their inner solutions. The zeroth-order temperature solution determines the zeroth-order inner solution of induced flow field; then the zeroth-order inner solution of induced flow affects the temperature field and interface shape, resulting in the first-order correction of the temperature field and leading-order correction of the interface shape. The outer solution of the flow field is totally passive, which is uniquely determined by the up-stream far-field condition and the matching conditions with the inner solution of flow field.

It is found that in the limiting case, $\mathrm{Pr} \gg 1$, the buoyancy effect does not affect the normalization parameter η_0^2. However, it does affect the shape of dendrite. Due to the presence of the buoyancy-induced flow, the shape function $\hat{h}_s$ becomes a function of $\bar{\sigma}$ involving the component of $L_1^{(1)}(\bar{\sigma})$. Hence, the interface is no longer a paraboloid. It becomes fatter with larger Peclet number, when the Grashof number $\mathrm{Gr} > 0$ increases. The change of Peclet number is proportional to the parameter, $\widehat{\mathrm{Gr}} = \frac{\mathrm{Gr}}{\mathrm{Pr}^2} = O(1)$.

Chapter 9

STABILITY AND SELECTION OF DENDRITIC GROWTH WITH CONVECTIVE FLOW

In the previous chapters, we have studied the basic steady state solutions for systems with different types of convective motion in the melt. Like the Ivantsov solutions, without taking into account the surface tension, the solutions obtained are not unique, under a given growth condition. Thus, the selection remains. To resolve the selection problem of dendritic growth with convective flow, one needs to take the surface tension into account and study the stability property of the steady state solutions. For this purpose, one may apply the approach of the *Interfacial Wave Theory* (IFW) (Xu, 1997) to find the neutrally stable mode solution and determine the stability and selection criterion ε_*. For the system under study, the criterion ε_* may be a function of the undercooling parameter T_∞ and a properly defined flow parameter $\mathcal{F}$, which measures the strength of convective flow.

With convective flow in the melt, the associated linear perturbed system for the stability analysis, in general, involves rather complicated non-uniform steady solutions of flow and temperature field. Therefore, the corresponding eigenvalue problem will be subject to a highly inhomogeneous dynamical system, which is very difficult to solve. However, the Interfacial Wave (IFW) theory has shown that unsteady perturbations around the basic state are restricted in a thin boundary layer near the interface, which we call the IW layer. As a consequence, the stability property of the system only depends on the nature of the basic state inside the IW layer, irrelevant to the nature of the basic state away from this layer. On the other hand, from the results obtained in the previous chapters, it is seen that, in some important practical cases, just in the inner region near the interface the form of the basic state solution is rather simple. In the paraboloidal coordinates system (ξ, η) the stream function of a flow field can be expressed in the form

$$\Psi_(\xi,\eta) \approx \xi^2 f(\eta), \tag{9.1}$$

whereas the temperature field can be described by a similar, or nearly similar solution and the interface shape is, or nearly, paraboloid. This situation allows us to derive a general theory for the stability of dendritic growth with various kinds of convective flows in the melt and the selection condition of the dendrite's tip velocity. In this chapter, we attempt to carry out such analysis.

The general mathematical formulation has been given in Chap. 3. The convective flow is assumed to be generated by different sources, characterized by the Grashof number Gr, enforced flow parameter $\overline{U}_\infty$ and density change parameter α. For convenience, we may denote all these parameters by the common symbol $\mathcal{F}$. The case with no convection corresponds to $\mathcal{F} = 0$.

1. Basic Steady State Solution

Some axi-symmetrical, basic state solutions of dendritic growth with zero surface tension,

$$\{\Psi = \Psi_{\mathrm{B}0};\ \zeta = \zeta_{\mathrm{B}0};\ T = T_{\mathrm{B}0};\ T_{\mathrm{S}} = T_{\mathrm{SB}0} = 0;\ \eta_{\mathrm{s}} = \eta_{\mathrm{B}0}\},$$

have been obtained in the previous chapters of this book. Some other types of steady state solutions can be found in the literature as well. Here, we summarize some of the typical solutions obtained in this book as follows.

1.1 Convection Flow Field Induced by Uniform External Flow

For the case: $\mathrm{Pr} \gg 1$, $\lambda_0 = (1+\overline{U}_\infty)$, $\mathrm{Gr} = 0$, we let $\epsilon_2 = \frac{1}{\ln \mathrm{Pr}}$,

$$\bar{\sigma} = \frac{\eta_0^2\lambda_0}{2\mathrm{Pr}}\xi^2, \qquad \hat{\tau} = \frac{\eta_0^2\lambda_0}{2}\eta^2\ , \tag{9.2}$$

and derive (see Chap. 6) that

$$\left\{\begin{array}{rl} \Psi = & \dfrac{1}{2}\eta_0^4\xi^2\eta^2 - \dfrac{\eta_0^2}{\ln\epsilon_2}\left(\dfrac{\lambda_0 - 1}{\lambda_0}\right)\xi^2\Big[\hat{\tau}\ln\hat{\tau} - \hat{\tau}(1+\ln\hat{\tau}_{0,0}) + \hat{\tau}_{0,0}\Big] \\ & +O\,(\mathrm{h.r.t})\,, \\ \zeta = & \dfrac{2}{\ln\epsilon_2}(\lambda_0 - 1) + O\,(\mathrm{h.r.t})\,, \end{array}\right. \tag{9.3}$$

$$T = T_\infty + \bar{I}_{0,0} E_1(\hat{\lambda}_0 \hat{\tau}) + \frac{1}{\ln \epsilon_2} \hat{D}_{1,0,0}(\hat{\tau}) + O\,(\text{h.r.t})\,, \tag{9.4}$$

where

$$T_\infty = -\bar{I}_{00} E_1(\hat{\lambda}_0 \hat{\tau}_{0,0}), \tag{9.5}$$

and

$$\bar{I}_{0,0} = \frac{\hat{\tau}_{0,0}}{\lambda_0} e^{\hat{\lambda}_0 \hat{\tau}_{0,0}}, \quad \hat{\lambda}_0 = \frac{2\lambda_0 - 1}{\lambda_0}. \tag{9.6}$$

Furthermore, it is derived that the interface shape function,

$$\eta_s = 1 + O\,(\text{h.r.t})\,, \tag{9.7}$$

the normalization parameter,

$$\eta_0^2 = \eta_{0,0}^2 \left[1 - \frac{1}{\ln \text{Pr}} \mathcal{Z}_e(\hat{\tau}_{0,0}, \lambda_0) + O\,(\text{h.r.t})\right] \tag{9.8}$$

where

$$\eta_{0,0} = \eta_{0,0}(\lambda_0, T_\infty) = \frac{2}{\lambda_0} \hat{\tau}_{0,0}, \tag{9.9}$$

and the Peclet number

$$\text{Pe} = \frac{\ell_t}{\ell_T} = \eta_{0,0}^2 \left[1 - \frac{1}{\ln \text{Pr}} \mathcal{Z}_e(\hat{\tau}_{0,0}, \lambda_0) + O\,(\text{h.r.t})\right]\,. \tag{9.10}$$

1.2 Convection Flow Field Induced by Buoyancy Effect

For the case: $\text{Pr} \gg 1$ $\lambda_0 = 1$, $\overline{\text{G}} = O(1)$. We let $\epsilon_2 = \frac{1}{\ln \text{Pr}}$,

$$\bar{\sigma} = \frac{\eta_0^2}{2\text{Pr}} \xi^2, \qquad \hat{\tau} = \frac{\eta_0^2}{2} \eta^2\,, \tag{9.11}$$

and derive (see Chap. 8) that

$$\begin{cases} \Psi = \dfrac{1}{2} \eta_0^4 \xi^2 \eta^2 - \dfrac{\epsilon_2^4 \eta_0^2}{2} \overline{\text{G}} I_{0,0}\; \xi^2 \Big[\widehat{Q}_0(\hat{\tau}) - \widehat{Q}_0(\hat{\tau}_{0,0}) - \widehat{Q}_0'(\hat{\tau}_{0,0})(\hat{\tau} - \hat{\tau}_{0,0})\Big] \\ \qquad \times \Big[2L_0^{(1)}(\bar{\sigma}) - L_1^{(1)}(\bar{\sigma})\Big] + O\,(\text{h.r.t})\,, \\ \zeta = -\dfrac{\epsilon_2^5 \eta_0^2}{2} \overline{\text{G}} I_{0,0}\; \xi^2 \hat{P}_0(\hat{\tau}) + O\,(\text{h.r.t})\,, \end{cases} \tag{9.12}$$

$$T = T_\infty + I_{0,0} E_1(\hat{\tau}) + \epsilon_2^4 \Big[\hat{D}_{1,0}(\hat{\tau}) + \hat{D}_{1,1}(\hat{\tau})(2 - \bar{\sigma})\Big] + O\,(\text{h.r.t})\,, \tag{9.13}$$

where

$$T_\infty = -I_{0,0}E_1(\hat{\tau}_{0,0}), \tag{9.14}$$

$$\eta_{0,0}^2 = 2\hat{\tau}_{0,0}, \qquad I_{0,0} = \hat{\tau}_{0,0}\mathrm{e}^{\hat{\tau}_{0,0}}. \tag{9.15}$$

Furthermore, it is derived that the interface shape function

$$\eta_s^2 = 1 - \frac{1}{2}\frac{\hat{\tau}_{0,0}\mathrm{e}^{\hat{\tau}_{0,0}}\tilde{h}_1(\hat{\tau}_{0,0})}{E_1(\hat{\tau}_{0,0})}\frac{\mathrm{Gr}}{\mathrm{Pr}^2}\xi^2 + O\,(\mathrm{h.r.t}), \tag{9.16}$$

the normalization parameter

$$\eta_0^2 = 2\hat{\tau}_0 = 2\hat{\tau}_{0,0}\Big[1 + O\,(\mathrm{h.r.t})\Big], \tag{9.17}$$

and the Peclet number

$$\mathrm{Pe} = \frac{\ell_t}{\ell_T} = \mathrm{Pe}_0\left[1 - \frac{\hat{\tau}_{0,0}\mathrm{e}^{\hat{\tau}_{0,0}}\tilde{h}_1(\hat{\tau}_{0,0})}{E_1(\hat{\tau}_{0,0})}\frac{\mathrm{Gr}}{\mathrm{Pr}^2}\right] + O\,(\mathrm{h.r.t}), \tag{9.18}$$

where $\mathrm{Pe}_0 = \eta_{0,0}^2 = 2\hat{\tau}_{0,0}$ is the Peclet number of dendritic growth in zero gravity.

1.3 Convection Motion Induced by Density Change During Phase Transition

The flow parameter for this case is $\mathcal{F} = \alpha$. With different far field conditions, one may obtain different steady solutions. Nevertheless, for this case the system allows an exact, similar solution (McFadden and Coriell, 1986, Xu, 1994b). It is obtained that

$$\begin{aligned}
&\Psi = \tfrac{1}{2}\eta_0^4\xi^2\eta^2 + \tfrac{1}{2}\alpha\eta_0^4\xi^2,\\
&T = T_\infty + \mathrm{e}^{\frac{\eta_0^2}{2}}\left(\tfrac{\eta_0^2}{2}\right)^{1+\alpha\eta_0^2/2}\Gamma\Big(-\tfrac{\alpha\eta_0^2}{2}, \tfrac{\eta_0^2\eta^2}{2}\Big),\\
&\eta_s = 1,
\end{aligned} \tag{9.19}$$

where $\Gamma(a, b)$ is the incomplete Gamma function. From here, we obtain

$$T_\infty = -\mathrm{e}^{\frac{\eta_0^2}{2}}\left(\frac{\eta_0^2}{2}\right)^{1+\alpha\eta_0^2/2}\Gamma\Big(-\frac{\alpha\eta_0^2}{2}, \frac{\eta_0^2}{2}\Big), \tag{9.20}$$

which yields the Peclet number $\mathrm{Pe} = \eta_0^2$, as a function of the parameters $\{\alpha, T_\infty\}$. Namely,

$$\mathrm{Pe} = \eta_0^2 = \mathcal{S}_d(\alpha, T_\infty). \tag{9.21}$$

1.4 More General Steady State Solutions with Nearly Paraboloid Interface

In addition to the above-listed steady state solutions, there are some other types of steady state solutions with convection flow in the literature, such as the model solutions obtained by Ananth and Gill (1989, 1991), and the local solution obtained by Canright and Davis (1991). For all these solutions with the inclusion of convection, the interface shape of the steady growing dendrite with zero surface tension is either paraboloidal, or nearly paraboloidal; the temperature field is described either by a similarity solution or by a nearly similarity solution. It can be presumed that these properties may be retained in a large class of steady state solutions of dendritic growth with convection.

Specifically, we assume that the basic state of system in this class involves some small factor $\nu_0(\epsilon_2) \ll 1$, where ϵ_2 is a physical parameter (for instance, one may set $\epsilon_2 = \frac{1}{\mathrm{Pr}}$, when the Prandtl number is large), accordingly, the normalization number, interface shape function can be expressed in the following form

$$\begin{aligned} \eta_0^2 &= \eta_{0,0}^2 + \nu_0(\epsilon_2)\eta_{0,1}, \\ \eta_{\mathrm{B}0}(\xi) &= 1 + \nu_0(\epsilon_2)\Delta_0(\xi). \end{aligned} \tag{9.22}$$

Furthermore, we assume that in the interfacial wave (IW) layer near the interface, the temperature and flow field can be expressed in the form

$$\begin{aligned} T_{\mathrm{B}0} &= T_\infty + \overline{T}_0(\eta) + \nu_0(\epsilon_2)\widehat{T}_1(\xi,\eta), \\ T_{\mathrm{BS}0} &= 0, \end{aligned} \tag{9.23}$$

and

$$\begin{aligned} \Psi_{\mathrm{B}0} &= \frac{\eta_0^4}{2}\xi^2\eta^2 + \nu_0(\epsilon_2)\eta_0^2\widehat{\Psi}_1(\xi,\eta), \\ \zeta_{\mathrm{B}0} &= \nu_0(\epsilon_2)\widehat{\zeta}_1(\xi,\eta). \end{aligned} \tag{9.24}$$

The modification terms $\widehat{T}_1(\xi,\eta)$, $\widehat{T}_{\mathrm{S}1}(\xi,\eta)$, $\widehat{\Psi}_1(\xi,\eta)$, $\widehat{\zeta}_1(\xi,\eta)$ may have different forms for different cases.

In this chapter, we attempt to explore the stability properties of such a class of systems in a uniform way. To investigate the interfacial stability of the system, we shall adopt interfacial stability parameter ε as the basic small parameter and look for the asymptotic solution in the limit of $\varepsilon \to 0$, by following the procedure as described in Chap. 2.

For a realistic dendritic growth system with non-zero surface tension, $\varepsilon > 0$, the interface shape will have an additional correction term of

$O(\varepsilon^2)$, so that we have

$$\eta_{\rm B} = 1 + \nu_0(\epsilon_2)\Delta_0(\xi) + \varepsilon^2\eta_1(\xi) + O(\text{h.o.t.}). \tag{9.25}$$

The basic state of dendritic growth with small surface tension can be defined in the same manner described in the IFW theory (Xu, 1997). Hence, in the region away from the root, the temperature in liquid and solid, the stream function and vorticity function in the IW layer can be expressed as:

$$\begin{aligned}
T_{\rm B} &= T_{\rm B0}(\xi,\eta) + O(\varepsilon^2),\\
T_{\rm SB} &= T_{\rm SB0}(\xi,\eta) + O(\varepsilon^2),\\
\Psi_{\rm B} &= \Psi_{\rm B0}(\xi,\eta) + O(\varepsilon^2),\\
\zeta_{\rm B} &= \zeta_{\rm B0}(\xi,\eta) + O(\varepsilon^2),
\end{aligned} \tag{9.26}$$

The present system now contains three small parameters: the interfacial stability parameter ε, which describes the effect of surface tension, and the parameters $\nu_0(\epsilon_2)$ and α, which are associated with the effect of convective flow. We denote the ratios of two parameters by

$$\Theta_0 = \frac{\nu_0(\epsilon_2)}{\varepsilon} = O(1), \tag{9.27}$$

and

$$\Theta_1 = \frac{\alpha}{\varepsilon} = O(1) \tag{9.28}$$

and shall consider Θ_0, Θ_1 as fixed constants. Thus, we may write the interface shape function in the form

$$\eta_{\rm B}(\xi) = 1 + \varepsilon\Theta_0\Delta_0(\xi) + \varepsilon^2\eta_1(\xi)\cdots. \tag{9.29}$$

With the solution (9.23), in many cases, one can derive that at $\eta = 1$,

$$\begin{aligned}
&\frac{\rm d}{{\rm d}\eta}T_{\rm B0}(1) = -\eta_0^2(1-\varepsilon\Theta_0\overline{\Lambda}_1) + O(\text{h.o.t.}),\\
&\frac{{\rm d}^2}{{\rm d}\eta^2}T_{\rm B0}(1) = \eta_0^2\overline{\Lambda}_2 + O(\text{h.o.t.}),\\
&\frac{\partial}{\partial\xi}T_{\rm B0}(\xi,1) \approx 0,
\end{aligned} \tag{9.30}$$

and

$$\begin{aligned}
\frac{1}{\eta_0^2}\frac{\partial\Psi_{\rm B0}}{\partial\xi} &= \eta_0^2\xi + \varepsilon\Theta_0 P_1\\
\frac{1}{\eta_0^2}\frac{\partial\Psi_{\rm B0}}{\partial\eta} &= \eta_0^2\xi^2 + \varepsilon\Theta_0 P_2\,.
\end{aligned} \tag{9.31}$$

Moreover, from (9.22)–(9.30), one derives that at the interface $\eta = 1$,

$$\begin{aligned} &\frac{\partial}{\partial \eta}(T_{\mathrm{B}} - T_{\mathrm{SB}}) = -\eta_0^2\left(1 - \varepsilon\Theta_0\overline{\Lambda}_1\right) + O(\varepsilon^2), \\ &\frac{\partial^2}{\partial \eta^2}(T_{\mathrm{B}} - T_{\mathrm{SB}}) = \eta_0^2\overline{\Lambda}_2 + O(\varepsilon^2), \\ &\frac{\partial}{\partial \xi}(T_{\mathrm{B}} - T_{\mathrm{SB}}) = o(\varepsilon^2). \end{aligned} \tag{9.32}$$

In the above, the functions $\overline{\Lambda}_1$, $\overline{\Lambda}_2$, P_1, P_2 are determined by the basic state considered.

2. Linear Perturbed System around the Basic Steady State Solution

To study the stability of the steady state solution $\{T_{\mathrm{B}}, T_{\mathrm{SB}}, \Psi_{\mathrm{B}}\}$, we consider the general unsteady state, that can be separated into two parts: the basic state (9.26) and the small perturbation $\{\tilde{T}, \tilde{T}_{\mathrm{S}}, \tilde{h}, \tilde{\Psi}, \tilde{\zeta}\}$, namely,

$$\begin{aligned} T(\xi,\eta,t) &= T_{\mathrm{B}} + \tilde{T}(\xi,\eta,t), \\ T_{\mathrm{S}}(\xi,\eta,t) &= T_{\mathrm{SB}} + \tilde{T}_{\mathrm{S}}(\xi,\eta,t), \\ \eta_{\mathrm{s}} &= \eta_{\mathrm{B}} + \tilde{h}(\xi,t)/\eta_0^2, \\ \Psi(\xi,\eta,t) &= \Psi_{\mathrm{B}} + \tilde{\Psi}(\xi,\eta,t), \\ \zeta(\xi,\eta,t) &= \zeta_{\mathrm{B}} + \tilde{\zeta}(\xi,\eta,t)\ . \end{aligned} \tag{9.33}$$

The perturbed states are subject to the following system:

$$\begin{aligned} \nabla^2\tilde{T} = {}& \eta_0^4(\xi^2+\eta^2)\frac{\partial \tilde{T}}{\partial t} + \frac{1}{\eta_0^2\xi\eta}\left(\frac{\partial \Psi_{\mathrm{B}}}{\partial \eta}\frac{\partial \tilde{T}}{\partial \xi} - \frac{\partial \tilde{\Psi}}{\partial \xi}\frac{\partial T_{\mathrm{B}}}{\partial \eta}\right) \\ & - \frac{1}{\eta_0^2\xi\eta}\left(\frac{\partial \Psi_{\mathrm{B}}}{\partial \xi}\frac{\partial \tilde{T}}{\partial \eta} - \frac{\partial \tilde{\Psi}}{\partial \eta}\frac{\partial T_{\mathrm{B}}}{\partial \xi}\right), \end{aligned} \tag{9.34}$$

$$D^2\tilde{\Psi} = -\eta_0^4(\xi^2+\eta^2)\tilde{\zeta}, \tag{9.35}$$

$$\begin{aligned} \mathrm{Pr} D^2\tilde{\zeta} = {}& \eta_0^2(\xi^2+\eta^2)\frac{\partial \tilde{\zeta}}{\partial t} + \frac{2\zeta_{\mathrm{B}}}{\eta_0^4\xi^2\eta^2}\frac{\partial(\tilde{\Psi}, \eta_0^2\xi\eta)}{\partial(\xi,\eta)} + \frac{2\tilde{\zeta}}{\eta_0^4\xi^2\eta^2}\frac{\partial(\Psi_{\mathrm{B}}, \eta_0^2\xi\eta)}{\partial(\xi,\eta)} \\ & - \frac{1}{\eta_0^2\xi\eta}\frac{\partial(\Psi_{\mathrm{B}}, \tilde{\zeta})}{\partial(\xi,\eta)} - \frac{1}{\eta_0^2\xi\eta}\frac{\partial(\tilde{\Psi}, \zeta_{\mathrm{B}})}{\partial(\xi,\eta)} \end{aligned}$$

$$+\frac{\eta_0^4 \mathrm{Gr}}{\mathrm{Pr}|T_\infty|}\xi\eta\left(\eta\frac{\partial\tilde{T}}{\partial\xi}+\xi\frac{\partial\tilde{T}}{\partial\eta}\right). \tag{9.36}$$

The boundary conditions are:

1. The up-stream far-field condition: As $\eta \to \infty$,

$$\tilde{T},\ \tilde{\Psi},\ \tilde{\zeta} \to 0. \tag{9.37}$$

2. The regular condition: As $\eta \to 0$,

$$\tilde{T}_{\mathrm{S}} \to 0. \tag{9.38}$$

3. The interface condition: At $\eta = \eta_{\mathrm{B}}$,

$$\tilde{T} = \tilde{T}_S - \frac{\partial(T_{\mathrm{B}} - T_{\mathrm{SB}})}{\partial\eta}\frac{\tilde{h}}{\eta_0^2} + (\mathrm{h.o.t}), \tag{9.39}$$

$$\begin{aligned}\tilde{T}_S &= \frac{\varepsilon^2}{\hat{S}}\left\{\frac{\partial^2\tilde{h}}{\partial\xi^2}+\frac{(1+2\xi^2)}{\xi\tilde{S}^2}\frac{\partial\tilde{h}}{\partial\xi}-\frac{1}{\tilde{S}^2}\tilde{h}\right\}\\ &\qquad -\frac{\partial T_{\mathrm{SB}}}{\partial\eta}\frac{\tilde{h}}{\eta_0^2}+(\mathrm{h.o.t}),\end{aligned} \tag{9.40}$$

$$\begin{aligned}&\frac{\partial(\tilde{T}-\tilde{T}_S)}{\partial\eta}+\left[\frac{1}{\eta_0^2}\frac{\partial(T_{\mathrm{B}}-T_{\mathrm{SB}})}{\partial\xi}+\xi\right]\frac{\partial\tilde{h}}{\partial\xi}+\eta_0^2\hat{S}^2\frac{\partial\tilde{h}}{\partial t}\\ &\quad +\tilde{h}+\frac{\partial^2(T_{\mathrm{B}}-T_{\mathrm{SB}})}{\partial\eta^2}\frac{\tilde{h}}{\eta_0^2}-\eta_{\mathrm{B}}'\frac{\partial}{\partial\xi}(\tilde{T}-\tilde{T}_{\mathrm{S}})\\ &\quad -\alpha\frac{\partial\tilde{T}_{\mathrm{S}}}{\partial\eta}-\frac{\alpha}{\eta_0^2}\left[\frac{\partial T_{\mathrm{SB}}}{\partial\xi}\frac{\partial\tilde{h}}{\partial\xi}+\frac{\partial^2 T_{\mathrm{SB}}}{\partial\eta^2}\tilde{h}\right]\\ &\quad +\alpha\eta_{\mathrm{s}}'\frac{\partial\tilde{T}_{\mathrm{S}}}{\partial\xi}+(\mathrm{h.o.t})=0,\end{aligned} \tag{9.41}$$

$$\begin{aligned}&\frac{\partial\tilde{\Psi}}{\partial\xi}+\left[\frac{1}{\eta_0^2}\frac{\partial\Psi_{\mathrm{B}}}{\partial\eta}-\eta_0^2\xi^2\eta_{\mathrm{B}}\right]\frac{\partial\tilde{h}}{\partial\xi}-\eta_0^2\xi(\eta_{\mathrm{B}}-\xi\eta_{\mathrm{B}}')\tilde{h}\\ &\qquad =(\mathrm{h.o.t}),\end{aligned} \tag{9.42}$$

$$\frac{\partial \tilde{\Psi}}{\partial \eta} - \left[\frac{1}{\eta_0^2}\frac{\partial \Psi_B}{\partial \xi} - \eta_0^2 \xi \eta_B^2\right]\frac{\partial \tilde{h}}{\partial \xi} + (2\eta_0^2 \xi \eta_B \eta_B' - \eta_0^2 \xi^2)\tilde{h}$$
$$= (\text{h.o.t}), \qquad (9.43)$$

where

$$\hat{S} = \sqrt{\eta_B^2 + \xi^2}\,. \qquad (9.44)$$

4. The tip smoothness condition: As $\xi \to 0$,

$$\tilde{h}(0) < \infty; \qquad \tilde{h}'(0) = 0\,. \qquad (9.45)$$

5. The down-stream far-field condition: As $\xi \to \infty$, the amplitude of the solution $\tilde{h}$ cannot be exponentially growing. It turns out that the solution $\tilde{h}$ must vanish, namely,

$$\tilde{h} \to 0, \qquad (9.46)$$

as $\xi \to \infty$, whose asymptotic form is to be determined by the local dispersion relationship of the system.

The coefficients of the above linear perturbation system contain the basic state solution $\{\Psi_B;\ \eta_B;\ T_B;\ T_{BS}\}$. As we have pointed out before and it will be seen in the next section, the unsteady perturbed states are restricted in the so-called interfacial wave (IW) layer at the interface with the thickness of $O(\varepsilon)$. Hence, these basic state solution in the above system can be approximately described by the inner solution of $\{\Psi_{B0};\ \eta_{B0};\ T_{B0};\ T_{BS0}\}$ in the inner region near the interface.

Our goal is to solve the linear eigenvalue problem associated with the above system, by following the procedure developed in Chap. 2: first solve the system (9.34)–(9.46) for any given $(\sigma, \eta_0^2, \varepsilon)$ disregarding the tip condition; then apply the tip condition (9.45) to the asymptotic solution obtained. Thus, the parameter σ can be solved as a function of ε and other physical parameters.

The same as the system with no convection, the present system has the singularity at the tip, $\xi = 0$, and some other points in the complex plane. To find the uniformly valid asymptotic solutions, one needs to separate the inner regions around these singular points from the remaining region, which is called the outer region.

3. Outer Expansion Solution

We solve the problem in the first step with the multiple variables expansion (MVE) method (Kevorkian and Cole, 1996) and use the same fast variables

$$
\begin{aligned}
\eta_{+} &= \frac{\eta - 1}{\varepsilon}, \\
\xi_{+}(\xi, \eta_{+}, \varepsilon) &= \bar{\xi}_{+} + \varepsilon \int_{0}^{\eta_{+}} K(\xi, \eta'_{+}, \varepsilon) \mathrm{d}\eta'_{+}, \\
t_{+} &= \frac{t}{\eta_0^2 \varepsilon}
\end{aligned}
\tag{9.47}
$$

as used in Chap. 2.

The converted system with the multiple variables is given as follows.

1. The thermal conduction equation in the liquid phase is

$$
\begin{aligned}
\left(k^2 \frac{\partial^2}{\partial \xi_{+}^2} + \frac{\partial^2}{\partial \eta_{+}^2}\right) \tilde{T} &= \varepsilon \eta_0^2 (\xi^2 + \eta^2) \frac{\partial \tilde{T}}{\partial t_{+}} \\
&- \varepsilon \left(2 \frac{\partial^2}{\partial \eta \partial \eta_{+}} + 2K \frac{\partial^2}{\partial \xi_{+} \partial \eta_{+}} + \frac{\partial K}{\partial \eta_{+}} \frac{\partial}{\partial \xi_{+}}\right) \tilde{T} \\
&- \varepsilon \left(\frac{k}{\xi} \frac{\partial}{\partial \xi_{+}} + \frac{1}{\eta} \frac{\partial}{\partial \eta_{+}}\right) \tilde{T} - \varepsilon \left(2k \frac{\partial^2}{\partial \xi \partial \xi_{+}} + k' \frac{\partial}{\partial \xi_{+}}\right) \tilde{T} \\
&+ \frac{\varepsilon}{\eta_0^2 \xi \eta} \frac{\partial \Psi_B}{\partial \eta} \left(k \frac{\partial}{\partial \xi_{+}} + \varepsilon \frac{\partial}{\partial \xi}\right) \tilde{T} \\
&- \frac{\varepsilon}{\eta_0^2 \xi \eta} \frac{\partial T_B}{\partial \eta} \left(k \frac{\partial}{\partial \xi_{+}} + \varepsilon \frac{\partial}{\partial \xi}\right) \tilde{\Psi} \\
&- \frac{\varepsilon}{\eta_0^2 \xi \eta} \frac{\partial \Psi_{\mathrm{B}}}{\partial \xi} \left(\frac{\partial}{\partial \eta_{+}} + \varepsilon \frac{\partial}{\partial \eta} + \varepsilon K \frac{\partial}{\partial \xi_{+}}\right) \tilde{T} \\
&- \frac{\varepsilon}{\eta_0^2 \xi \eta} \frac{\partial T_{\mathrm{B}}}{\partial \xi} \left(\frac{\partial}{\partial \eta_{+}} + \varepsilon \frac{\partial}{\partial \eta} + \varepsilon K \frac{\partial}{\partial \xi_{+}}\right) \tilde{\Psi} + O(\varepsilon^2).
\end{aligned}
\tag{9.48}
$$

2. The thermal conduction equation in the solid phase is

$$
\left(k^2 \frac{\partial^2}{\partial \xi_{+}^2} + \frac{\partial^2}{\partial \eta_{+}^2}\right) \tilde{T}_{\mathrm{S}} = \varepsilon \eta_0^2 (\xi^2 + \eta^2) \frac{\partial \tilde{T}_{\mathrm{S}}}{\partial t_{+}}
$$

$$+\varepsilon\eta_0^2\Big(\xi k\frac{\partial}{\partial\xi_+}-\eta\frac{\partial}{\partial\eta_+}\Big)\tilde{T}_{\rm S}$$

$$-\varepsilon\Big(\frac{k}{\xi}\frac{\partial}{\partial\xi_+}+\frac{1}{\eta}\frac{\partial}{\partial\eta_+}\Big)\tilde{T}_{\rm S}-\varepsilon\Big(2k\frac{\partial^2}{\partial\xi\partial\xi_+}+k'\frac{\partial}{\partial\xi_+}\Big)\tilde{T}_{\rm S}$$

$$-\varepsilon\Big(2\frac{\partial^2}{\partial\eta\partial\eta_+}+2K_{\rm S}\frac{\partial^2}{\partial\xi_+\partial\eta_+}+\frac{\partial K_{\rm S}}{\partial\eta_+}\frac{\partial}{\partial\xi_+}\Big)\tilde{T}_{\rm S}$$

$$+O(\varepsilon^2). \tag{9.49}$$

3. The kinematic equation is

$$\left(k^2\frac{\partial^2}{\partial\xi_+^2}+\frac{\partial^2}{\partial\eta_+^2}\right)\tilde{\Psi}=-\varepsilon^2\eta_0^4(\xi^2+\eta^2)\tilde{\zeta}$$

$$-\varepsilon\left(2\frac{\partial^2}{\partial\eta\partial\eta_+}+2K\frac{\partial^2}{\partial\xi_+\partial\eta_+}+\frac{\partial K}{\partial\eta_+}\frac{\partial}{\partial\xi_+}\right)\tilde{\Psi}$$

$$+\varepsilon\left(\frac{k}{\xi}\frac{\partial}{\partial\xi_+}+\frac{1}{\eta}\frac{\partial}{\partial\eta_+}\right)\tilde{\Psi}-\varepsilon\left(2k\frac{\partial^2}{\partial\xi\partial\xi_+}+k'\frac{\partial}{\partial\xi_+}\right)\tilde{\Psi}$$

$$+O(\varepsilon^2). \tag{9.50}$$

4. The vorticity equation is

$${\rm Pr}\left(k^2\frac{\partial^2}{\partial\xi_+^2}+\frac{\partial^2}{\partial\eta_+^2}\right)\tilde{\zeta}=\varepsilon\eta_0^2(\xi^2+\eta^2)\frac{\partial\tilde{\zeta}}{\partial t_+}$$

$$-\varepsilon{\rm Pr}\left(2\frac{\partial^2}{\partial\eta\partial\eta_+}+2K\frac{\partial^2}{\partial\xi_+\partial\eta_+}+\frac{\partial K}{\partial\eta_+}\frac{\partial}{\partial\xi_+}\right)\tilde{\zeta}$$

$$+\varepsilon{\rm Pr}\left(\frac{k}{\xi}\frac{\partial}{\partial\xi_+}+\frac{1}{\eta}\frac{\partial}{\partial\eta_+}\right)\tilde{\zeta}-\varepsilon{\rm Pr}\left(2k\frac{\partial^2}{\partial\xi\partial\xi_+}+k'\frac{\partial}{\partial\xi_+}\right)\tilde{\zeta}$$

$$+\varepsilon\frac{\eta_0^4{\rm Gr}}{{\rm Pr}|T_\infty|}\xi\eta\left(\eta k\frac{\partial\tilde{T}}{\partial\xi_+}+\xi\frac{\partial\tilde{T}}{\partial\eta_+}\right)+O(\varepsilon^2). \tag{9.51}$$

The boundary conditions are:

1. The up-stream far-field condition: away from the IW layer in the liquid phase, as $\eta_+ \to \infty$,

$$\tilde{T},\ \tilde{\Psi},\ \tilde{\zeta} \to 0. \tag{9.52}$$

2. The regular condition: away from the IW layer in the solid phase, as $\eta_+ \to -\infty$,

$$\tilde{T}_S \to 0. \tag{9.53}$$

3. For the interface conditions, (9.39)–(9.43), we may further make a Taylor expansion around the interface $\eta = 1$. The resultant interface conditions will still be homogeneous. The multiple variable form of the interface conditions are listed as follows: at $\eta = 1$, $\eta_+ = \Theta_0\Delta_0$,

$$\tilde{T} = \tilde{T}_S + (1 - \varepsilon\Theta_0\overline{\Lambda}_1)\tilde{h} - (\Theta_0\Delta_0 + \varepsilon\eta_1)\frac{\partial(\tilde{T} - \tilde{T}_S)}{\partial\eta_+} + O(\text{h.o.t}), \tag{9.54}$$

$$\begin{aligned} \tilde{T}_S = {} & \frac{1}{S}\left\{\left(k^2\frac{\partial^2}{\partial\bar{\xi}_+^2} + 2\varepsilon k\frac{\partial^2}{\partial\bar{\xi}_+\partial\xi} + \varepsilon\frac{\partial k}{\partial\xi}\frac{\partial}{\partial\bar{\xi}_+} + \varepsilon^2\frac{\partial^2}{\partial\xi^2}\right)\right. \\ & \left. +\varepsilon\left(\frac{1}{\xi} + \frac{\xi}{S^2}\right)\left(k\frac{\partial}{\partial\bar{\xi}_+} + \varepsilon\frac{\partial}{\partial\xi}\right) - \frac{\varepsilon^2}{S^2}\right\}\tilde{h} \\ & -(\Theta_0\Delta_0 + \varepsilon\eta_1)\frac{\partial\tilde{T}_S}{\partial\eta_+} + O(\text{h.o.t}), \end{aligned} \tag{9.55}$$

$$\begin{aligned} & \left(\frac{\partial\tilde{T}}{\partial\eta_+} - \frac{\partial\tilde{T}_S}{\partial\eta_+}\right) + (\Theta_0\Delta_0 + \varepsilon\eta_1)\frac{\partial^2(\tilde{T} - \tilde{T}_S)}{\partial\eta_+^2} + \varepsilon\left(\frac{\partial\tilde{T}}{\partial\eta} - \frac{\partial\tilde{T}_S}{\partial\eta}\right) \\ & +\varepsilon\left(\hat{K}\frac{\partial\tilde{T}}{\partial\bar{\xi}_+} - \hat{K}_S\frac{\partial\tilde{T}_S}{\partial\bar{\xi}_+}\right) + S^2\frac{\partial\tilde{h}}{\partial t_+} + \xi\left(k\frac{\partial\tilde{h}}{\partial\bar{\xi}_+} + \varepsilon\frac{\partial\tilde{h}}{\partial\xi}\right) \\ & -\varepsilon\Theta_1\frac{\partial\tilde{T}_S}{\partial\eta_+} + \varepsilon(1 + \overline{\Lambda}_2)\tilde{h} + O(\text{h.o.t}) = 0\,, \end{aligned} \tag{9.56}$$

$$\begin{aligned} & \left(k\frac{\partial\tilde{\Psi}}{\partial\bar{\xi}_+} + \varepsilon\frac{\partial\tilde{\Psi}}{\partial\xi}\right) + \varepsilon\Theta_0\left(P_1 - \eta_{0,0}^2\xi^2\Delta_0\right)\left(k\frac{\partial\tilde{h}}{\partial\bar{\xi}_+} + \varepsilon\frac{\partial\tilde{h}}{\partial\xi}\right) \\ & -\varepsilon\eta_{0,0}^2\xi\tilde{h} + (\Theta_0\Delta_0 + \varepsilon\eta_1)\frac{\partial^2\tilde{\Psi}}{\partial\eta_+\partial\xi} = O(\text{h.o.t}), \end{aligned} \tag{9.57}$$

$$\left(\frac{\partial\tilde{\Psi}}{\partial\eta_+}+\varepsilon\frac{\partial\tilde{\Psi}}{\partial\eta}+\varepsilon\hat{K}\frac{\partial\tilde{\Psi}}{\partial\bar{\xi}_+}\right)-\varepsilon\Big[\Theta_1\eta_{0,0}^2\xi+\Theta_0(P_2+\eta_{0,0}^2\xi\Delta_0)\Big]$$
$$\times\left(k\frac{\partial\tilde{h}}{\partial\bar{\xi}_+}+\varepsilon\frac{\partial\tilde{h}}{\partial\xi}\right)+(\Theta_0\Delta_0+\varepsilon\eta_1)\frac{\partial^2\tilde{\Psi}}{\partial\eta_+^2}-\varepsilon\eta_{0,0}^2\xi^2\tilde{h}$$
$$=O(\text{h.o.t}). \tag{9.58}$$

In the above,

$$S=\sqrt{1+\xi^2}. \tag{9.59}$$

In terms of the multiple variables, $(\xi,\eta,\xi_+,\eta_+,t_+)$, we make the following MVE for perturbed states:

$$\begin{aligned}
\tilde{T}&=\tilde{\mu}_0(\varepsilon)\left\{\tilde{T}_0(\xi,\eta,\xi_+,\eta_+)+\varepsilon\tilde{T}_1(\xi,\eta,\xi_+,\eta_+)+\cdots\right\}e^{\sigma t_+},\\
\tilde{T}_{\mathrm{S}}&=\tilde{\mu}_0(\varepsilon)\left\{\tilde{T}_{\mathrm{S}0}(\xi,\eta,\xi_+,\eta_+)+\varepsilon\tilde{T}_{\mathrm{S}1}(\xi,\eta,\xi_+,\eta_+)+\cdots\right\}e^{\sigma t_+},\\
\tilde{h}&=\tilde{\mu}_0(\varepsilon)\left\{\tilde{h}_0(\xi,\bar{\xi}_+)+\varepsilon\tilde{h}_1(\xi,\bar{\xi}_+)+\cdots\right\}e^{\sigma t_+},\\
\tilde{\Psi}&=\tilde{\nu}_0(\varepsilon)\left\{\tilde{\Psi}_0(\xi,\eta,\xi_+,\eta_+)+\varepsilon\tilde{\Psi}_1(\xi,\eta,\xi_+,\eta_+)+\cdots\right\}e^{\sigma t_+},
\end{aligned} \tag{9.60}$$

with

$$\begin{aligned}
k(\xi,\varepsilon)&=k_0(\xi)+\varepsilon k_1(\xi)+\cdots,\\
K(\xi,\eta_+,\varepsilon)&=K_0(\xi,\eta_+)+\varepsilon K_1(\xi,\eta_+)+\cdots,\\
K_{\mathrm{S}}(\xi,\eta_+,\varepsilon)&=K_{\mathrm{S}0}(\xi,\eta_+)+\varepsilon K_{\mathrm{S}1}(\xi,\eta_+)+\cdots,\\
\sigma&=\sigma_0+\varepsilon\sigma_1+\varepsilon^2\sigma_2+\cdots.
\end{aligned} \tag{9.61}$$

In the above, the eigenvalue $\sigma=\sigma_{\mathrm{R}}-\mathrm{i}\omega$ $(\omega\geq 0)$ is, generally, a complex number. Note that the eigenvalue σ introduced here should not be confused with one of the independent variables (σ,τ) used in (4.11) for the basic state solution. In the first step, we set σ_0 as an arbitrary constant.

From the multiple variables form of the system (9.48)–(9.58), one can successively derive each order approximation solution. Before proceeding, we note from the boundary conditions (9.57)– (9.58) that the interplay between perturbed flow field and temperature field is through the perturbation of the interface $\tilde{h}$, and one must set

$$\tilde{\nu}_0(\varepsilon)=\varepsilon\tilde{\mu}_0(\varepsilon). \tag{9.62}$$

3.1 Zeroth-Order Multiple Variables Expansion (MVE) Solutions

In the zeroth-order approximation, we obtain the following system:

$$\begin{aligned} \left(k_0^2\frac{\partial^2}{\partial \xi_+^2}+\frac{\partial^2}{\partial \eta_+^2}\right)\tilde{T}_0 = 0, \\ \left(k_0^2\frac{\partial^2}{\partial \xi_+^2}+\frac{\partial^2}{\partial \eta_+^2}\right)\tilde{T}_{S0} = 0, \end{aligned} \tag{9.63}$$

and

$$\begin{aligned} \left(k_0^2\frac{\partial^2}{\partial \xi_+^2}+\frac{\partial^2}{\partial \eta_+^2}\right)\tilde{\Psi}_0 = \eta_{0,0}^4(\xi^2+\eta^2)\tilde{\zeta}_0, \\ \left(k_0^2\frac{\partial^2}{\partial \xi_+^2}+\frac{\partial^2}{\partial \eta_+^2}\right)\tilde{\zeta}_0 = 0\ . \end{aligned} \tag{9.64}$$

The boundary conditions are:

1. As $\eta_+ \to \infty$,

$$\tilde{T}_0,\ \tilde{\Psi}_0,\ \tilde{\zeta}_0 \to 0. \tag{9.65}$$

2. As $\eta_+ \to -\infty$,

$$\tilde{T}_{S0} \to 0. \tag{9.66}$$

3. At $\eta = 1,\ \eta_+ = \Theta_0\Delta_0,$

$$\tilde{T}_0 = \tilde{T}_{S0} + \tilde{h}_0 - \Theta_0\Delta_0\frac{\partial(\tilde{T}_0-\tilde{T}_{S0})}{\partial \eta_+}, \tag{9.67}$$

$$\tilde{T}_{S0} = \frac{k_0^2}{S}\frac{\partial^2\tilde{h}_0}{\partial \xi_+^2} - \Theta_0\Delta_0\frac{\partial \tilde{T}_{S0}}{\partial \eta_+}, \tag{9.68}$$

$$\frac{\partial\left(\tilde{T}_0-\tilde{T}_{S0}\right)}{\partial \eta_+} + \Theta_0\Delta_0\frac{\partial^2(\tilde{T}_0-\tilde{T}_{S0})}{\partial \eta_+^2} + S^2\sigma_0\tilde{h}_0 + k_0\xi\frac{\partial \tilde{h}_0}{\partial \xi_+} = 0, \tag{9.69}$$

$$\frac{\partial \tilde{\Psi}_0}{\partial \xi_+} + k_0\Theta_0\Delta_0\frac{\partial^2\tilde{\Psi}_0}{\partial \eta_+\partial \xi_+} + \Theta_0 k_0(P_1-\eta_{0,0}^2\xi^2\Delta_0)\tilde{h}_0' - \eta_{0,0}^2\xi\tilde{h}_0 = 0, \tag{9.70}$$

$$\frac{\partial \tilde{\Psi}_0}{\partial \eta_+} + \Theta_0 \Delta_0 \frac{\partial^2 \tilde{\Psi}_0}{\partial \eta_+^2} - \Big[\Theta_1 \eta_{0,0}^2 \xi + \Theta_0 (P_2 + \eta_{0,0}^2 \xi \Delta_0)\Big] \tilde{h}_0'$$
$$- \eta_{0,0}^2 \xi^2 \tilde{h}_0 = 0. \quad (9.71)$$

From (9.67)–(9.71), one can see that in the leading-order approximation, perturbations of the temperature field are subject to the Laplace equation; the effects of convective flow, as well as growth on the perturbed temperature field are negligible. The perturbed flow field can be considered as inviscid. Moreover, the perturbations of temperature field can be solved first without being affected by the perturbed flow field. On the other hand, the perturbed flow field is passive, which may be affected by the perturbed temperature field through the perturbation of interface shape $\tilde{h}_0$. It is also weaker one order of magnitude than the perturbed temperature field.

The zeroth-order approximation yields the normal mode solution:

$$\begin{cases} \tilde{T}_0 = A_0(\xi,\eta)\ \exp\{\mathrm{i}\,\xi_+ - k_0(\eta_+ - \Theta_0\Delta_0)\}\,, \\ \tilde{T}_{S0} = A_{S0}(\xi,\eta)\ \exp\{\mathrm{i}\,\xi_+ + k_0(\eta_+ - \Theta_0\Delta_0)\}\,, \\ \tilde{\Psi}_0 = \Big[C_0(\xi,\eta) + \eta_+ E_0(\xi,\eta)\Big]\ \exp\{\mathrm{i}\,\xi_+ - k_0(\eta_+ - \Theta_0\Delta_0)\}\,, \\ \tilde{h}_0 = \hat{D}_0\ \exp\{\mathrm{i}\,\xi_+\}\ . \end{cases} \quad (9.72)$$

By setting

$$\begin{cases} \hat{A}_0(\xi) = A_0(\xi,1)\,, & \hat{A}_{S0}(\xi) = A_{S0}(\xi,1)\,, \\ \hat{C}_0(\xi) = C_0(\xi,1)\,, & \hat{E}_0(\xi) = E_0(\xi,1)\,, \end{cases} \quad (9.73)$$

from (9.67)–(9.71), we derive that

$$(1 - k_0\Theta_0\Delta_0)\hat{A}_0 - (1 + k_0\Theta_0\Delta_0)\hat{A}_{S0} - \hat{D}_0 = 0, \quad (9.74)$$

$$(1 + k_0\Theta_0\Delta_0)\hat{A}_{S0} + \frac{k_0^2}{S}\hat{D}_0 = 0, \quad (9.75)$$

$$-k_0(1 - k_0\Theta_0\Delta_0)\hat{A}_0 - k_0(1 + k_0\Theta_0\Delta_0)\hat{A}_{S0}$$
$$+ S^2\sigma_0\hat{D}_0 + \mathrm{i}\xi k_0\hat{D}_0 = 0, \quad (9.76)$$

$$\mathrm{i}(\hat{C}_0 + \Theta_0\Delta_0\hat{E}_0) + \mathrm{i}k_0\Big[\hat{E}_0(1 - k_0\Theta_0\Delta_0) - k_0\hat{C}_0\Big]$$
$$+ \Big[\mathrm{i}\Theta_0 k_0(P_1 - \eta_{0,0}^2\xi^2\Delta_0) - \eta_{0,0}^2\xi\Big]\hat{D}_0 = 0, \quad (9.77)$$

$$\Big[\hat{E}_0(1-k_0\Theta_0\Delta_0)-k_0\hat{C}_0\Big]+\Theta_0\Delta_0\Big[\hat{E}_0(k_0^2\Theta_0\Delta_0-2k_0)+k_0^2\hat{C}_0\Big]$$

$$-\Big\{\mathrm{i}\Big[\Theta_1\eta_{0,0}^2\xi+\Theta_0(P_2+\eta_{0,0}^2\xi\Delta_0)\Big]+\eta_{0,0}^2\xi^2\Big\}\hat{D}_0=0. \qquad (9.78)$$

From the first three conditions, (9.74)–(9.76) we solve

$$\begin{cases} \hat{A}_0(\xi)=\dfrac{1}{1-k_0\Theta_0\Delta_0}\Big(1-\dfrac{k_0^2}{S}\Big)\hat{D}_0, \\ \hat{A}_{S0}(\xi)=-\dfrac{1}{1+k_0\Theta_0\Delta_0}\dfrac{k_0^2}{S}\hat{D}_0, \end{cases} \qquad (9.79)$$

and the local dispersion formula

$$\sigma_0=\Sigma(\xi,k_0)=\frac{k_0}{S^2}\Big(1-\frac{2k_0^2}{S}\Big)-\mathrm{i}\frac{\xi}{S^2}k_0 \qquad (9.80)$$

where $\hat{D}_0$ is an arbitrary constant. From (9.77)–(9.78), one may solve $\hat{C}_0(\xi)$ and $\hat{E}_0(\xi)$ in terms of $\hat{D}_0$.

The above local dispersion formula is the same as that for dendritic growth without convective flow. It shows that, in the leading-order approximation, the convective flow does not affect the spectrum of the eigenvalues, it only changes the profiles of the eigenfunctions. Therefore, with the inclusion of convection, the instability mechanism of the system remains unchanged in the leading order.

The determination of σ_0 follows completely the same procedure as for the system without convection in Chap. 2 (Xu, 1997). In what follows, we shall only list the results without detailed derivations. For any given constant σ_0, one can find three roots: $\{k_0^{(1)}(\xi);\, k_0^{(2)}(\xi);\, k_0^{(3)}(\xi)\}$. Among them, only $k_0^{(1)}(\xi)$, $k_0^{(3)}(\xi)$ are meaningful. Thus, the general solution in the outer region is:

$$\tilde{h}=\Big\{D_0^{(1)}\ \exp\Big\{\frac{\mathrm{i}}{\varepsilon}\int_0^{\bar{\xi}}(k_0^{(1)}+\varepsilon k_1^{(1)}+\cdots)\mathrm{d}\xi_1\Big\}$$

$$+D_0^{(3)}\ \exp\Big\{\frac{\mathrm{i}}{\varepsilon}\int_0^{\bar{\xi}}(k_0^{(3)}+\varepsilon k_1^{(3)}+\cdots)\mathrm{d}\xi_1\Big\}+\cdots\Big\}e^{\sigma t_+}, \qquad (9.81)$$

where the coefficients $(D_0^{(1)}; D_0^{(2)})$ remain unknown. It will be seen from the first-order approximation solution to be derived later that the above MVE solution has singularity at the following points:

- The tip point of the dendrite $\xi=0$.

- The critical points $\xi = \xi_c$, which are the roots of the equation

$$\frac{\partial \Sigma(\xi, k_0)}{\partial k_0} = 0. \tag{9.82}$$

Combining (9.80) with (9.82), one finds that the singular points ξ_c are the roots of the equation

$$\sigma_0 = \sqrt{\frac{2}{27}} \left(\frac{1 - \mathrm{i}\xi}{S} \right)^{3/2}_{\xi=\xi_c}. \tag{9.83}$$

Hence, the MVE solution obtained above is valid only in the region away from these singular points. Therefore in order to derive a uniformly valid asymptotic expansion for the solution, one must divide the whole complex ξ-plane into three sub-regions: the outer region, inner region near the tip and inner region near the singular point ξ_c, then derive the asymptotic expansion of the solution in each sub-region separately, and finally match all these asymptotic expansion solutions in the intermediate regions. After the uniformly valid asymptotic solution is obtained, as the second step, one may apply the smooth tip condition (9.46) and derive the quantization condition for the eigenvalues σ_0. Such procedure is completely the same as shown in (Xu, 1997). We shall not give the detailed derivation, but only list the final results below. The quantization condition for the eigenvalues is

$$\begin{aligned} \frac{1}{\varepsilon} \int_0^{\xi_c} \left(k_0^{(1)} - k_0^{(3)} \right) d\xi &= (2n + 1 + \frac{2}{3} + \frac{\theta_0}{2})\pi - \frac{i}{2} \log \alpha_0 \\ n &= (0, \pm 1, \pm 2, \pm 3, \ldots) \end{aligned} \tag{9.84}$$

where

$$\alpha_0\, e^{i\theta_0 \pi} = k_0^{(1)}(0) \big/ k_0^{(3)}(0)\,. \tag{9.85}$$

For any given small stability parameter ε, the above quantization condition determines a discrete set of the complex eigenvalues $\{\sigma_0 = \sigma_n^{(0)}\}$ $(n = 0, 1, 2, \ldots)$ and the corresponding global modes. As in the previous works, we call these global modes the Global Trapped Wave (GTW) Modes.

As we stated before, the leading-order approximation of eigenvalue σ_0 does not reflect the effect of convective flow. To see the effect of convection on the eigenvalue, one must look into the first-order MVE solution.

3.2 First-Order Approximation

The first-order MVE solution will determine the amplitude functions $A_0(\xi,\eta)$, $A_{S0}(\xi,\eta)$, the functions $k_1(\xi)$, and σ_1. At this level, the perturbed temperature field is still de-coupled from the perturbed flow field and the equation of perturbed temperature field is not affected by the perturbed flow field. As the form of the equation of perturbed temperature field is the same as that for the system without convection derived in Chap. 2, the following formulas still hold:

$$\begin{aligned} Q_0 &= \frac{1}{\hat{D}_0}\frac{\partial}{\partial\eta}(A_0 - A_{S0})\Big|_{\eta=1} \\ &= -2\mathrm{i}\frac{\partial}{\partial\xi}\left(\frac{k_0^2}{S}\right) - \left(1 - \frac{2}{S}k_0^2\right)\left[\frac{\eta_{0,0}^2}{2k_0}\sigma_0 S^2 + k_1\right. \\ &\quad \left. + \frac{\mathrm{i}}{2}\left(\xi\eta_{0,0}^2 - \frac{1}{\xi} - \frac{k_0'}{k_0}\right)\right] - \left(\frac{1+\eta_{0,0}^2}{2}\right), \end{aligned} \tag{9.86}$$

and

$$\begin{aligned} \hat{K}_0 &= K_0(\xi,0) = -\frac{k_0'}{2k_0}, \\ \hat{K}_{S0} &= K_{S0}(\xi,0) = \frac{k_0'}{2k_0}. \end{aligned} \tag{9.87}$$

Moreover, for the first-order approximation, one has the solutions

$$\begin{aligned} \tilde{T}_1 &= A_1(\xi,\eta)\exp\{\mathrm{i}\xi_+ - k_0(\eta_+ - \Theta_0\Delta_0)\}, \\ \tilde{T}_{S1} &= A_{S1}(\xi,\eta)\exp\{\mathrm{i}\xi_+ + k_0(\eta_+ - \Theta_0\Delta_0)\}, \\ \tilde{h}_1 &= \hat{D}_1\exp\{\mathrm{i}\xi_+\} \end{aligned} \tag{9.88}$$

and the system:

$$(1 - k_0\Theta_0\Delta_0)\hat{A}_1 - (1 + k_0\Theta_0\Delta_0)\hat{A}_{S1} - \hat{D}_1 = I_1\hat{D}_0, \tag{9.89}$$

$$(1 + k_0\Theta_0\Delta_0)\hat{A}_{S1} + \frac{k_0^2}{S}\hat{D}_1 = I_2\hat{D}_0, \tag{9.90}$$

$$\begin{aligned} -k_0(1 - k_0\Theta_0\Delta_0)\hat{A}_1 - k_0(1 + k_0\Theta_0\Delta_0)\hat{A}_{S1} \\ + S^2\sigma_0\hat{D}_1 + \mathrm{i}\xi k_0\hat{D}_1 = I_3\hat{D}_0, \end{aligned} \tag{9.91}$$

where we denoted

$$\hat{A}_1(\xi) = A_1(\xi,1)\,, \qquad \hat{A}_{S1}(\xi) = A_{S1}(\xi,1). \tag{9.92}$$

The solvability condition is then derived as follows:

$$\det \begin{pmatrix} 1 & -1 & I_1 \\ 0 & 1 & I_2 \\ -k_0 & -k_0 & I_3 \end{pmatrix} = 0, \tag{9.93}$$

which leads to:

$$I_3 + 2k_0 I_2 + I_1 k_0 = 0. \tag{9.94}$$

With the inclusion of the effect of convection, we now have

$$\begin{aligned} I_1 &= -\Theta_0 \overline{\Lambda}_1 + k_0 \eta_1 \tfrac{(1-2k_0^2/S)+k_0\Theta_0\Delta_0}{1-k_0^2\Theta_0^2\Delta_0^2}, \\ I_2 &= \tfrac{1}{S}\left\{ \mathrm{i}k_0' + \mathrm{i}k_0\left(\tfrac{1}{\xi} + \tfrac{\xi}{S^2}\right) - 2k_0 k_1 \right\} + \tfrac{k_0^3/S}{1+k_0\Theta_0\Delta_0}\eta_1, \\ I_3 &= -(\sigma_1 S^2 + \mathrm{i}\xi k_1) - Q_0 - (1+\overline{\Lambda}_2) - \tfrac{\Theta_1 k_0^3}{S(1+k_0\Theta_0\Delta_0)} \\ &\quad + \mathrm{i}\tfrac{k_0'}{2k_0} \tfrac{(1-2k_0^2/S)+k_0\Theta_0\Delta_0}{1-k_0^2\Theta_0^2\Delta_0^2} + \eta_1 k_0^2 \tfrac{1+k_0\Theta_0\Delta_0(1-2k_0^2/S)}{1-k_0^2\Theta_0^2\Delta_0^2}. \end{aligned}$$

From (9.94), we obtain

$$\begin{aligned} k_1 F(\xi) - \frac{\mathrm{i}}{2\xi}F(\xi) - S^2\sigma_1 - \overline{\Lambda}_2 + \frac{\eta_{0,0}^2}{2} + \frac{\eta_{0,0}^2}{2}\left(1 - \frac{2k_0^2}{S}\right)^2 \\ + \mathrm{i}\frac{k_0'}{k_0}\left[\frac{6k_0^2}{S} + \frac{1}{2}k_0\Theta_0\Delta_0 \frac{1+(1-2k_0^2/S)k_0\Theta_0\Delta_0}{1-k_0^2\Theta_0^2\Delta_0^2}\right] \\ - \frac{\Theta_1 k_0^3}{S(1+k_0\Theta_0\Delta_0)} - k_0\Theta_0(\overline{\Lambda}_1 + \Delta_0\overline{\Lambda}_2) = 0 \end{aligned} \tag{9.95}$$

or

$$k_1 = \frac{\mathrm{i}}{2\xi} + \frac{R_1(\xi)}{F(\xi)} - \mathrm{i}\frac{R_2(\xi)}{F(\xi)}\frac{k_0'}{k_0}, \tag{9.96}$$

where we denote

$$\begin{aligned} F(\xi) &= 1 - \mathrm{i}\xi - \frac{6k_0^2}{S}, \\ R_1(\xi) &= S^2\sigma_1 + \left(\overline{\Lambda}_2 - \frac{1}{2}\eta_{0,0}^2\right) - \frac{\eta_{0,0}^2}{2}\left(1 - \frac{2k_0^2}{S}\right)^2 \\ &\quad + \frac{\Theta_1 k_0^3}{S(1+k_0\Theta_0\Delta_0)} + k_0\Theta_0\overline{\Lambda}_1, \\ R_2(\xi) &= \frac{6k_0^2}{S} + \frac{1}{2}k_0\Theta_0\Delta_0\frac{1+(1-2k_0^2/S)k_0\Theta_0\Delta_0}{1-k_0^2\Theta_0^2\Delta_0^2}. \end{aligned} \tag{9.97}$$

From (9.96), the same as in Chap. 2, we derive that

$$k_1 \sim \frac{\mathrm{i}}{2\xi} \tag{9.98}$$

as $\xi \to 0$, and

$$k_1 \sim \frac{R_1(\xi_c)}{2^{\frac{1}{2}} m_0^{\frac{1}{2}} (\xi - \xi_c)^{\frac{1}{2}}} + \frac{\mathrm{i}}{4(\xi - \xi_c)} + O(1) \tag{9.99}$$

as $\xi \to \xi_c$. Finally, because $k_1(\xi)$ can only allow a pole singularity at the isolated singular point ξ_c, we deduce that

$$R_1(\xi_c) = 0\,. \tag{9.100}$$

Then it follows that

$$\begin{aligned}\sigma_1 = & -\frac{1}{(1+\xi_c^2)}\left[1 + \frac{\eta_{0,0}^2}{2} - \frac{\eta_{0,0}^2}{18}\left(2 + \mathrm{i}\xi_c\right)^2 + \frac{\Theta_1 k_0^3}{S(1 + k_0\Theta_0\Delta_0)}\right. \\ & \left. +(\overline{\Lambda}_2 - 1 - \eta_{0,0}^2) + k_0\Theta_0\overline{\Lambda}_1\right].\end{aligned} \tag{9.101}$$

In the above, all functions are evaluated at $\xi = \xi_c$, and we have

$$k_0^2(\xi_c) = \frac{1}{6}(1 - \mathrm{i}\xi_c)\sqrt{1 + \xi_c^2}. \tag{9.102}$$

Again, it is seen that the up to the first order approximation, $O(\varepsilon)$, the interface shape correction term for the Basic state caused by the isotropic surface tension,which is $\varepsilon^2\eta_1(\xi)$, have no influence on the eigenvalues and eigenfunctions of perturbed states.

Given the stability parameter ε, from quantization condition (9.84), one first computes the discrete set of the complex eigenvalues $\{\sigma_0 = \sigma_n^{(0)}\}$ $(n = 0, 1, 2, \ldots)$. Then, in terms of formula (9.101), one can calculate corresponding σ_1. From the results obtained, we may write the eigenvalue σ in the form

$$\sigma = \sigma_{f,0} + \sigma_{f,1} + \sigma_{f,2} + \sigma_{f,3} + \cdots. \tag{9.103}$$

Here, $\sigma_{f,0}$ is the eigenvalue of the system without convective flow, while

$$\sigma_{f,1} = -\nu_0(\epsilon_2)\frac{k_0\overline{\Lambda}_1}{(1 + \xi_c^2)}, \tag{9.104}$$

$$\sigma_{f,2} = -\varepsilon\frac{\overline{\Lambda}_2 - 1 - \eta_{0,0}^2}{(1 + \xi_c^2)}, \tag{9.105}$$

and

$$\sigma_{f,3} = -\alpha \frac{k_0^3}{S(1+\xi_c^2)(1+k_0\Theta_0\Delta_0)}, \tag{9.106}$$

represent the effects of convective flow. The first term, $\sigma_{f,1}$ is caused by the change of temperature gradient of the basic state, the second term, $\sigma_{f,2}$ is caused by the change of second derivative of the temperature profile of the basic state, whereas the third term, $\sigma_{f,3}$ is caused by the effect of density change. For the validity of the asymptotic expansion solution, evidently, one needs to require that all these higher-order correction terms are not bigger than the leading-order term, namely,

$$\left\{|\sigma_{f,1}|, |\sigma_{f,1}|, |\sigma_{f,1}|\right\} < |\sigma_{f,0}|. \tag{9.107}$$

4. Stability Criterion and Selection Condition of Tip Velocity

The Global Neutrally Stable (GNS) state of the system is mathematically expressed as

$$\sigma_R(\varepsilon_*) = 0, \quad (n = 0). \tag{9.108}$$

The stability criterion is

$$\begin{cases} \varepsilon \geq \varepsilon_* & \text{(stable)}, \\ \varepsilon < \varepsilon_* & \text{(unstable)}. \end{cases} \tag{9.109}$$

The selection condition of the dendrite's tip velocity can be expressed as

$$\varepsilon = \varepsilon_*. \tag{9.110}$$

If one uses ℓ_c as the length scale, while using $(\frac{\kappa_T}{\ell_c})$ as the velocity scale, then the selected non-dimensional tip velocity and tip radius can re-written in the form

$$\begin{cases} U_{\text{tip}} = \eta_0^4 \varepsilon_*^2, \\ R_{\text{tip}} = \frac{\text{Pe}}{\eta_0^4 \varepsilon_*^2}. \end{cases} \tag{9.111}$$

Furthermore the non-dimensional frequency of oscillation can be written as

$$\Omega_* = \frac{\omega_*}{\varepsilon_* \eta_0^2}. \tag{9.112}$$

From the results obtained it is seen that, in general, the convective flow will not change the velocity profile in the IW layer. So, it cannot affect the perturbed system directly. Convective flow, via the following two ways, affects the stability properties and the selection of the tip velocity U_{tip}, tip radius R_{tip} and frequency of oscillation Ω_*. The first way is that the convective flow may change the interface shape of the basic steady state and the local temperature profile near the interface, hence changing the parameters $\overline{\Lambda}_1$, $\overline{\Lambda}_2$ and Δ_0. As a consequence, it affects the eigenvalue σ and then affects the selection criterion ε_*, and the quantities: U_{tip}, R_{tip} and Ω_*. However, these effects only appear at higher-order approximations, not in the leading-order approximation. The second way is that the convective flow may affect the normalization parameter η_0^2 and the Peclet number Pe of the basic state. The change of the parameters η_0^2 and Pe subsequently affects the quantities: U_{tip}, R_{tip} and Ω_*.

5. Some Special Cases

5.1 Convection Motion Induced by Uniform External Flow with Pr $\gg$ 1

In this case, we have $\text{Gr} = \alpha = 0$. With the solution described in section 1.1, we have $\epsilon_2 = \frac{1}{\text{Pr}}$, and $\nu_0(\epsilon_2) = \frac{1}{\ln \epsilon_2}$. One derives that at $\eta = 1$,

$$\begin{aligned}
\frac{d\bar{T}_0}{d\eta} &= -\eta_{0,0}^2\Big[1 - \nu_0(\epsilon_2)\hat{\lambda}_0\hat{\tau}_{0,1}\Big] - \eta_0^2\Big[1 - \nu_0(\epsilon_2)\Big(\hat{\lambda}_0 + \tfrac{1}{\hat{\tau}_{0,0}}\Big)\hat{\tau}_{0,1}\Big] \\
\frac{d^2\bar{T}_0}{d\eta^2} &= \eta_{0,0}^2\Big[1 + 2\eta_{0,0}^2\Big(\lambda_0 - \tfrac{1}{2}\Big)\Big], \\
\frac{\partial \bar{T}_1}{\partial \eta} &= -\eta_0^2\Big(\hat{\lambda}_0 + \tfrac{1}{\hat{\tau}_{0,0}}\Big)\hat{\tau}_{0,1}
\end{aligned} \tag{9.113}$$

It follows that

$$\begin{aligned}
&\frac{d}{d\eta}\overline{T}_{B0}(1) = -\eta_0^2, \\
&\frac{d^2}{d\eta^2}\overline{T}_{B0}(1) = \eta_0^2\left[1 + 2\eta_{0,0}^2\Big(\lambda_0 - \frac{1}{2}\Big) - \nu_0(\epsilon_2)\frac{\hat{\tau}_{0,1}}{\hat{\tau}_{0,0}}\right].
\end{aligned} \tag{9.114}$$

Hence, we have

$$\begin{aligned}
&\Delta_0 = \overline{\Lambda}_1 = 0; \\
&\overline{\Lambda}_2 = \Big[1 + 2\eta_{0,0}^2\Big(\lambda_0 - \tfrac{1}{2}\Big) + \tfrac{1}{\ln \text{Pr}}\mathcal{Z}_e(\hat{\tau}_{0,0}, \lambda_0)\Big].
\end{aligned} \tag{9.115}$$

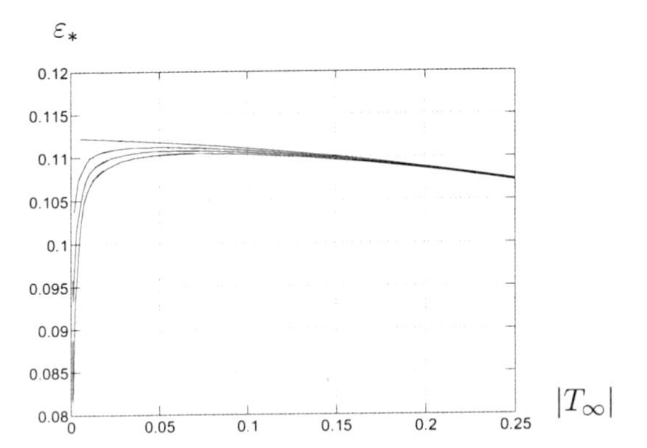

Figure 9.1. The variation of stability criterion ε_* versus T_∞ for the cases: $\mathrm{Pr} = 22$ and the Reynolds number of enforced flow, $\mathrm{Re_c} = 0.0, 1.0 \times 10^{-7}, 2.0 \times 10^{-7}, 3.0 \times 10^{-7}$ from top to bottom

With the above results, we derive

$$\begin{aligned} \sigma_{f,1} &= \sigma_{f,3} = 0, \\ \sigma_{f,2} &= -\varepsilon \frac{\Lambda_2 - 1 - \eta_{0,0}^2}{(1 + \xi_c^2)}, \end{aligned} \tag{9.116}$$

It is seen that the eigenvalue σ, is a function of the parameters: λ_0, T_∞ and Pr. From the stability criterion, one can derive the critical number ε_* as the function of the parameters λ_0, T_∞ and Pr, and accordingly obtain

$$\begin{cases} U_{\text{tip}} = \varepsilon_*^2 \eta_{0,0}^4 (1 - \frac{1}{\ln \mathrm{Pr}} \mathcal{Z}_e)^2, \\ R_{\text{tip}} = \dfrac{1}{\varepsilon_*^2 \eta_{0,0}^2 (1 - \frac{1}{\ln \mathrm{Pr}} \mathcal{Z}_e)}, \\ \Omega_* = \dfrac{\omega_*}{\varepsilon_* \eta_{0,0}^2 (1 - \frac{1}{\ln \mathrm{Pr}} \mathcal{Z}_e)}. \end{cases} \tag{9.117}$$

Note that in this case the parameter $\eta_{0,0}^2$ is the function of both the undercooling temperature T_∞ and the flow parameter λ_0 through the formula:

$$|T_\infty| = \frac{\eta_{0,0}^2}{2} \mathrm{e}^{(\lambda_0 - \frac{1}{2})\eta_{0,0}^2} E_1 \left[\left(\lambda_0 - \frac{1}{2} \right) \eta_{0,0}^2 \right]. \tag{9.118}$$

In practice, the flow parameter $\lambda_0 = 1 + \frac{U_\infty}{U}$ cannot be prescribed, which depends on both the flow velocity U_∞ and the tip velocity U. To measure the strength of the external flow, it is better to use the following

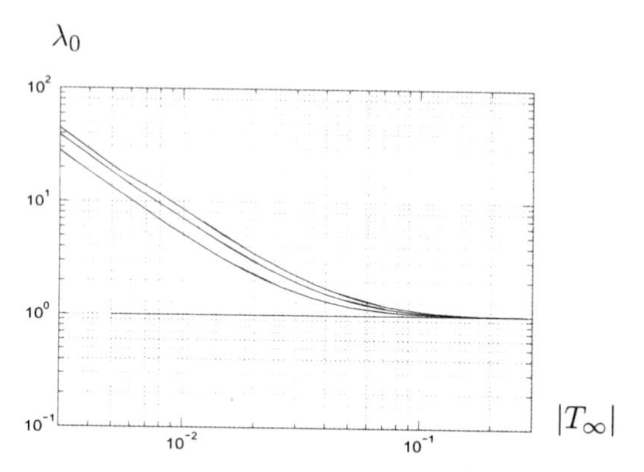

Figure 9.2. The variation of flow parameter λ_0 versus T_∞ for the cases: Pr = 22 and the Reynolds number of enforced flow, $\mathrm{Re_c} = 0.0, 1.0 \times 10^{-7}, 2.0 \times 10^{-7}, 3.0 \times 10^{-7}$ from top to bottom

Reynolds number $(\mathrm{Re})_\mathrm{c}$ based on the capillary length ℓ_c and the velocity of external flow U_∞:

$$\mathrm{Re_c} = \frac{(U_\infty)\ell_\mathrm{c}}{\nu}. \tag{9.119}$$

Thus, the selection condition should be given by the parameters: $T_\infty, (\mathrm{Re})_\mathrm{c}$ and Pr. It is derived that

$$\lambda_0 = 1 + \mathrm{Pr}\frac{\mathrm{Re_c}}{U_{\mathrm{tip}}} = 1 + \frac{\mathrm{PrRe_c}}{\eta_0^4 \varepsilon^2}, \tag{9.120}$$

where

$$\eta_0^2 = \eta_{0,0}^2 \left[1 - \frac{1}{\ln \mathrm{Pr}} \mathcal{Z}_e(\lambda_0, \eta_{0,0}^2)\right]^2. \tag{9.121}$$

(9.120)–(9.121) yields a quite complicated function relationship between the flow parameter λ_0 and the independent parameters: $T_\infty, (\mathrm{Re})_\mathrm{c}$ and Pr. To simplify the calculation, one may assume $\mathrm{Pr} = \infty$, so that, the factor $\frac{1}{\ln \mathrm{Pr}} \mathcal{Z}_e$ is negligible. With this approximation, one may have

$$\eta_0^2 \approx \eta_{0,0}^2, \quad \lambda_0 \approx 1 + \frac{\mathrm{PrRe_c}}{\eta_{0,0}^4 \varepsilon^2}, \tag{9.122}$$

and

$$\overline{\Lambda}_2 = \left[1 + 2\eta_{0,0}^2\left(\lambda_0 - \frac{1}{2}\right)\right]. \tag{9.123}$$

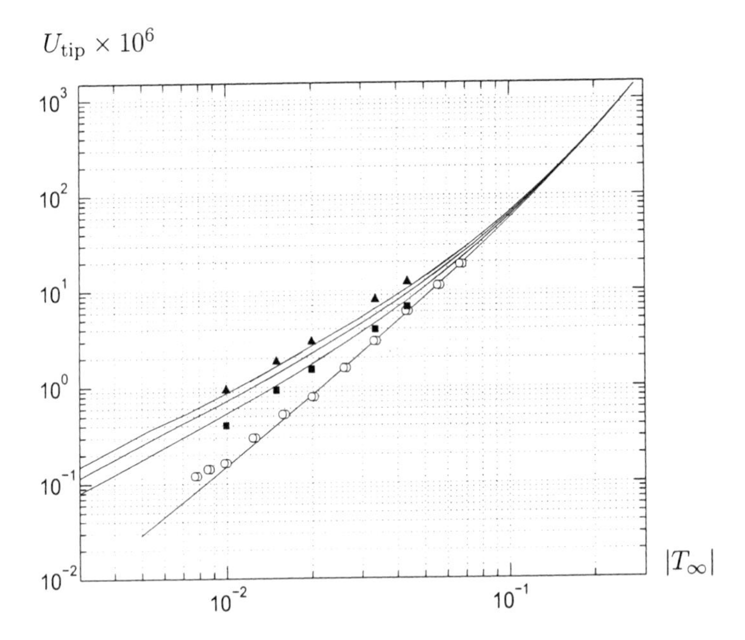

Figure 9.3. The variation of tip velocity U_{tip} versus $|T_\infty|$ for the cases: $\text{Pr} = 22$ and the Reynolds number of the enforced flow, $\text{Re}_c = 0.0, 1.0 \times 10^{-7}, 2.0 \times 10^{-7}, 3.0 \times 10^{-7}$ from bottom to top. The circles are the micro-gravity experimental data of the material SCN with $\text{Re}_c = 0$, obtained by Glicksman et. al. The squares are the experimental data with $\text{Re}_c = 1.11 \times 10^{-6}$, while the triangles are the experimental data with $\text{Re}_c = 1.11 \times 10^{-5}$. Both experiments with external flow are conducted on the ground by Gill et. al, by using the material SCN

Accordingly,

$$\begin{cases} U_{\text{tip}} = \varepsilon_*^2 \eta_{0,0}^4, \\ R_{\text{tip}} = \dfrac{1}{\varepsilon_*^2 \eta_{0,0}^2}, \\ \Omega_* = \dfrac{\omega_*}{\varepsilon_* \eta_{0,0}^2}. \end{cases} \tag{9.124}$$

The numerical results presented below are obtained with this $\text{Pr} = \infty$ simplification. In Fig 9.1, we show ε_* versus $|T_\infty|$ for the cases of $\text{Pr} = 22$ and $\text{Re}_c = 0.0, 1.0 \times 10^{-7}, 2.0 \times 10^{-7}, 3.0 \times 10^{-7}$. It is shown that some noticeable differences between these curves can be seen in the low undercooling region. However, these curves are quite close to each other in the region $|T_\infty| > 0.1$ that they are hardly seen. This implies that the effect of external flow on the stability criterion ε_* is very insensitive if the undercooling temperature is not too low.

In Fig 9.2 and Fig 9.3, we respectively show the flow parameter λ_0 and tip growth velocity $U_{\rm tip}$ versus $|T_\infty|$ for the same cases. It is seen that the effect of external flow is important in the small undercooling regime. In Fig 9.3, we also include the micro-gravity experimental data of the material SCN with no external flow obtained by Glicksman, et. al, as well as the ground experimental data of SCN with external flow obtained by Gill, et. al. The tendency of the theoretical curves are in excellent agreement with the experimental data. The corresponding Reynolds numbers $\mathrm{Re_c}$, however, are quite different. The causes for such an inconsistency are not clear yet. One of the possible causes might be that our asymptotic expansion solution derived in the limit $\mathrm{Pr} \to \infty$ with $\lambda_0 = O(1)$, numerically, was not accurate enough for the cases under discussion, some higher order terms might need to include; another cause might be that the experimental data involved some physical effects that we missed.

5.2 Convection Motion Induced by Buoyancy Effect with Pr ≫ 1

For this case, we have $\lambda_0 = 1, \alpha = 0$. With the solution described in section 1.2, we have $\epsilon_2 = \frac{1}{\mathrm{Pr}}$, $\nu_0(\epsilon_2) = \epsilon_2^4$ and $\hat{\tau}_{0,1} = 0$. One derives that at $\eta = 1$,

$$\frac{d\bar{T}_0}{d\eta} = -\eta_{0,0}^2 = -\eta_0^2 \tag{9.125}$$

$$\frac{d^2\bar{T}_0}{d\eta^2} = \eta_{0,0}^2(1 + \eta_{0,0}^2) + O(\epsilon_2^4). \tag{9.126}$$

Furthermore, with the results

$$\begin{aligned} \hat{D}'_{1,0}(\hat{\tau}_{0,0}) &= (4 + 2\hat{\tau}_{0,0})\frac{\hat{h}_{0,1}}{\hat{\tau}_{0,0}} \\ \hat{D}'_{1,1}(\hat{\tau}_{0,0}) &= -(2 + \hat{\tau}_{0,0})\frac{\hat{h}_{0,1}}{\hat{\tau}_{0,0}}, \end{aligned} \tag{9.127}$$

one derives that at $\eta = 1$,

$$\frac{\partial \widehat{T}_1}{\partial \eta} = \frac{\hat{h}_{0,1}}{\hat{\tau}_{0,0}}(2 + \hat{\tau}_{0,0})\bar{\sigma}. \tag{9.128}$$

and

$$\frac{\partial T_{\rm B0}}{\partial \eta} = -\eta_0^2\Big[1 - \epsilon_2^4\frac{\hat{h}_{0,1}}{\hat{\tau}_{0,0}}(2 + \hat{\tau}_{0,0})\bar{\sigma}\Big]. \tag{9.129}$$

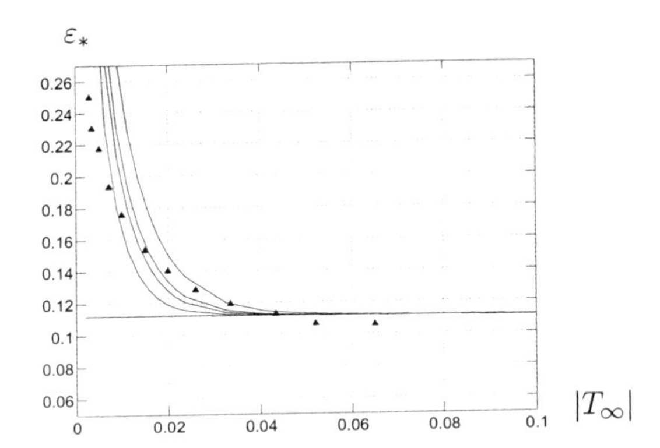

Figure 9.4. The variation of stability criterion ε_* versus T_∞ for the cases: $\mathrm{Pr} = 22$ and the fixed buoyancy parameter $\mathrm{Gr_c} = 2.0 \times 10^{-14}, 6.0 \times 10^{-14}, 1.0 \times 10^{-13}, 3.0 \times 10^{-13}$ from bottom to top. The circles are the micro-gravity experimental data of material SCN obtained by Glicksman et. al, while the triangles are the ground experimental data of material SCN with $\mathrm{Gr_c} = 3.218 \times 10^{-13}$ obtained by Gill, et. al

It follows that

$$\begin{aligned}
\Delta_0 &= -\frac{1}{2}\frac{\hat{h}_{0,1}}{\hat{\tau}_{0,0}}\bar{\sigma} = -\frac{\hat{\tau}_{0,0}\mathrm{e}^{\hat{\tau}_{0,0}}\tilde{h}_1(\hat{\tau}_{0,0})}{E_1(\hat{\tau}_{0,0})}\mathrm{Gr}\mathrm{Pr}^2\xi^2, \\
\overline{\Lambda}_1 &= \frac{\hat{h}_{0,1}}{\hat{\tau}_{0,0}}(2+\hat{\tau}_{0,0})\bar{\sigma} \\
&= (2+\hat{\tau}_{0,0})\frac{\hat{\tau}_{0,0}^2\mathrm{e}^{2\hat{\tau}_{0,0}}\tilde{h}_1(\hat{\tau}_{0,0})\overline{\mathrm{G}}}{2\mathrm{Pr}}\xi^2 \\
\overline{\Lambda}_2 &= 1+\eta_{0,0}^2.
\end{aligned} \tag{9.130}$$

Hence we obtain

$$\begin{aligned}
\sigma_{f,2} &= \sigma_{f,3} = 0, \\
\sigma_{f,1} &= -\overline{\mathrm{G}}\tilde{h}_1(\hat{\tau}_{0,0})\frac{\hat{\tau}_{0,0}^2\mathrm{e}^{2\hat{\tau}_{0,0}}(2+\hat{\tau}_{0,0})}{2\mathrm{Pr}^5}\frac{k_0(\xi_c)\xi_c^2}{(1+\xi_c^2)}.
\end{aligned} \tag{9.131}$$

Noting

$$\begin{aligned}
\eta_0^2 &= \eta_{0,0}^2 \\
\mathrm{Pe} &= \eta_{0,0}^2\left[1-\frac{\hat{\tau}_{0,0}^2\mathrm{e}^{2\hat{\tau}_{0,0}}\tilde{h}_1(\hat{\tau}_{0,0})}{2}\frac{\overline{\mathrm{G}}}{\mathrm{Pr}^5}\right] \\
&= \eta_{0,0}^2\left[1-\frac{\hat{\tau}_{0,0}\mathrm{e}^{\hat{\tau}_{0,0}}\tilde{h}_1(\hat{\tau}_{0,0})}{E_1(\hat{\tau}_{0,0})}\frac{\mathrm{Gr}}{\mathrm{Pr}^2}\right],
\end{aligned} \tag{9.132}$$

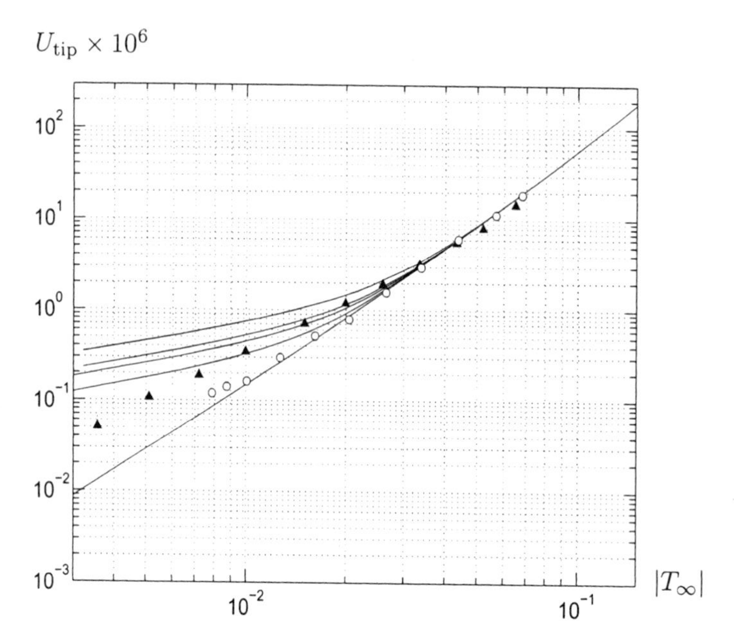

Figure 9.5. The variation of tip velocity U_{tip} versus $|T_\infty|$ for the cases: $\text{Pr} = 22$ and the fixed buoyancy parameter $\text{Gr}_c = 0.0, 2.0 \times 10^{-14}, 6.0 \times 10^{-14}, 1.0 \times 10^{-13}, 3.0 \times 10^{-13}$ from bottom to top. The circles are the micro-gravity experimental data of material SCN obtained by Glicksman et. al, while the triangles are the ground experimental data of material SCN with $\text{Gr}_c = 3.218 \times 10^{-13}$ obtained by Gill, et. al

From the stability criterion, we can derive the critical number ε_* as the function of the given parameters: T_∞, Gr and Pr:

$$\begin{cases} U_{\text{tip}} = \eta_{0,0}^4 \varepsilon_*^2, \\ R_{\text{tip}} = \dfrac{1}{\eta_{0,0}^2 \varepsilon_*^2} \left[1 - \dfrac{\hat{\tau}_{0,0} e^{\hat{\tau}_{0,0}} \tilde{h}_1(\hat{\tau}_{0,0})}{E_1(\hat{\tau}_{0,0})} \dfrac{\text{Gr}}{\text{Pr}^2} \right], \\ \Omega_* = \dfrac{\omega_*}{\varepsilon_* \eta_{0,0}^2}. \end{cases} \tag{9.133}$$

Note that the parameter Gr depends on the tip velocity U and the undercooling temperature. Hence, it is not suitable to be used as an independent growth condition. In the practice, to measure the strength of buoyancy effect, it is better to introduce the following buoyancy pa-

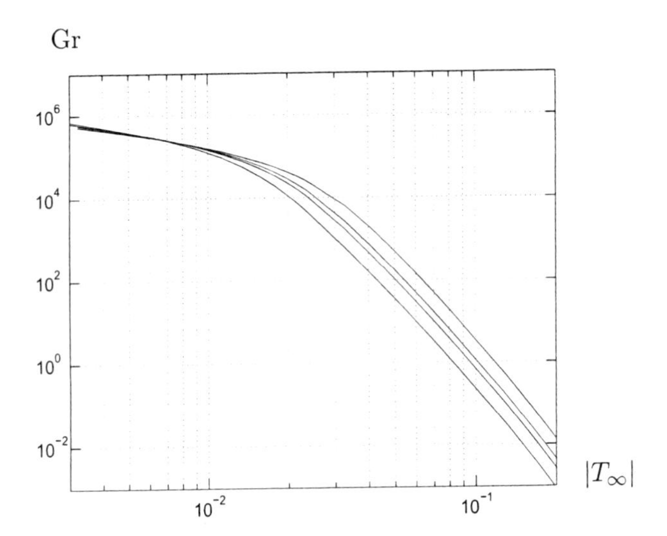

Figure 9.6. The variation of Grashof number Gr versus $|T_\infty|$ for the cases: $\mathrm{Pr} = 22$ and the fixed buoyancy parameter $\mathrm{Gr_c} = 0.0, 2.0\times10^{-14}, 6.0\times10^{-14}, 1.0\times10^{-13}, 3.0\times10^{-13}$ from bottom to top

rameter based on the length scale ℓ_c and the temperature scale $\left(\frac{\Delta H}{c_p \rho}\right)$:

$$\mathrm{Gr_c} = \frac{g\beta\ell_c^3}{\kappa_T^2}\frac{\Delta H}{c_p\rho}. \tag{9.134}$$

One can derive

$$\mathrm{Gr} = \frac{\mathrm{PrGr_c}|T_\infty|}{\eta_0^{12}\varepsilon^6} \approx \frac{\mathrm{PrGr_c}|T_\infty|}{\eta_{0,0}^{12}\varepsilon^6}. \tag{9.135}$$

Thus, we one use $T_\infty, \mathrm{Gr_c}$ and Pr as the independent parameters to describe the growth condition. In doing so, one can re-write

$$\sigma_{f,1} = -\frac{\mathrm{Gr_c}\tilde{h}_1(\hat{\tau}_{0,0})}{\mathrm{Pr}}\frac{e^{2\hat{\tau}_{0,0}}(2+\hat{\tau}_{0,0})}{\eta_{0,0}^8\varepsilon^6}\frac{k_0(\xi_c)\xi_c^2}{(1+\xi_c^2)}, \tag{9.136}$$

and calculate the selected physical quantities as the functions of the parameters of $T_\infty, \mathrm{Gr_c}$ and Pr. In viewing the condition (9.107), to ensure the validity of the asymptotic solution, we need to require

$$\left|\frac{\tilde{h}_1\mathrm{Gr_c}}{\eta_0^{10}\varepsilon^6}\right| < 1. \tag{9.137}$$

In Fig 9.4, we show ε_* versus $|T_\infty|$ for the cases of $\mathrm{Pr} = 22$ and the buoyancy parameter $\mathrm{Gr_c} = 2.0 \times 10^{-14}, 6.0 \times 10^{-14}, 1.0 \times 10^{-13}, 3.0 \times 10^{-13}$. It is seen that the effect of buoyancy-induced convection on ε_* is significant in the small undercooling regime, while it is negligible in the regime of $|T_\infty| > 0.04$ for the above cases. In Fig 9.5, we show the tip growth velocity U_{tip} versus $|T_\infty|$ for the same cases. It is seen that, affected by the buoyancy-induced convection, in the small undercooling regime the tip growth velocity increases remarkably. Such increases are due to the changes of stability criterion ε_*, but not due to the changes of Peclet number Pe, nor the change of the normalization parameter η_0^2. In Fig 9.6, we show the Grashof number Gr versus $|T_\infty|$ for these cases. It is seen that in the low undercooling regime, the Grashof number is quite large for the given buoyancy parameter $\mathrm{Gr_c}$.

In Fig 9.4 and Fig 9.5, as the comparisons, we also show the microgravity experimental data of the material SCN obtained by Glicksman et. al, as well as the ground experimental data of the material SCN with the buoyancy parameter $\mathrm{Gr_c} = 3.218 \times 10^{-13}$ obtained by Gill et. al. It is seen that the theoretical predictions are in reasonably good agreements with the experimental data, though there is certain degree of discrepancies in the regime of very low undercooling temperature. This is a surprising result, owing to the fact that for the cases under discussion the modified Grashof numbers $\overline{\mathrm{G}}$ are extremely large.

5.3 Convection Motion Induced by Density Change During Phase Transition

In this case, we have $\lambda_0 = 1, \mathrm{Gr} = 0$, $\alpha \ll 1$. It is derived that

$$\Delta_0 = \Lambda_1 = 0, \qquad \Lambda_2 - 1 - \eta_{0,0}^2 = 0, \tag{9.138}$$

$$\mathrm{Pe} = \eta_0^2 = \eta_{0,0}^2 = \mathcal{S}_d(\alpha, T_\infty), \tag{9.139}$$

$$\sigma_{f,3} = -\alpha \frac{k_0^3}{S^3}, \tag{9.140}$$

and

$$\sigma_{f,1} = \sigma_{f,2} = 0. \tag{9.141}$$

Thus, we can calculate the selection criterion ε_* as the function of the parameters (α, η_0^2), subsequently determine the selected physical quantities: $U_{\mathrm{tip}}, R_{\mathrm{tip}}, \Omega_*$.

In Fig. 9.7, we show the variation of stability criterion ε_* with the undercooling temperature $|T_\infty|$ for the cases of density parameter $\alpha =$

$-0.2, -0.1, 0.0, 0.1, 0.2$. It is seen that differing from the effect of natural convection or external flow, the effect of convection induced by the density change on the selection criterion ε_* is uniform in the entire range of the undercooling temperature $|T_\infty|$. For the same cases, in Fig. 9.8, we show the variation of the Peclet number with the undercooling temperature $|T_\infty|$, while in Fig. 9.9 and Fig. 9.10, we show the variation of the tip velocity $U_{\rm tip}$ with the undercooling temperature $|T_\infty|$.

It is seen from Fig. 9.8 that The Peclet number increases with the increasing density change parameter α; the larger undercooling temperature T_∞, the stronger the effect of the density change parameter α on Pe. This can be explained as the following. As the parameter $\alpha > 0$, the liquid is pulled towards the dendrite's interface during dendrite growth. Such convective flow will play a similar role as an external flow against the growing dendrite, which strengthen the transport process of carrying latent heat away from the interface through the liquid phase. Hence the Peclet number becomes bigger, compared with the case of $\alpha = 0$. On the other hand, the convective flow induced by the density change during the phase transition is stronger in the region of large undercooling, where the growth rate is larger, compared with the small undercooling region. Consequently, the increase of the Peclet number will be larger in the region of large undercooling region than small undercooling region. As the parameter $\alpha < 0$, the situation is just opposite.

From Fig. 9.10 one sees that with the increasing density change parameter α, the tip velocity $U_{\rm tip}$ decreases in the small undercooling region, while it increases in the large undercooling region. This is because that in the small undercooling region the decrease of the selection criterion ε_* dominates the small increase of the Peclet number, whereas in the large undercooling region, the large increase of the Peclet number, dominates the decrease of the selection criterion ε_*. However, from Fig. 9.10, it is seen that, generally speaking, the magnitude of the change of tip velocity affected by the convection induced by the density change is rather small in the entire region of the undercooling temperature $|T_\infty|$, provide the parameter $|\alpha| < 0.1$.

6. A Summary

In the present chapter, we studied the interfacial stability of dendritic growth from a melt for a class of systems with convection flow, whose steady state solutions with zero surface tension, in the IW layer near the interface, can be expressed in the form (9.22)–(9.24). The convection flow is generally measured by some flow parameter $\mathcal{F}$, which may have different definitions for different cases. The convection may be in-

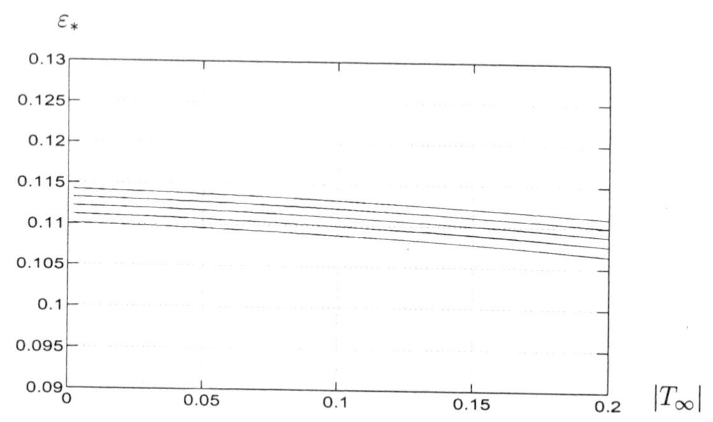

Figure 9.7. The variation of stability criterion ε_* versus the undercooling temperature $|T_\infty|$ for the cases of the density parameter $\alpha = -0.2, -0.1, 0, 0.1, 0.2$ from top to bottom

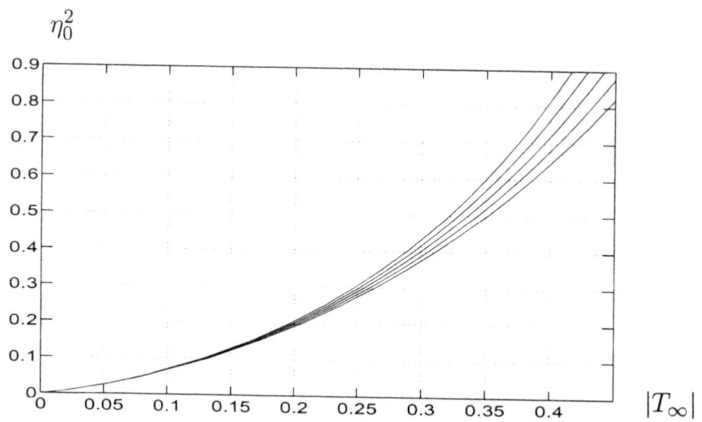

Figure 9.8. The variation of the Peclet number $\text{Pe} = \eta_0^2$ versus the undercooling temperature $|T_\infty|$ for the cases of the density parameter $\alpha = -0.2, -0.1, 0, 0.1, 0.2$, from bottom to top

duced by a uniform enforced flow at the far field, or by buoyancy in the gravitational field, or by a density change during phase transition.

On the basis of stability analysis for the basic state solution, we established a general theory for the selection of the tip growth velocity. The approach of the interfacial wave (IFW) theory is adopted in the present investigation, and the Global Travelling Wave (GTW) instability mechanism previously discovered in the system without convection is found

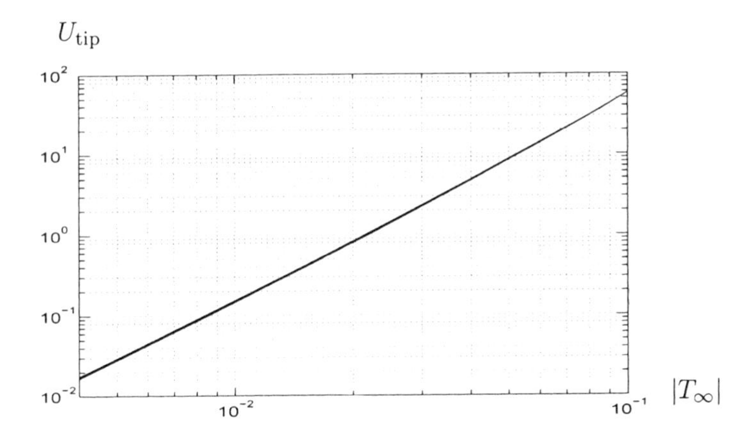

Figure 9.9. The variation of tip velocity U_{tip} versus the undercooling temperature $|T_\infty|$ for the cases of the density parameter $\alpha = -0.2, -0.1, 0, 0.1, 0.2$. The differences between the curves are invisible in the logarithmic graph

to remain unchanged. The major conclusions drawn from this study are as follows:

1. The isotropic surface tension is still the determined parameter for the stability of dendritic growth and the selection of tip velocity. This global neutrally stable mode (GTW) is selected at the later stage of dendritic growth. The tip velocity U_{tip}, tip radius R_{tip} and the eigenfrequency of the interface Ω_* are uniquely determined by the selection criterion ε_* and other parameters, such as η_0^2, Pe and T_∞.

2. In general, convective flow may affect the selection of dendritic growth in two ways. First, it may affect the eigenvalues σ, and the selection criterion ε_* through the parameters Δ_0, α, $\overline{\Lambda}_1$ and $\overline{\Lambda}_2$, which are determined by the modifications of the interface shape and the local temperature gradient at the interface of the basic state, respectively. However, by this way, convection flow only affects the first-order approximation solutions of the eigenvalues σ and the selection criterion ε_*, in the limit process $\varepsilon \to 0$. In the leading-order approximation, the selection criterion ε_* is the same as the system without convection, and only the parameter Δ_0 can affect the eigenfunctions. Furthermore, the convective flow may affect the normalization parameter η_0^2 and Peclet number Pe of the basic state. The changes of these two parameters on one hand affect the eigenvalues σ and selection criterion ε_*, on the other hand, they may directly affect the tip velocity U_{tip}, tip radius R_{tip} and frequency of oscillation Ω_*.

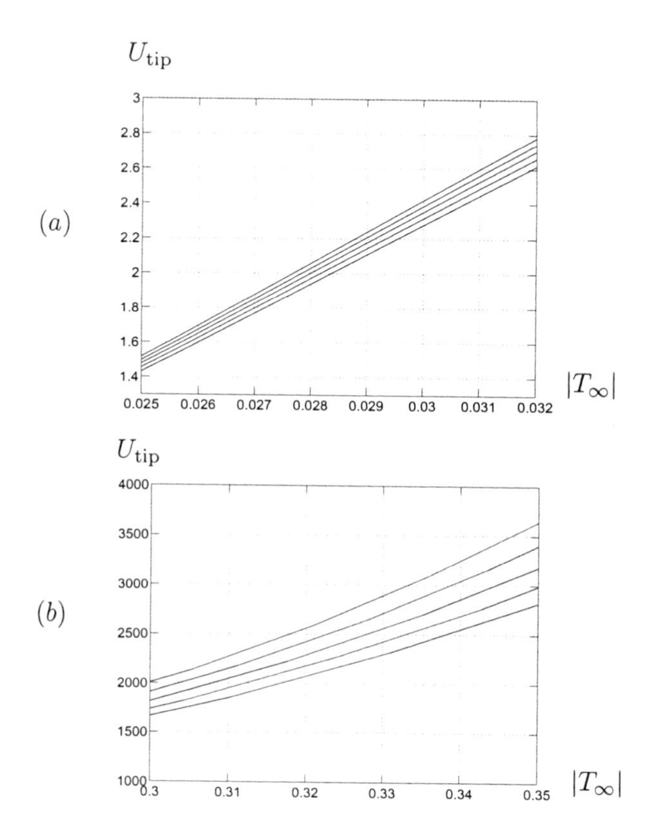

Figure 9.10. The variation of tip velocity $U_{\rm tip}$ versus the undercooling temperature $|T_\infty|$ for the cases of the density parameter $\alpha = -0.2, -0.1, 0, 0.1, 0.2$, (a) in the low undercooling region, $U_{\rm tip}$ decreases with increasing α; (b) in the high undercooling region, $U_{\rm tip}$ increases with increasing α

3. The effect of external flow on the selection of dendritic growth is mainly from the changes of Peclet number and the normalization parameter η_0^2. The effect of external flow becomes noticeable only in the low undercooling region. With a fixed undercooling temperature, the tip velocity increases with the Reynolds number $\mathrm{Re_c}$ of external flow.

4. The effect of natural convection on the selection of dendritic growth is also mainly from the changes of the selection criterion ε_* and the Peclet number. It does not affect the normalization parameter η_0^2. The tip velocity increases with the buoyancy number $\mathrm{Gr_c}$ and becomes noticeable also only in the low undercooling region.

5. Finally, convection induced by density change affects dendritic growth via the both ways. It changes the stability criterion ε_*, as well as the Peclet number Pe and the normalization parameter η_0^2. In this case, the Peclet number Pe and the normalization parameter η_0^2 are equal. For a fixed the undercooling temperature $|T_\infty|$, the selection criterion ε_* decreases with increasing density change parameter α uniformly, while the Peclet number Pe increases with increasing density change parameter α. The effect of convection induced by density change on the tip velocity U_{tip} is through both the Peclet number Pe and selection criterion parameter ε_*. With increasing α, U_{tip} decreases in the small undercooling region, while U_{tip} increases in the large undercooling region. However, when $|\alpha| < 0.1$, such effect is rather small.

Chapter 10

CONCLUDING REMARK

In this monograph, I reviewed some recent research works on dynamics of dendritic growth interacting with convective flow. My attention is focused on the results obtained by a systematic, analytical approach in the framework of the IFW theory, some relevant numerical simulation works on this subject are left out. Moreover, as mentioned at the beginning of the book, dendritic growth is a complex nonlinear phenomenon occurring in a variety of crystal growth systems, hence, the studies presented in this book can only reveal a portion of the nature of this phenomenon. Many significant problems in this field are open. Among these unsolved issues, in my opinion, the following are particularly of interest and should be further explored in the near future:

- Dendritic growth interacting with more complicated flow field in melt;
- Non-axi-symmetric dendritic growth, caused by anisotropic surface tension, or non-axi-symmetric buoyancy effect.
- Dendritic growth affected by external forces other than gravity.
- Dendritic growth with the effect of anisotropic kinetic attachment.
- Stability, selection and formation of an array of dendritic growth under various growth conditions. Especially, the stability and transition of cellular growth to an array of dendritic growth in directional solidification. This long-standing, challenging problem in the area of materials science is still not well resolved.

References

Abramovitz, M. and Stegun, I.A. (1964). *Handbook of Mathematical Functions.* National Bureau of Standards.

Amar, M. Ben, Bouissou, Ph., and Pelce, P. (1988). An exact solution for the shape of a crystal growing in a forced flow. *Journal of Crystal Growth*, 92:97–100.

Ananth, R. and Gill, W. N. (1989). Dendritic growth of an elliptical paraboloid with forced convection in the melt. *J. Fluid Mech.*, 208:575–593.

Ananth, R. and Gill, W. N. (1991). Self-consistent theory of dendritic growth with convection. *J. Crystal Growth*, 108:173–189.

Bender, C. M. and Orszag, S. A. (1978). *Advanced Mathematical Methods for Scientists and Engineers.* McGraw–Hill, New York.

Bouissou, P., Chiffaudel, A., Perrin, B., and Tabelling, P. (1990). Dendritic side-branching forced by an external flow. *Europhys. Lett.*, 13:89–94.

Bouissou, P., Perrin, B., and Pelce, P. (1989). Influence of an external flow on dendritic growth. *Phys. Rev., A*, 40:509–512.

Bouissou, Ph. and Pelce, P. (1989). Effect of a forced flow on dendritic growth. *Phys. Rev. A*, 40:6637–6680.

Canright, D. and Davis, S. H. (1991). Buoyancy effect of a growing, isolated dendrite. *J. Crystal Growth*, 114:153–185.

Davis, R.T. (1972). Numerical solution of the Navier–Stokes equation for symmetric laminar incompressible flow past a parabola. *J. Fluid Mech.*, 51:417–433.

Davis, R.T. and Werle, M.J. (1972). Numerical solution for laminar incompressible flow past a paraboloid of revolution. *AIAA J.*, 10:1224–1230.

Davis, S. H. (2001). *Theory of Solidification.* Cambridge University Press, Cambridge.

Dennis, S.C.R. and Warsh, J.D. (1971). Numerical solution for steady symmetric viscous flow past a parabolic cylinder in a uniform stream. *J. Fluid Mech.*, 48:771–789.

Dougherty, A. and Gollub, J. P. (1988). Steady-state dendritic growth of nh_4br from solution. *Phys. Rev. A*, 38:3043–3053.

Van Dyke, Milton (1975). *Perturbation Methods in Fluid Mechanics.* The Parabolic Press, Stanford, California.

Emsellem, V. and Tabeling, P. (1995). Experimental study of dendritic growth with an external flow. *J. Crystal Growth*, 156:285–295.

Emsellem, V. and Tabeling, P. (1996). On the role of convection in free dendritic growth: Experimental measurement of the concentration field. *J. Crystal Growth*, 166:251–255.

Glicksman, M. E., Koss, M. B., Bushnell, L. T., Lacombe, J. C., and Winsa, E. A. (1995). Dendritic growth of succinonitrile in terrestrial and microgravity conditions as a test of theory. *ISIJ International*, 35, No: 6:604–610.

Glicksman, M. E., Koss, M. B., and Winsa, E. A. (1994). Dendritic growth velocities in microgravity. *Phy. Rev. Let.*, 73, No: 4:573–576.

Glicksman, M. E., Schaefer, R. J., and Ayers, J. D. (1975). High-confidence measurement of solid/liquid surface energy in a pure material. *Philosophical Magazine*, 32:725–743.

Glicksman, M. E., Schaefer, R. J., and Ayers, J. D. (1976). Dendrite growth — a test of theory. *Metall. Trans.*, 7A:1747–1757.

Gradshteyn, I.S. and Ryzhik, I.M. (1980). *Table of Integrals, Series, and Products (Corrected and Enlarged Edition)*, Academic Press, New York.

Horvay, G. and Cahn, J. W. (1961). Dendritic and spheroidal growth. *Acta Metall.*, 9:695–705.

Huang, S. C. and Glicksman, M. E. (1981). Fundamentals of dendritic solidification — (i). steady–state tip growth; (ii). development of sidebranch structure. *Acta Metall.*, 29:701–734.

Hurle, D. T. J., editor (1993). *Handbook of Crystal Growth, Vol. 1: Fundamentals, Part B: Transport and Stability*. Elsevier Science, North–Holland, Amsterdam.

Ivantsov, G. P. (1947). Temperature field around a spheroidal, cylindrical and acicular crystal growing in a supercooled melt. *Dokl. Akad. Nauk, SSSR.*, 58, No: 4:567–569.

Kessler, D. A., Koplik, J., and Levine, H. (1986). Pattern formation far from equilibrium: the free space dendritic crystal growth. In *Patterns, Defects and Microstructures in Non-equilibrium Systems*. NATO A.R.W., Austin, TX.

Kevorkian, J. and Cole, J. D. (1996). *Multiple Scale and Singular Perturbation Methods, Applied Mathematical Sciences, Vol. 114*. Springer, Berlin, Heidelberg.

Kobayashi, T. and Furukawa, Y. (1991). *Snow crystals*. Snow Crystal Museum Asahikawa, Hokkaido.

Koo, K. K., Ananth, R., and Gill, W. N. (1992). Thermal convection, morphological stability and the dendritic growth of crystals. *AIChE Journal*, 38, No: 6:945–954.

Kruskal, M. and Segur, H. (1991). Asymptotics beyond all orders in a model of crystal growth. *Stud. in Appl. Math*, No: 85:129–181.

Langer, J. S. (1980). Instability and pattern formation in crystal growth. *Rev. Mod. Phys*, 25:1–28.

Langer, J. S. (1986). *Lectures in the Theory of Pattern Formation, USMG NATO AS Les Houches Session XLVI 1986 — Le hasard et la matiere / chance and matter. Ed. by J. Souletie, J. Vannimenus and R. Stora.* Elsevier Science, Amsterdam.

Langer, J. S. (1992). Issues and opportunities in materials research. *Physics Today*, pages 24–31.

Langer, J. S. and Müller-Krumbhaar, H. (1978). Theory of dendritic growth — (i). elements of a stability analysis; (ii). instabilities in the limit of vanishing surface tension; (iii). effects of surface tension. *Acta Metall*, 26:1681–1708.

Lee, Y-W. (1991). *Pattern Formation and Convective Heat Transfer during Dendritic Crystal Growth.* PhD dissertation, Rensselaer Polytechnic Institute, Troy, NY.

Lee, Y-W., Gill, W., and Ananth, R. (1992). Forced convection heat transfer during dendritic crystal growth: Local solution of Navier–Stokes equations. *Chem. Eng. Comm.*, 116:193–200.

Lee, Y-W., Smith, R. N., Glicksman, M. E., and Koss, M. B. (1996). Effects of buoyancy on the growth of dendritic crystals. *Annual Review of Heat Transfer*, 7:59–139.

Lin, C. C. (1955). *The Theory of Hydrodynamic Stability.* Cambridge University Press, Cambridge.

Lin, C. C. and Lau, Y. Y. (1979). Density wave theory of spiral structure of galaxies. *Studies In Applied Mathematics*, 60:97–163.

McFadden, G.B. and Coriell, S.R. (1986). The effect of fluid flow due to the crystal-melt density change on the growth of a parabolic isothermal dendrite. *Journal of Crystal Growth*, 74:507–512.

Mullins, W. W. and Sekerka, R. F. (1963). Morphological stability of a particle growing by diffusion or heat flow. *J. Appl. Phys.*, 34:323–329.

Mullins, W. W. and Sekerka, R. F. (1964). Stability of a planar interface during solidification of a dilute binary alloy. *J. Appl. Phys.*, 345:444–451.

Nash, G. E. and Glicksman, M. E. (1974). Capillarity–limited steady-state dendritic growth (i). theoretical development. *Acta Metall.*, 22:1283–1299.

Oseen, C.W. (1910). Ueber die stokes'sche formel und über eine verwandie aufgabe in der hydrodynamik. *Ark. Matj. Astronom. Fys.*, 6:No: 29.

Pelce, P. (1988). *Dynamics of Curved Front.* Academic Press, New York.

Pines, V., Chait, A., and Zlatkowski, M. (1996). Anomaly in dendritic growth data – effect of density change upon solidification. *J. Crystal Growth*, 169:798–802.

Saville, D. A. and Beaghton, P. J. (1988). Growth of needle-shaped crystals in the presence of convection. *Phys. Rev.A*, 37:3423–3430.

Sekerka, R. F., Coriell, S. R., and McFadden, G. B. (1995). Stagnation film model of the effect of natural convection on dendritic operating state. *J. Crystal Growth*, 154:370–376.

Sekerka, R. F., Coriell, S. R., and McFadden, G. B. (1996). The effect of container size on dendritic growth in microgravity. *J. Crystal Growth*, 171:303–306.

Stokes, G. G. (1851). On the effect of internal friction of fluids on motion of pendulums. *Trans. Camb. Phil. Soc. (Reprinted: Math. and Phys. Papers 3 pp. 1–141, Cambridge Univ. Press)*, 9, Part II:8–106.

Takahashi, K., Furukawa, Y., and Takahashi, Y. (1995). *Story of snow crystals.* Koudansya, Tokyo.

Verdman, A. E. P. (1973). The numerical solution of the Navier–Stokes equation for laminar incompressible flow past a paraboloid of revolution. *Computers and Fluids*, 1:251–271.

Wilkinson, J. (1955). A note on the Oseen approximation for a paraboloid in a uniform stream parallel to its axis. *Quart. Journ. Mech. and Applied Math.*, VIII, Pt. 4:415–421.

Xu, J. J. (1987). Global asymptotic solution for axi-symmetric dendrite growth with small undercooling. In *Structure and Dynamics of Partially Solidified System*, pages 97–109. Ed. by D.E. Loper, NATO ASI Series E. No. 125.

Xu, J. J. (1989). Interfacial wave theory for dendritic structure of a growing needle crystal (i): Local instability mechanism; (ii): Wave-emission mechanism at the turning point. *Phys. Rev. A*, 40, No. 3:1599–1614.

Xu, J. J. (1990a). Asymptotic theory of steady axisymmetric needle-like crystal growth. *Studies in Applied Mathematics*, 82:71–91.

Xu, J. J. (1990b). Global neutral stable state and selection condition of tip growth velocity. *J. Crystal Growth*, 100:481–490.

Xu, J. J. (1991a). Interfacial wave theory of solidification — dendritic pattern formation and selection of tip velocity. *Phys. Rev. A15*, 43, No: 2:930–947.

Xu, J. J. (1991b). Two-dimensional dendritic growth with anisotropy of surface tension. *Physics (D)*, 51:579–595.

Xu, J. J. (1994a). Dendritic growth from a melt in an external flow: Uniformly valid asymptotic solution for the steady state. *J. Fluid Mech.*, 263:227–243.

Xu, J. J. (1994b). Effect of convection motion in melt induced by density-change on dendritic solidification. *Canad. J. Phys.*, 72, No: 3 & 4:120–125.

Xu, J. J. (1996). Generalized needle solutions, interfacial instabilities and pattern formations. *Phys. Rev. E*, 53 No: 5:5051–5062.

Xu, J. J. (1997). *Interfacial Wave Theory of Pattern Formation: Selection of Dendrite Growth and Viscous Fingering in a Hele–Shaw Flow.* Springer–Verlag, New York.

Xu, J. J. and Yu, D. S. (1998). Regular perturbation expansion solution for generalized needle crystal growth. *Journal of Crystal Growth*, 187:314–326.

Xu, J. J. and Yu, D. S. (2001a). Further examinations of dendrite growth theories. *J. of Crystal Growth*, 222:399–413.

Xu, J. J. and Yu, D. S. (2001b). Selection and resonance of dendrite growth with interference of oscillatory external sources. *J. of Crystal Growth*, 226:378–392.

Yu, D. S. and Xu, J. J. (1999). Dendrite growth in external flow: The selection of tip velocity. *J. of Crystal Growth*, 198, No: 49:49–55.

Index